1-3-12

JOHNNIE MAE BERRY LIBRARY
CINCINNATI STATE
3520 CENTRAL PARKWAY
CINCINNATI, OH 45223-2690

D1158134

HVAC Design Sourcebook

697
A581
2012

HVAC Design
Sourcebook

W. Larsen Angel, P.E., LEED AP

New York   Chicago   San Francisco
Lisbon   London   Madrid   Mexico City
Milan   New Delhi   San Juan
Seoul   Singapore   Sydney   Toronto

The **McGraw·Hill** Companies

Copyright © 2012 by The McGraw-Hill Companies, Inc. All rights reserved. Printed in the United States of America. Except as permitted under the United States Copyright Act of 1976, no part of this publication may be reproduced or distributed in any form or by any means, or stored in a data base or retrieval system, without the prior written permission of the publisher.

1 2 3 4 5 6 7 8 9 0   DOC/DOC   1 7 6 5 4 3 2 1

ISBN 978-0-07-175303-6
MHID 0-07-175303-6

Sponsoring Editor	**Project Manager**	**Indexer**
Joy Evangeline Bramble	Anupriya Tyagi, Cenveo Publisher Services	Robert Swanson
Editing Supervisor		**Art Director, Cover**
Stephen M. Smith	**Copy Editor**	Jeff Weeks
Production Supervisor	Mary Kay Kozyra	**Composition**
Richard C. Ruzycka	**Proofreader**	Cenveo Publisher Services
Acquisitions Coordinator	Manish Tiwari, Cenveo Publisher Services	
Molly T. Wyand		

Printed and bound by RR Donnelley.

McGraw-Hill books are available at special quantity discounts to use as premiums and sales promotions, or for use in corporate training programs. To contact a representative, please e-mail us at bulksales@mcgraw-hill.com.

This book is printed on acid-free paper.

Information contained in this work has been obtained by The McGraw-Hill Companies, Inc. ("McGraw-Hill") from sources believed to be reliable. However, neither McGraw-Hill nor its authors guarantee the accuracy or completeness of any information published herein, and neither McGraw-Hill nor its authors shall be responsible for any errors, omissions, or damages arising out of use of this information. This work is published with the understanding that McGraw-Hill and its authors are supplying information but are not attempting to render engineering or other professional services. If such services are required, the assistance of an appropriate professional should be sought.

To Lisa, my dear wife whom I love, whose constant encouragement and insight helped to make this book a reality.

"… with God all things are possible."—Matthew 19:26

About the Author

W. Larsen Angel, P.E., LEED AP, is a principal in the MEP consulting engineering firm Green Building Energy Engineers. He has worked in the MEP consulting engineering industry for more than 20 years. Mr. Angel has contributed to the development of design standards and continues to find new ways to streamline the HVAC system design process. He is a Member of the American Society of Heating, Refrigerating and Air-Conditioning Engineers (ASHRAE) and is certified by ASHRAE as a Commissioning Process Management Professional (CPMP).

Contents

Preface

A re you searching for a practical handbook that will assist you in the process of designing heating, ventilating, and air-conditioning (HVAC) systems for commercial buildings? The *HVAC Design Sourcebook* is the tool you need to quickly become a valuable member of your design team.

The typical approach to training junior employees in the process of designing HVAC systems is to have them try to learn the skills they need to work as effective members of a design team from the senior HVAC engineers they are working under. Unfortunately, the knowledge the senior engineers are expected to impart has taken them years to develop and, without a practical training resource, the process of teaching junior HVAC system designers the essentials of HVAC system design becomes time-consuming, ineffective, and costly.

The *HVAC Design Sourcebook* fills the void in the industry for a practical resource to assist in the process of training junior HVAC system designers in the basics of HVAC system design. Essential design concepts are clearly explained and illustrated with photographs of actual HVAC systems installations and graphical conventions used in the preparation of construction drawings. Codes and standards are referenced frequently to emphasize the need for HVAC systems to be designed in accordance with the requirements of the regulating authorities. Other topics such as the overall design process, HVAC systems and equipment, piping and ductwork distribution systems, noise and vibration control, and automatic temperature controls are presented in a manner that can be understood and applied by the junior HVAC system designer. The ultimate goal of preparing complete, well-coordinated HVAC system construction drawings is consistently in view throughout the book.

The *HVAC Design Sourcebook* is the essential resource for individuals who are considering or pursuing a career in the field of HVAC system design.

W. Larsen Angel, P.E., LEED AP

Acknowledgments

Special thanks to my two sons for their understanding during this busy time.

Thanks also to Joe Podson, Executive Director of B'nai B'rith Homecrest House located in Silver Spring, Maryland, for allowing me to use photographs of the Homecrest House facility in this book.

HVAC Design Sourcebook

CHAPTER 1

What Is HVAC?

The term *HVAC* stands for heating, ventilating, and air-conditioning. It describes the field that is concerned with heating, ventilating, and air-conditioning the indoor environment in order to meet the comfort, health, and safety needs of building occupants and the environmental needs of indoor equipment or processes. Although HVAC systems are required for airplanes, ships, automobiles, and other special applications, this book will focus on HVAC systems for commercial buildings.

Heating and ventilating systems for buildings have been in existence for centuries. Fireplaces and windows, the earliest forms of indoor heating and ventilating, remained the primary means of heating and ventilating buildings into the late nineteenth century. It was in the nineteenth century that engineers began to use steam heating systems, which consisted of coal-fired boilers, pipes, and radiators, to heat buildings. Steam heating systems are still widely used today, although natural gas and fuel oil have replaced coal as the primary fuel source.

In the Middle Ages, people made the connection between "bad air" in overcrowded or smoky rooms and disease. In the eighteenth and nineteenth centuries, scientists and physicians began to study the sources of indoor air contaminants and the effects these contaminants had on human health. As a result, in 1895 the American Society of Heating and Ventilating Engineers (ASHVE) adopted a minimum ventilation rate of 30 cubic feet per minute (cfm) of outdoor air per occupant as a ventilation standard for public buildings. It was understood at the time that this ventilation rate was sufficient to dilute the indoor air contaminants to a level that was acceptable for human occupancy. Outdoor air ventilation rates that are required to produce acceptable indoor air quality for various occupancies continue to be studied by the American Society of Heating, Refrigerating, and Air-Conditioning Engineers (ASHRAE), the successor of ASHVE. *ANSI/ASHRAE Standard 62.1-2007—Ventilation for Acceptable Indoor Air Quality* is devoted to the subject of indoor air quality. The guidelines of this standard have largely been incorporated into the various mechanical codes applied to building construction, such as the *International Mechanical Code* (IMC), published by the International Code Council, Inc. (ICC).

Mechanical cooling (air-conditioning) systems are a relatively recent development. The first central building air-conditioning system, designed for the Missouri State Building, was demonstrated to the public during the 1904 World's Fair held in St. Louis. Since that time, a great deal of research has been performed and a wide variety of air-conditioning equipment has been developed to meet the diverse air-conditioning needs of modern buildings.

In addition to heating, ventilating, and air-conditioning the indoor environment to meet the comfort and health needs of the building occupants, modern HVAC systems are frequently required to protect the safety of the occupants or, in industrial applications, to provide a clean environment for the processes performed within the building. For example, smoke control systems may be required to minimize the spread of smoke within a building during a fire. Also, the HVAC systems may be required to maintain air pressure relationships between adjacent spaces where hazardous materials are handled and to signal the building operator or activate emergency ventilation systems if these pressure relationships are not maintained. HVAC systems may also be required to provide a high level of air filtration in order to maintain a clean indoor environment for such processes as semiconductor fabrication.

With the understanding of what HVAC systems are and what they are expected to accomplish, it is the role of the HVAC system designer to design HVAC systems to meet project needs. In order to do this, the HVAC system designer must first understand the project requirements. The designer must then use this information, along with a knowledge of the potential HVAC system options, to design the HVAC systems (in accordance with the applicable codes) that are appropriate for the project. Factors governing the HVAC system selection and the ultimate HVAC system design include:

- HVAC system types that are available to meet the project needs
- Building owner's preferences or standards
- Building owner's budget
- Installed cost, operating cost, and maintenance cost of the potential HVAC system options
- Space limitations, both indoors and outdoors, and coordination with other building elements such as the architectural, structural, and electrical systems

After all factors have been considered and the final HVAC system configuration developed, the HVAC system designer must present the HVAC system design in a clear and concise way through the use of construction documents. Construction documents are the drawings and specifications for a project that are used by the installing contractor to construct the HVAC systems. The construction documents are also used by the building maintenance personnel as a resource in the ongoing operation and maintenance of the HVAC systems.

In this book, we will discuss the HVAC system design process from concept to completion of the construction documents (Chap. 2); piping, valves, and specialties, which are an integral part of HVAC systems (Chap. 3); the central plant, which is where fuel sources are converted to heating and cooling energy (Chap. 4); air systems, which circulate air within the building (Chap. 5); piping and ductwork distribution systems, which are used to distribute the heating and cooling energy from the central plant to the air systems and terminal equipment and eventually to the spaces within the building (Chap. 6); terminal equipment, which is often used in the distribution of the heating and cooling energy to the spaces within the building (Chap. 7); noise and vibration control, which is a critical component of a successful HVAC system design (Chap. 8); automatic temperature controls, without which HVAC systems cannot function properly (Chap. 9); and finally the preparation of construction drawings, including some important drafting and computer-aided design concepts (Chap. 10).

CHAPTER 2

The Design Process

HVAC Load Calculations

HVAC load calculations are the foundation upon which the HVAC system design is built. Therefore, it is imperative that the HVAC system designer accurately calculate the peak heating and cooling loads for the project in order to properly design the HVAC systems. The most accurate method for calculating the HVAC loads of a commercial building is the heat balance method. This method is described in detail in Chap. 18 of the 2009 *ASHRAE Handbook—Fundamentals*. The HVAC system designer must have a good understanding of this method in order to understand how a building's geometry, orientation, and internal functions affect the HVAC loads within the different areas of the building.

Because the process of calculating HVAC loads is quite involved, commercially available HVAC load calculation software is used almost exclusively for commercial projects. This section provides an overview of the major considerations associated with HVAC load calculations for commercial buildings. The details of how building information is entered into the HVAC load calculation program will vary from one program to another. The HVAC system designer should consult the software user's manual for detailed instructions on how to set up the load calculations.

Buildings are affected by heating and cooling loads both external to and internal to the building. External loads include heat gains or losses from exterior walls (above or below grade), windows, roofs, skylights, doors, floors, partitions (walls, floors, or ceilings internal to a building that separate conditioned spaces from unconditioned spaces), and outdoor air leakage (referred to as infiltration). External loads vary with outdoor air temperature and relative humidity, the intensity and position of the sun, wind speed, and the temperature of the ground. External loads are also dependent upon the geographical location of the project. Internal loads include heat gains from people, lighting, and equipment. These loads also vary and depend upon the occupancy of the various spaces within the building and equipment usage.

In most situations, the goal of proper HVAC system design is to maintain a constant indoor air temperature year-round, regardless of the outdoor conditions or internal functions. In some cases, it is desirable to maintain a constant indoor air relative humidity. However, for the sake of simplicity, we will assume that only indoor air temperature is being controlled.

In order to maintain a constant indoor air temperature, the HVAC systems serving a building must be sized to offset the heat that is lost from the various spaces within the building (when the heat losses exceed the heat gains) and offset the heat that is added to the various spaces within the building (when the heat gains exceed the heat losses).

Terms

Before we continue with the discussion of HVAC load calculations for commercial buildings, it is necessary to define some terms:

- Space: The smallest area defined in the HVAC load calculation, usually consists of a single room.

- Zone: Typically a collection of spaces, all of which have similar HVAC loading characteristics. HVAC loading characteristics are defined as the manner in which the HVAC loads vary within a space. For example, three offices located on the south side of a building would normally have similar HVAC loading characteristics because the heat gains and losses through the exterior building components would vary similarly for all of the spaces. Therefore, these spaces would commonly be grouped into one HVAC zone. The space temperature of these three offices would be controlled by a single thermostat[1] located in one of the offices. However, single-space zones in a building are common where temperature control of that one space is critical. An example would be a conference room. It would not be desirable for the space temperature in a conference room to be controlled by a thermostat located in a nearby office. If that office was unoccupied, the thermostat would call for minimum cooling. However, if the conference room was fully occupied at the same time, it would require maximum cooling. In this case, the thermostat would not adequately satisfy the cooling requirement of the conference room because the conference room has different HVAC loading characteristics than the office. Another example of a single-space zone would be an office located in the corner of a building. Because this office would have two exterior walls with exposures that are at right angles to each other, its HVAC loading characteristics would be different from any other space on that floor.

- Terminal Equipment: The equipment that delivers the heating or cooling energy to the HVAC zones in response to the zone thermostats. An example of terminal equipment for a commercial building is a variable air volume (VAV) terminal unit. Multiple VAV terminal units are normally served by a single VAV air system (see air system description below). Each VAV terminal unit receives conditioned supply air from the air system, modulates the supply airflow, and may add heat to the supply airflow in response to the zone thermostat controlling the VAV terminal unit. Another example of terminal equipment for a commercial building is a finned-tube radiator that is sometimes used to provide radiant heat for zones having external loads.

- Air System: The HVAC equipment that conditions the air supplied to the HVAC zones. An air system also returns air from the HVAC zones and provides outdoor air ventilation when required. An air system may serve multiple zones (as in the case of the VAV terminal units described above) or it may serve only one zone. In the case of the single-zone air system, the zone thermostat controls the heating and cooling capacity of the air system, and there is no terminal equipment. An example of a single-zone air system that most people are familiar with is the fan-coil unit that provides heating and cooling for a home in response to the zone thermostat.

  Air can be conditioned in an air system by heating, cooling, humidifying (adding moisture), or dehumidifying (removing moisture). The components of an air system that are relevant to HVAC load calculations are the supply fan and return fan (if applicable), which circulate the air through the heating and/or cooling coils in the unit; the heating and/or cooling coils, which transfer the necessary heating and/or cooling energy that is required by the zones to the airstream; and the outdoor air ventilation, which is usually introduced at a mixing point upstream of the heating and cooling coils. The combination of return air and outdoor air is called mixed air.

Other components of an air system that are relevant to HVAC load calculations but are not as widely used include a humidifier, which is used to add moisture to the supply air; a reheat coil, which is used to reheat the supply air after it has been cooled (often used as a means to maintain the relative humidity of the zones at a maximum level); and an energy recovery coil, which is used to exchange energy from an exhaust airstream to the outdoor air ventilation airstream.

- Central Plant: Buildings require heating and cooling energy to offset the heat gains and heat losses and to condition the outdoor air ventilation for the building. The central plant refers to the equipment that generates the heating and cooling energy utilized by the building. This equipment can be either centralized or decentralized. In a centralized system the central plant equipment is remote from the air systems and terminal equipment. An example of a centralized system would be a central heating and cooling plant for a building where the plant is remote from the air systems and contains boilers that provide heating water or steam to the air system heating coils and chillers that provide chilled water to the air system cooling coils. Heating and cooling energy may also be supplied by the central plant to various types of heating and/or cooling terminal equipment in the building.

  In a decentralized system, the central plant equipment is an integral part of each air system. An example of a decentralized system would be multiple rooftop units serving a building where each rooftop unit contains a gas-fired furnace that provides heating energy to the airstream through a heat exchanger and a complete refrigeration system that provides cooling energy to the airstream through a cooling coil. This type of unit is referred to as a packaged, or self-contained, unit because all of the necessary heating and cooling equipment is contained within one complete package. In this example of a decentralized system, each air system (rooftop unit) contains the central heating and cooling plant equipment.

Geographical Location

Now that the terms for calculating HVAC loads have been defined, we will discuss the process of setting up the HVAC load calculations, assuming the calculation will be performed with commercially available HVAC load calculation software. The first step in the process is to define the inputs to the program, starting with the building's geographical location. Once the location has been selected, the program will utilize that area's database of annual weather data (contained within the program) to simulate the outdoor conditions, which include air temperature and relative humidity, wind speed and direction, intensity and position of the sun, and ground temperature. The database contains weather data for 365 typical (not actual) 24-hour days, totaling 8,760 hours of weather data for that location.

Building Materials

Opaque Materials

All materials conduct heat to some degree. The conductance, or U-Value, of a material, expressed in terms of British thermal units per hour per square foot per Fahrenheit degree (Btu/h·ft²·°F), is a measure of how well the material conducts heat. The higher the U-Value, the better the material conducts heat and vice versa. For HVAC load calculations, the U-Value for each type of wall, roof, and partition needs to be calculated. This is done by examining the wall, roof, and partition sections in the architectural drawings for the building. The properties of common construction materials are listed in Chap. 26 of the 2009 *ASHRAE Handbook—Fundamentals* and are also included in the building materials database of some HVAC load calculation programs.

In order to calculate the U-Value for a wall, roof, or partition type, it is necessary to first sum the resistances of all of the components for each wall, roof, or partition that are shown on the architectural drawings. The resistance of a material is the inverse of the U-Value and is given in terms of hour square foot Fahrenheit degree per British thermal unit (h·ft²·°F/Btu). Most people are familiar with this term because it is used to describe the insulating value of fiberglass batt insulation (e.g., 3½ in. of batt fiberglass insulation has an R-Value of 11, which is typically denoted as R-11). Once the total R-Value of the wall, roof, or partition has been determined, the reciprocal of this total R-Value will be the U-Value.

Figure 2-1 provides an example of how the U-Value is determined for a typical wall that consists of the following components (from inside to outside): gypsum wall board, batt insulation, vegetable board sheathing, air space, and face brick. The U-Value for this wall section is the reciprocal of the R-Value, or 0.066 Btu/h·ft²·°F.

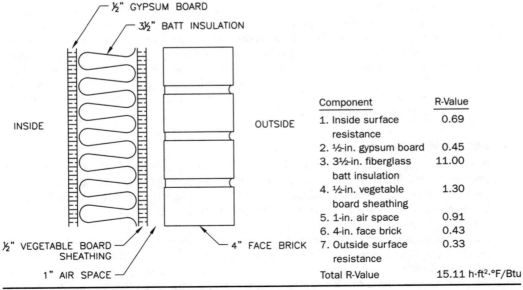

Component	R-Value
1. Inside surface resistance	0.69
2. ½-in. gypsum board	0.45
3. 3½-in. fiberglass batt insulation	11.00
4. ½-in. vegetable board sheathing	1.30
5. 1-in. air space	0.91
6. 4-in. face brick	0.43
7. Outside surface resistance	0.33
Total R-Value	15.11 h·ft²·°F/Btu

FIGURE 2-1 Typical architectural wall section.

In addition to the U-Value of the building envelope[2] materials, the color (light, medium, dark) and the weight (light, medium, heavy) of these materials have to be entered into the HVAC load calculation program because these factors affect the heat absorption and transmittance of these materials. For example, lighter-colored building materials exposed to the sun reflect more of the sun's radiant energy than darker-colored building materials do and, therefore, do not absorb as much of the sun's radiant energy. Second, lighter-weight building materials transmit the energy that they absorb from the sun to the interior of the building more quickly than heavier-weight building materials do. As a result, the peak cooling load of a lightweight building occurs shortly after the outdoor air temperature and intensity of the sun reach their peak; the peak cooling load of a heavyweight building will occur at a longer time interval after the outdoor conditions peak. Lightweight buildings will also cool off more quickly at night in the summer than heavyweight buildings will. In short, heavyweight buildings have more thermal mass than lightweight buildings and, as a result, transmit changes in the outdoor conditions more slowly to the indoor environment.

Fenestration

Fenestrations (windows, skylights, and doors) in a building also have a U-Value associated with them. However, in addition to defining the U-Value for the glazing (glass component only—no frame) contained within the windows, skylights, and doors, the solar heat gain coefficient (SHGC) needs to be determined as well. The SHGC is dimensionless and represents the percentage of the sun's radiant energy that is transferred through the glazing to the space. The SHGC coefficient decreases with added panes of glazing, tinting, or low-e (low-emittance) coatings. For example, SHGCs for various types of glazing are as follows:

Glazing	SHGC
1/8-in. clear, uncoated, single-pane	0.86
1/8-in. clear, uncoated, double-pane	0.76
1/8-in. gray, uncoated, double-pane	0.60
1/8-in. gray, low-e, double-pane	0.54

It is best to obtain the U-Value and SHGC for the various types of fenestration proposed for a project from the fenestration manufacturers' product data because these values vary considerably from one product to another and from one manufacturer to another. It is also necessary to determine if the window frames are thermally broken from the walls in which they installed, that is, if the frame is insulated from the wall. This can be determined by reviewing the details of the various window types in the architectural drawings for the building.

Buildings will sometimes incorporate elements that provide an external shading of the fenestration components, such as a roof overhang or shading above the top or along the sides of windows. The dimensions and positions relative to the fenestration components of all external shades must be entered into the description of each type of fenestration because they will have a significant impact on the percentage of the sun's radiant energy that is transmitted through the fenestration to the building's spaces. Internal shades such as drapes or venetian blinds may also be used, although it is recommended that they be omitted from the description of the fenestration types because their use will vary from one space to another. Omitting internal shades from the description of the fenestration types will result in a more conservative calculation of the cooling load for the affected spaces and will provide a bit of a safety factor for the cooling airflow calculation if the internal shades are actually used within a particular space.

Unconditioned Spaces

Once the components of the building envelope have been defined, the unconditioned spaces within the building should be determined so that the partitions (walls, floors, or ceilings) separating the conditioned spaces from the unconditioned spaces can be identified, their U-Values calculated, and the partition areas (square feet) entered into the HVAC load calculation. Although an unconditioned space is referred to as unconditioned, it can also mean that the space is partially conditioned to a certain degree. For example, equipment rooms in buildings are usually heated to at least 60°F in the winter and are ventilated during the summer with outdoor air. Therefore, the minimum space temperature of equipment rooms will typically be 60°F in the winter and the maximum space temperature will be 100 to 105°F in the summer. The variable temperature in the unconditioned space will need to be entered into the HVAC load calculation program so that the heat losses (in the winter) and heat gains (in the summer) through the partition wall are accounted for in the calculation of peak heating and cooling loads for the conditioned spaces adjacent to the unconditioned spaces.

Conditioned Spaces

Once the unconditioned spaces within the building have been determined, the remaining spaces in the building will be heated during the winter and cooled during the summer by the HVAC systems and their external and internal loads must be accounted for. Spaces having at least one component that separates the space from the outdoors are referred to as perimeter spaces. Perimeter spaces for open office areas can also be defined as the first 15 ft of space from the exterior walls. Spaces having no components that separate them from the outdoors are referred to as interior spaces. For perimeter spaces, the area and orientation (N, NE, E, SE, S, SW, W, NW) of all the exterior walls, windows, and doors will be entered into the space input menu of the HVAC load calculation program. The areas and slopes of the roofs and skylights also need to be entered into the space input menu.

Spaces with Floors at or Below Grade

If the floor elevation of the space is at or below grade, the exterior perimeter of the floor as well as the depth of the floor below grade must be entered into the space input menu. This will enable the HVAC load calculation program to account for the heat losses through the edges of the floor slab and walls below grade during the winter. It is common for floor slabs on grade to be insulated around the perimeter in order to provide a thermal break between the floor slab and the ground outdoors. The thickness and R-Value per inch of insulation of the floor slab perimeter insulation will need to be entered into the space input menu. Also, for floors below grade, it is common for the exterior wall to be insulated underground from the grade elevation to several feet below grade. Once again, the thickness, R-Value per inch, and depth of this insulation below grade will have to be entered into the space input menu. Perimeter floor slab insulation and insulation of walls below grade will be shown in the wall sections on the architectural drawings for the building.

Infiltration

An estimate of infiltration (air leakage through cracks in the building envelope) may be considered for perimeter spaces above grade. However, this is not necessary if the building will be sufficiently pressurized with outdoor air during occupied periods. Positive building pressurization results when the outdoor air introduced to the building through the air systems to provide ventilation exceeds the air that is mechanically exhausted from the building. This is the case for most buildings because it is undesirable for a building to be under negative pressurization. Negative pressurization can cause unfiltered, unconditioned outdoor air to be introduced directly to the spaces within the building, resulting in drafts and other undesirable conditions. However, it is recommended that infiltration be accounted for in spaces having exterior doors that will be opened frequently, such as vestibules. Outdoor air infiltration through exterior doors should be estimated according to the procedure described in Chap. 16 of the 2009 *ASHRAE Handbook—Fundamentals*. For an exterior door that serves approximately 75 people per hour, the outdoor air infiltration can be estimated at about 50 cfm per 3-ft × 7-ft door, based on a 0.10-in. water column (w.c.) pressure difference across the door.

Internal Loads

For all spaces (perimeter and interior) within the building, it is necessary to account for all the internal loads, including heat gains from people, lighting, and equipment. The number of people who will occupy a space can generally be obtained from the furniture plan that is part of the architectural drawings for the project. If this information is not available, each space will have to be categorized as to its use (i.e., office, conference room, auditorium, etc.) and the minimum number of people assigned to each space in accordance with the applicable mechanical code. Table 6-1 in *ANSI/ASHRAE Standard 62.1-2007—Ventilation for Acceptable Indoor Air Quality*, which also lists the default occupant density (number of people per 1,000 ft²) for various occupancy categories, can be referred to if the actual occupant density is unknown. Next, the rate of heat gain per person (both sensible and latent heat gains) needs to be assigned for each space. The rate of heat gain per person for various degrees of activity is given in Chap. 18, Table 1, of the 2009 *ASHRAE Handbook—Fundamentals*.

Lighting power densities (Chap. 18, Table 2, 2009 *ASHRAE Handbook—Fundamentals*) for various space types can be used for preliminary HVAC load calculations. However, it is recommended that the actual lighting power for each space be obtained from the lighting plan that is part of the electrical drawings for the project. The quantity of lighting fixtures and power (watts) per fixture should be tabulated for each space and this information entered into the space input menu for the final HVAC load calculations.

Finally, heat gains from equipment used in each space need to be accounted for. It is best to request a list of equipment that will be used in each space from the building owner and coordinate this information with the architect. Some commonsense estimates can be made for typical spaces, such as offices, where the HVAC system designer would estimate that there will be one computer per occupant. Heat gains from common equipment and appliances are listed in various tables in Chap. 18 of the 2009 *ASHRAE Handbook—Fundamentals*. However, it is best to obtain the manufacturer's product data for large pieces of equipment in order to accurately estimate the heat gain from this equipment. Also, a diversity factor, which represents the percentage of time the equipment is actually operating, should be applied to equipment that does not run continuously, like copiers or printers. The tables in the 2009 *ASHRAE Handbook— Fundamentals* list the average power use in watts for common equipment and appliances. However, the HVAC system designer will have to determine a diversity factor for equipment and appliances that do not fall into the listed categories. The best way to do this is through a discussion with the building owner and architect.

Schedules

Schedules are used in HVAC load calculation programs to vary the percentage (on an hourly basis) of an internal load's heat gain to the space. For example, a schedule that has 100% values for the hours of 7:00 a.m. to 5:00 p.m. and 0% values for the hours of 6:00 p.m. to 7:00 a.m. may be used to vary the lighting heat gain for an office space. Schedules are also used to identify periods of the day when the space temperature will be maintained at the occupied setpoint or the unoccupied setpoint.

Zones

If an air system will serve multiple zones, as in the case of a VAV air handling unit that serves multiple VAV terminal units, each zone will need to be defined in the HVAC load program in terms of the spaces that it serves, occupied and unoccupied space temperature setpoints, and the type of terminal equipment that will be used.

Typical occupied space setpoints are 75°F for cooling and 70°F for heating. A typical unoccupied space setpoint for heating is 60°F. It is common for cooling not to be provided for commercial buildings during unoccupied periods. In this case, the temperature within the building will rise in proportion to the thermal energy stored within the building over the course of the day and the nighttime outdoor air temperature.

Baseboard radiators and VAV terminal units (with or without fans or heating coils) are examples of terminal equipment that may be used. Other information describing the terminal equipment will also need to be entered in the zone input menu as required by the HVAC load calculation software. The user's manual for the software should be consulted for detailed instructions.

Air Systems

Air systems in HVAC load calculations represent those systems that provide the necessary heating, cooling, and outdoor air ventilation required by the various zones and spaces that they serve. For each air system, it will be necessary to determine the air system type (i.e., constant air volume, VAV, etc.) and the various components of the air system that are relevant to the HVAC load calculations, such as the amount of outdoor air ventilation required, as well as specific information on the heating coils, cooling coils, supply fan, and other components.[3] The air system definition will also include the relevant information for the terminal equipment, such as finned-tube radiators and the VAV terminal units for VAV air systems. The heating and cooling supply air temperatures must also be determined so that the heating and cooling airflows can be calculated based on the heating and cooling loads of the areas served by each air system. A common heating supply air temperature for most occupancies is 85°F, which is 15°F above a normal space heating setpoint of 70°F. If the supply air used for heating is cooler than 85°F, the spaces will feel drafty. If the supply air used for heating is warmer than 85°F, the heating air supplied to the spaces will stratify, meaning that it will stay near the ceiling and not reach the occupied zone, which is from 0 to 6 ft above the floor.

A common cooling supply air temperature for most occupancies is 55 to 58°F. Cooling air in this temperature range has a low enough dew point[4] to achieve a space-relative humidity for most occupancies that is between 40 and 60% at a normal space cooling setpoint of 75°F.

Central Plant

Finally, it will be necessary to define the central plant equipment for each building. This is the equipment that provides the necessary heating and cooling energy required by all of the air systems and terminal equipment. For the purposes of determining the peak heating and cooling load of the building, it is not necessary to describe the central plant equipment in detail. Rather, the generic heating and cooling plants that are available within the HVAC load calculation program should be used instead. The HVAC load calculation program will sum the heating and cooling loads for each air system and piece of terminal equipment every hour of the year and will identify the peak heating and cooling loads and the month and hour that these peak loads occur. These peak heating and cooling loads, plus any additional capacity for future expansion, will be used to size the central plant heating and cooling equipment.

Codes and Standards

Codes

Many counties, cities, and towns have building codes that govern the construction of buildings within their jurisdictions. For smaller cities, towns, and counties that do not have building codes, the state's building codes apply to construction within these areas. The codes that are in effect are enforced by the local and state code officials who are frequently referred to as the authorities having jurisdiction (AHJs). The AHJs enforce the codes that are in effect for their jurisdictions through the permitting process.

The permitting process starts when the contractor, building owner, or member of the design team completes an application for a building permit and submits the application along with a fee and the required number of complete sets of drawings to the permitting department of the local or state jurisdiction where the project is located. These plans have to be stamped and signed by the architect and professional engineers on the design team who are registered in the state in which the project is located. The plans are reviewed by plan reviewers for each discipline; reviewers frequently issue a list of comments to the design team for their responses. Sometimes the building permit is granted pending the satisfactory responses of the design team; sometimes the permit is withheld until the responses are incorporated into revised plans, which are resubmitted to the permitting department.

Once the building permit has been issued, the trade permits can be applied for by the project mechanical, plumbing, fire protection, and electrical contractors. During construction, the inspectors for the local or state jurisdiction will perform rough-in and final inspections for the work of the various trades, citing concerns that need to be resolved. Upon resolution of the concerns, the inspectors will issue their approvals for the various trades at these project milestones. Once all final inspections are complete and the project has been approved by the state and local code officials, a use and occupancy permit is granted for the project and the owner is permitted to use and occupy the premises.

Since a project must be designed in accordance with the applicable codes, each member of the design team must know what codes apply for each project. The codes that apply to building construction within a local or state jurisdiction can be obtained by contacting the permit department of the appropriate jurisdiction.

In most cases, the local or state jurisdictions will adopt a published building code, mechanical code, plumbing code, fire prevention code, electric code, and other codes by reference and will write amendments to each code, the sum of which represents the codes for the local or state jurisdiction. Examples of published codes are the *International Mechanical Code* (IMC) published by the International Code Council, Inc. (ICC), which would apply to the HVAC systems, and the *National Electric Code* (NEC), published by the National Fire Protection Association (NFPA), which would apply to the electrical systems. Codes are normally updated every 3 years.

Standards

Standards are published by professional societies and associations as guidelines for the design of various building systems, with the goal of protecting the safety and health of building occupants. Standards are not codes and are therefore not enforced by the AHJs. However, the recommendations of standards are frequently incorporated into published codes and are thereby indirectly enforced by the AHJs as code requirements. An example is *ASHRAE Standard 62.1-2007—Ventilation for Acceptable Indoor Air Quality*. The recommendations of this standard have largely been incorporated into the 2009 version of the IMC. In other words, standards usually take the lead on issues related to safety and health and are later incorporated into the applicable codes. It is prudent for the HVAC system designer to not only be familiar with the applicable codes but also with the latest standards.

System Selection

The types of HVAC systems that will be utilized for projects will normally be determined by the senior HVAC engineer. However, it is helpful for the junior HVAC system designer to be familiar with some of the factors that govern proper HVAC system selection.

A leading factor in determining the HVAC systems for a project is whether the project is a renovation within an existing building where some of the existing HVAC systems will be retained or a new project where the HVAC systems will be designed from scratch. For a renovation project, the HVAC system options are usually limited because the new portions of the HVAC systems will often have to coordinate with the existing HVAC system components that will remain. For example, if the project consists of the renovation of a single floor within a 20-year-old, three-story office building where each floor is conditioned with a VAV air handling unit serving fan-powered VAV terminal units with hot water heating coils, most likely the air handling unit and all associated ductwork, piping, and VAV terminal units on that floor will be replaced. Furthermore, if the scope of the project is limited to that one floor, the existing heating and cooling plant, which (in this example) consists of hot water boilers, chillers, and pumping systems, would remain. Therefore, the HVAC system selection for this project would consist of a replacement of the HVAC systems serving that floor with similar systems. The selection of the heating and cooling coils within the VAV air handling unit serving the floor would have to coordinate with the operating parameters of the existing central heating and cooling plant that is to remain. If the central heating and cooling plant is designed to deliver 180°F heating water with a 20°F delta T (difference between the heating water supply temperature from the heating plant and the heating water return temperature to the heating plant) and deliver 45°F chilled water with a 10°F delta T (difference between the chilled water supply temperature from the cooling plant and the chilled water return temperature to the cooling plant), the hot water heating and chilled water cooling coils within the replacement air handling unit would have to be selected based on these criteria.

Although it is typical for the HVAC systems serving renovated areas of a building to be similar to the existing HVAC systems, there will be times when it is desirable to change the HVAC systems, especially when the usage of the project area has changed from its original use or if the existing HVAC system is obsolete or inefficient.

For new projects, the HVAC system options are more numerous than they are for renovation projects. The final HVAC system selection will depend upon a number of factors that should be discussed with the building owner and architect. The pros and cons of each option should be explored before finalizing the HVAC system selection. Some of the factors that affect the HVAC system selection include:

- The use and occupancy schedule of the various spaces within the building
- The building owner's preferences or standards
- The building owner's budget
- The presence or absence of a central heating and cooling plant to serve the HVAC systems
- Installed costs of the HVAC system options
- Energy efficiencies of the HVAC system options
- Maintenance costs of the HVAC system options
- Type of building construction
- Available interstitial space above ceilings
- Flexibility of the HVAC system options for future expansion
- Other project-specific requirements

Design Team Members

Before we begin a discussion of the design submissions, it is necessary to describe the various disciplines that are involved in the building design.

First, the architect for the project is usually the design team leader. In addition to designing the building envelope and core interior components that are essential to the building, the architect is responsible for coordinating all of the design disciplines, as well as being the main design team contact with the building owner. Starting from the outside and working to the interior of the building, the other disciplines that are commonly involved in a building design are:

- Civil Engineer: Responsible for the design of the site on which the building will be located, which includes the design of the grading, paving, and utilities.
- Environmental Engineer: May be required if there are hazardous materials located on the site or if it is necessary to remediate wetlands or deal with other environmentally sensitive issues on the project site.

- Landscape Architect: Responsible for designing the plantings, walkways, irrigation, and water features on the project site.
- Architect: Responsible for designing the building envelope and core interior components.
- Structural Engineer: Responsible for designing the foundations and major structural elements of the building, including floor slabs, roofs, columns (vertical members), and beams or joists (horizontal or angled members).
- Interior Designer: Responsible for designing the interior partitions (nonload-bearing walls), doors, and ceilings in the noncore areas and for specifying the finishes for the interior components of the building such as the walls, ceilings, floors, and doors. The interior designer is also responsible for designing the furniture, fixtures, and major equipment that will be utilized within the building.
- Mechanical, Electrical, and Plumbing (MEP) Engineer: Responsible for designing the HVAC, electrical power and lighting, and plumbing systems within the building.
- Fire Protection (FP) Engineer: May be involved in the design of the sprinkler and fire alarm systems within the building. Other times, the MEP engineer will design the major components of the sprinkler system and provide a specification for the sprinkler system, which will be designed and built by the sprinkler contractor. It is also common for the MEP engineer to design the fire alarm system for the building.
- Conveying Systems Engineer: Responsible for designing the conveying systems (elevators, dumb waiters, and escalators) in the building.
- Other Consultants: May be part of the design team for certain building projects and will design specialized systems such as audiovisual (AV), security, telephone, and data systems.

Design Submissions

Schematic Design

The schematic design (SD) is normally the first design submission for medium- and large-sized projects (projects larger than 10,000 ft^2) where the design team will issue preliminary design submissions prior to the final design submission. The SD submission represents an approximately 15 to 25% level of completion for the project. The purpose of the SD submission is to present the preliminary architectural and engineering design concepts for the building to the building owner. The SD submission will normally include preliminary architectural floor plans[5] and narrative descriptions of the proposed building systems and site improvements. It is customary for the SD submission to be followed by a design meeting where the design team discusses the major elements of the building design with the building owner to refine the design concepts and meet the owner's needs.

For the HVAC system design, the elements described in the following paragraphs should be included in the SD submission.

Block HVAC Load Calculations

Block HVAC load calculations are those in which the building is subdivided into areas that will be served by individual air systems. For each air system, the areas served will be considered as one large space, even though these areas will be further subdivided into more spaces and zones later in the development of the design. The heating, cooling, and outdoor air ventilation requirements for each air system are determined from the block load calculations. If the air systems will be served by a central heating and cooling plant, the heating and cooling capacity of the central plant equipment can also be determined by the HVAC load calculation software from the block HVAC load calculations.

HVAC System Options

In the SD submission, several HVAC system options should be presented based on the block HVAC load calculations and the other requirements of the project. A brief description of each system's advantages and disadvantages should also be presented in order to provide the building owner with sufficient information to make an informed decision on the HVAC system approach for the project. The HVAC system designer should keep in mind that all but one of the HVAC system options will be discarded, so a minimal amount of effort should be expended explaining each option. The chosen HVAC system option will be further developed in the subsequent design submissions.

Single-Line Layout of the Recommended HVAC System

Sometimes a single-line layout of the recommended HVAC system option is helpful to illustrate the HVAC system components for the owner and architect, enabling them to begin to conceptualize the major components of the likely HVAC system configuration. With a single-line HVAC layout, the central HVAC equipment is shown on the architectural floor plans as rectangular blocks having the approximate dimensions of the actual equipment. The equipment maintenance clearances and space required for ductwork and piping connections are also allowed for in the layout in order to give the architect a rough approximation of the size of the equipment rooms that may be required to house the indoor HVAC equipment. Approximate locations and sizes of the ductwork and piping mains should be shown on the architectural floor plans to illustrate the coordination that will be required between the HVAC systems and the other components of the building design. Outdoor HVAC equipment, located either on the building roof or on grade, should also be shown on the single-line HVAC drawings.

Design Development

Once the HVAC system configuration has been selected based on the various options presented in the SD submission, the project will move into the design development (DD) phase where the HVAC system design will be further developed. The DD submission represents an approximately 35 to 50% level of completion for the project. The purpose of the DD submission is to present a well-developed and coordinated design concept for the building to the building owner.

Zone-by-Zone HVAC Load Calculations

During the DD phase, the block HVAC calculations begun during the SD phase are further developed by subdividing the areas served by each air system into (single-space) zones even though these zones will be further subdivided into more spaces and possibly more zones later in the final phase of the design. Space-by-space HVAC load calculations are not appropriate at this phase of the design because it is likely that the architectural design will change. Also, the level of accuracy that is achieved through space-by-space HVAC calculations is not required at the DD phase.

Preliminary Equipment Selections

Preliminary selections of major equipment should be made for the selected HVAC systems based on the zone-by-zone HVAC load calculations. These equipment selections are used to provide preliminary information to the architect for space requirements, electrical engineer for the electrical loads, and structural engineer for the loading on the building structural elements.

Preliminary Sizing of Equipment Rooms

Once the major HVAC equipment has been selected, this information is used to prepare preliminary floor plans of the equipment rooms. The purpose of the preliminary equipment room floor plans is to provide the architect with an estimate of the floor space required for the equipment rooms within the building. Preferred areas within the building for the equipment rooms should also be communicated to the architect so that these areas can be worked into the architect's overall building design.

At this stage in design development, the HVAC equipment should still be represented by rectangular blocks of the approximate size of the equipment. The areas surrounding the HVAC equipment required for proper maintenance, ductwork and piping connections, and equipment replacement should be identified as well. Other equipment within the equipment rooms, such as electrical panelboards, should be identified so that adequate space is allowed for all equipment within the equipment rooms. Major building structural components, such as building columns, and vertical shafts, should be identified, and the major pieces of HVAC equipment should be coordinated with these elements. Finally, the locations and sizes of the doors for the equipment rooms should be shown so that the architect can coordinate the layout of the spaces surrounding the equipment rooms. The architect will determine the direction that the doors must swing based on the requirements of the building code. If they must swing into the equipment room, sufficient floor space must be allowed within the room for the door swings. Figure 2-2 shows a sample DD-level equipment room layout.

Obviously, at this stage in the design, some compromise is required between the architect and the HVAC system designer. The HVAC system designer should always keep in mind that the HVAC systems for a building, though essential, are supplemental to the building's main function. From the perspective of the building owner and architect, maximizing the useable space within a building and incorporating the desired architectural features are the most important issues. However, these goals should not be achieved by designing poorly functioning HVAC systems that are not allowed sufficient space for maintenance and equipment replacement.

At the DD level of the HVAC system design, the HVAC system designer should be somewhat conservative in estimating the space requirements for the equipment rooms. It is much easier to give space back to the architect than it is to gain more mechanical equipment space after the architect has laid out the areas surrounding the equipment rooms and the owner has agreed to the amount of mechanical space that will be allotted.

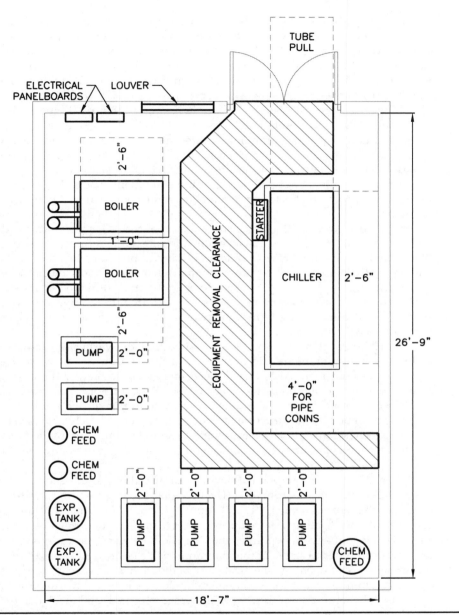

FIGURE 2-2   DD-level equipment room layout.

Preliminary Sizing of Vertical Shafts

In addition to the floor space required for equipment rooms, the architect and structural engineer will also need to know the sizes and locations of vertical shafts that will be required within the building. Vertical shafts are designed within buildings to allow the vertical ductwork and piping elements (referred to as risers) to be routed from one floor to the next. Vertical shafts are also required for the plumbing and electrical system risers. Plumbing system risers include sanitary, vent, storm water, and domestic cold and hot water piping. Electrical system risers include conduits that enclose the electrical power and telephone/data systems that are required to be routed between floors. Sometimes the plumbing and electrical system risers will be routed within the mechanical shafts that enclose the ductwork and piping systems. However, it is common for the plumbing and electrical system risers to be routed within dedicated shafts near the toilet rooms, electrical closets, and telephone/data closets on each floor.

Mechanical shafts are commonly located near the equipment rooms because the ductwork and piping within these shafts often have their source within the equipment rooms. Therefore, the sizes and locations of mechanical shafts need to be developed concurrent with the layout and sizing of the equipment rooms. The geometry of the vertical shafts will be affected by the architectural floor plan and building structural elements.

In the same way that floor space for equipment rooms is reluctantly yielded by the architect, so is the floor space that is required for the mechanical shafts. For this reason, it is incumbent upon the HVAC system designer to provide a layout of the ductwork and piping that will be located within the shafts in order to justify the need for the shaft space. This ductwork and piping needs to be configured within the shafts in a way that enables the ductwork and piping to enter the shaft from the equipment room (or other source) and also enables the branch ductwork and piping on each floor to be connected to the risers without conflict.

The HVAC system designer must realize that the vertical structural members (columns) of a building are laid out on a grid that coordinates with the needs of the architectural floor plan. Also, it is difficult and costly to modify the basic structural column grid to accommodate the needs of the HVAC or other building systems. It is much more reasonable for the building HVAC systems to coordinate with the structural columns designed by the structural engineer. However, the architect normally has the ability to reconfigure various rooms within the building core to provide floor space for shafts if this need is clearly communicated by the HVAC system designer early in the design process.

Figure 2-3 shows a sample shaft layout that would be provided to the architect and structural engineer for coordination. The floor plan layout shows the 24×10 exhaust air duct, 24×32 return air duct, 14×30 supply air duct, 4-in. chilled water supply and return, and 3-in. heating water supply and return risers that must be located in the shaft. This layout also incorporates the clearances between the various risers and the shaft walls in order to arrive at the minimum inside clear dimensions of the shaft and its associated floor opening. The inside face of the shaft wall will coincide with the floor opening. A minimum of 6 in. of clearance should be allowed between piping and ductwork risers to accommodate ductwork flanges (if required) and insulation on the ductwork and piping. A minimum clearance of 3 in. should be allowed between the piping and ductwork risers and the inside face of the shaft wall to accommodate the ductwork flanges and insulation. It may also be desirable to allow some additional space within the shaft for future services. The amount of space allowed for future services will depend upon the current and future finish-out plans for the building and should be discussed with the building owner and architect. The architect and structural engineer need to know the dimensions of the floor slab openings on each floor in order to accommodate the required mechanical shafts into the architectural and structural drawings.

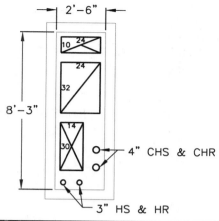

FIGURE 2-3 DD-level plan of vertical shaft.

Preliminary Space Required Above Ceilings

Equally as important as communicating the floor space requirements for equipment rooms and shafts to the architect and structural engineer is communicating the vertical clear space required above the ceilings (and below the horizontal structural members) to conceal the HVAC ductwork, piping, and equipment that will be located above the ceilings. The best way to communicate this need is to prepare partial floor plans and sections[6] in various locations of the building that show the preliminary dimensions and layout of the HVAC systems at each location. In addition to showing the HVAC systems on the sections, allowances should be shown for other building systems, such as conduits, sprinkler piping, and plumbing piping, in order to give the architect and structural engineer a reasonable understanding of systems that need to be concealed above the ceilings. The clear space requirements above ceilings are typically the greatest near the interior core area of each floor where the ductwork, piping, and electrical distribution systems are the largest. Because the need for clear space above the ceilings differs based upon the location on the floors, the HVAC system designer should select representative areas near the building core and building perimeter that will require the greatest clear space and closest coordination between disciplines when developing the sections.

When preparing the section drawings, it is easiest to think of the vertical space above the ceiling in terms of planes. For example, the first 6 to 8 in. above the finished ceiling[7] should be considered as the lighting plane if recessed lighting fixtures are used. Above the lighting plane will be the sprinkler piping plane, which requires between 2 and 4 in. of vertical space. Above the lighting and sprinkler planes will be the HVAC and plumbing systems plane. Most HVAC systems that are concealed above ceilings can fit within 12 to 18 in. of vertical space above the lighting and sprinkler planes. The plumbing systems can usually be coordinated within the HVAC systems plane as well, but it is necessary to coordinate the pitched horizontal components of the sanitary piping systems with the HVAC systems. Steam, steam condensate, and air-conditioning condensate drainage systems also have pitched piping, which requires additional coordination because the elevation of this piping changes along the piping run. Therefore, the vertical clear space from the finished ceiling to the underside of the building structure required to accommodate the HVAC, plumbing, sprinkler, and electrical systems

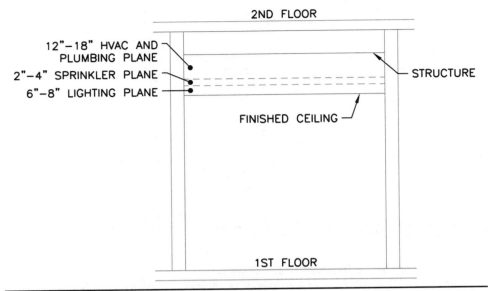

FIGURE **2-4** Coordination of the space above ceilings.

will normally be between 20 and 30 in. Figure 2-4 illustrates the coordination of the space above ceilings.

In general, all work above the ceilings must fit between the ceiling and the underside of the building structure. The building structure can be constructed of steel or concrete. In either case, the structural engineer may, on rare occasions, design holes within the webs of wide flange steel beams[8] or design framed openings in concrete beams to allow ductwork, piping, or other services to pass through the beams. However, if given notice early in the design process, the structural engineer can usually design some beams to have a lower height, which will allow more clear space below these beams for the building systems that will be concealed above the ceilings. Or, the structural engineer may choose to use bar joists[9] in lieu of beams in certain areas to make it possible for small pipes and ducts to be routed through the web of the bar joists if necessary. Clear communication early in the design process is key to obtaining the necessary clear space above the ceilings for the HVAC and other building systems.

Single-Line Layout of HVAC Systems

Once the equipment rooms have been laid out, the vertical shafts have been located, and HVAC plans and sections of various areas of the floors have been developed, a single-line layout of the HVAC systems serving the occupied spaces should be developed for the DD submission. Sizes of the main ducts and the locations of the terminal HVAC equipment should be shown on the single-line HVAC layout, as well as the locations of the supply, return, and exhaust air devices (air inlets and outlets) in the ceilings. All of this information is helpful not only to the architect, structural engineer, and electrical engineer who must coordinate the design of the ceilings, structure, and lighting systems with the HVAC systems, but also for the HVAC system designer. Developing a

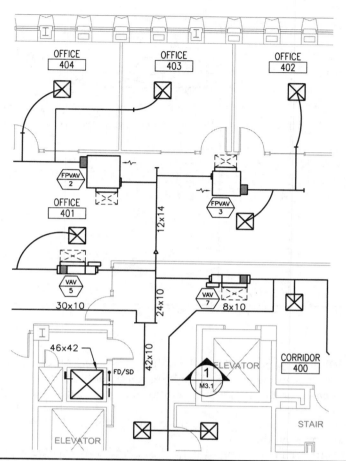

FIGURE 2-5   DD-level single-line HVAC layout.

single-line HVAC layout will help the HVAC system designer identify the logical routes for the main ducts and HVAC piping systems, identify areas where ducts and pipes need to cross over one another, and identify locations where close coordination with the architectural and structural components will be required. Figures 2-5 and 2-6 illustrate a DD-level single-line HVAC layout and a DD-level section, respectively. The section shows the combination fire/smoke damper that is required in the shaft wall and the associated access door, the 42 × 10 supply air duct, the 8 × 10 supply air duct, as well as the lighting, sprinkler, finished ceiling, structure, elevation above the finished floor to the finished ceiling, and the clear space required between the finished ceiling and the underside of the structure.

Preliminary Electrical Loads
At the DD level of the HVAC system design, it is necessary to give the electrical engineer an estimate of the electrical loads for the major pieces of HVAC equipment. The electrical load of HVAC equipment is measured in terms of power, which has units of

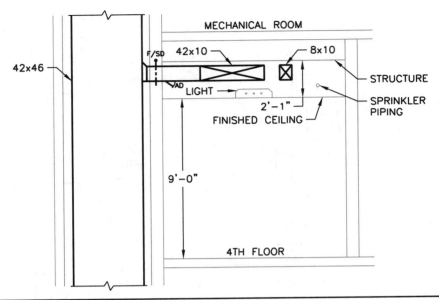

Figure 2-6 DD-level section.

watts (W) or horsepower (hp). The voltage that the HVAC equipment will utilize is very important to the electrical engineer and should be determined by the electrical engineer based on the design of the electrical power distribution system in the building and the available voltages that the HVAC equipment can utilize. The available voltages that the HVAC equipment can utilize can be obtained from the HVAC equipment manufacturer's product data.

For commercial buildings, the local electrical utility will normally provide a three-phase electrical service. From this three-phase service, single-phase power is derived within the building, and both three-phase and single-phase power is distributed throughout the building for use by the HVAC equipment, lighting, receptacles, and other equipment. Generally, three-phase power is used by large equipment and motors that are ½ hp and larger. Single-phase power is used for receptacles, lighting, small equipment, and motors that are smaller than ½ hp.

A list of the electrical loads for the major pieces of HVAC equipment can be used by the electrical engineer to add these loads to the lighting, receptacle, and other equipment loads in the building and to calculate the connected electrical load for the building. The electrical engineer will then apply appropriate demand factors to the connected loads (depending upon the type of load) to arrive at an estimated electrical demand for the building. The electrical demand will be used to size the main electrical distribution equipment and will also be supplied to the electrical utility for sizing of the electrical service and transformer (if necessary), which supply electrical energy to the building. The electrical demand of the HVAC systems in a building is a significant percentage of the total electrical demand of the building. Therefore, it is important that the HVAC system designer give the electrical engineer an accurate estimate of the electrical load of the major pieces of HVAC equipment during the DD phase.

Preliminary Structural Loads

The structural engineer is responsible for designing the foundation, floors, roofs, and supporting structure of the building. Therefore, it is necessary to communicate the size and weight of major pieces of HVAC equipment to the structural engineer during the DD phase. For HVAC equipment that will be mounted on the building floors or roofs, the structural engineer will need to know the plan dimensions (or footprint) of the HVAC equipment, as well as its operating weight. The operating weight of HVAC equipment, given in the manufacturer's product data, represents the weight of the equipment during operation (e.g., for air handling units containing water coils, the operating weight would include the weight of the coils filled with water). For major pieces of HVAC equipment that will be suspended from the building structure, the structural engineer will need to know the weight of the equipment, as well as the number of points of connection to the building structure.

It is not necessary to provide the structural engineer with the weights, dimensions, and support points for minor pieces of HVAC equipment, such as small pumps or VAV terminal units. The structural engineer will account for these minor loads, as well as the weight of ductwork, piping, and other systems supported from the building structure, in the standard estimate of the collateral load on the building structure. The collateral load is a uniform dead load allowance (in addition to the dead weight of the building structure) assessed by the structural engineer to account for the weight of all of the MEP systems [i.e., ductwork, piping (HVAC, plumbing, and sprinkler), lighting, and conduits] and sometimes includes the suspended ceiling. Collateral loads are typically set between 10 and 30 pounds per square foot (psf) applied uniformly as an average area load. Minor point loads less than 200 pounds are typically accommodated in this average loading, but highly concentrated point loads (dense duct configuration, pipe risers, or cable trays, etc.) might require special loading consideration or locally increased collateral load allowances.

Preliminary Openings in the Building Envelope

Openings in the building envelope are required for a building's HVAC systems in order to provide outdoor air ventilation, exhaust, and relief air from the building for indoor air handling units. They may also be required for supply and return air ductwork associated with outdoor air handling units. The sizes and locations of these openings need to be coordinated with the architect and structural engineer. Also, the architect may have a preference for the locations of these openings because they will affect the building's appearance.

Wall Openings Louvers are commonly used to protect openings in the exterior building walls from water penetration. The performance of louvers differs based on type. For example, louvers with drainable blades are able to limit water penetration at higher free area velocities[10] than louvers without drainable blades. For this reason, the required sizes of louvers will differ based on the type of louver selected. However, a conservative method for sizing louvers is to limit the face velocity[11] for intake louvers to 250 feet per minute (fpm) and limit the face velocity for exhaust or relief louvers to 500 fpm. For example, the louver area required for 1,000 cubic feet per minute (cfm) of outdoor air intake would be: 1,000 cfm ÷ 250 fpm = 4 square feet (ft²). The louver area required for 1,000 cfm of relief or exhaust air would be: 1,000 cfm ÷ 500 fpm = 2 ft². Louvers are available in many standard sizes, and custom louvers can also be fabricated for special

Figure 2-7 Typical wall louver.

applications. The standard sizes and associated rough opening requirements for louvers can be obtained from the manufacturer's product data. In addition to the dimensions (width and height) of the rough openings required, the architect and structural engineer also need to know the elevations of the openings within the walls required for the louvers. This information is typically given as an elevation above grade (or above finished floor) to the head (top) of the louver. Figure 2-7 is a photo of a typical wall louver, and Fig. 2-8 lists the wall opening information that needs to be communicated to the architect and structural engineer.

Louvers are available in many different colors. Consequently, it is important to have the architect select the proper color for each louver. Louvers should also be specified with a bird screen, which is a ½-in. × ½-in. wire mesh, installed on the inside face of the louver to prevent birds from entering the building or duct systems through the louvers.

Roof Openings In the same way that openings in the exterior building walls need to be coordinated with the architect and structural engineer, so do openings in the building roofs. Openings in the building roofs may be required for outdoor air ventilation, exhaust, relief, or for supply and return air ductwork associated with roof-mounted air handling units. The architect and structural engineer need to show the sizes of the rough openings on their plans, and the structural engineer must design the appropriate structural reinforcement for the openings. Typically, the architect

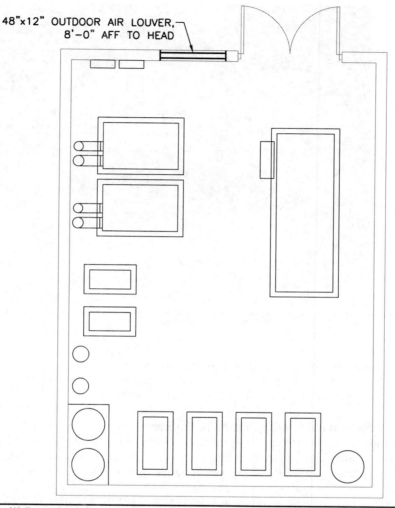

48"x12" OUTDOOR AIR LOUVER,
8'-0" AFF TO HEAD

FIGURE 2-8   Wall opening information for architectural/structural coordination.

will add notes to the architectural plans requiring the general contractor to coordinate with the mechanical contractor in order to provide properly sized roof openings for the HVAC systems. Openings in the building roof are commonly supported by structural members below the roof deck and are surrounded by roof curbs above the roof deck. Roof curbs provide a base for roof-mounted equipment and protect the openings from water penetration. Various types of equipment may be mounted on top of the roof curbs, including intake/exhaust/relief air hoods, ductwork goosenecks, air handling units, and exhaust fans. Figure 2-9 shows a typical roof-mounted hood, which can be used for outdoor air intake, exhaust, or relief air. Manufacturers' product data should be used for actual roof-mounted hood selections. However, a conservative method that can be used during the DD phase is to size the openings

FIGURE 2-9   Typical intake/exhaust/relief air hood.

required for roof-mounted intake/exhaust/relief air hoods so as to limit the throat velocity to 500 fpm. Figure 2-10 lists the roof opening information that needs to be communicated to the architect and structural engineer.

Coordination With Underground Site Utilities

Although it is more common for the plumbing engineer to coordinate with underground utilities located on the building site (i.e., sanitary, storm water, and domestic water services), the HVAC system designer may also have to coordinate the building HVAC systems with underground site utilities if the building is served by heating and cooling utilities whose source is located outside of the building in which the project is located. Frequently, multiple-building campuses such as universities or hospitals generate steam, heating water, and/or chilled water in a remote central heating and cooling plant and distribute these services to the various buildings on campus through underground piping systems. Also, some cities have companies that own central heating and cooling plants and offer heating and cooling utilities to building owners within the limits of their district heating and cooling systems.

Typically, the civil engineer for the project is responsible for designing all underground utilities on site. The line of demarcation between the underground site utilities and building systems is usually 5 ft outside of the exterior building wall. The HVAC system designer is responsible for all piping within the building to a point 5 ft outside of the exterior building wall; the civil engineer is responsible for all underground utilities beyond that point. For this reason, it is necessary for the HVAC system designer to provide the civil engineer with the size, location, and elevation for all HVAC utilities that will penetrate the exterior building walls below grade and continue underground on the project site.

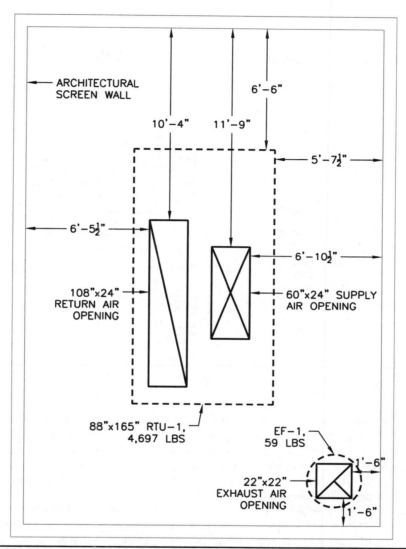

FIGURE 2-10   Roof opening information for architectural/structural coordination.

Because all elevations on the civil drawings are referenced to sea level, elevations of the HVAC utilities that penetrate the exterior building wall should also be provided to the civil engineer in terms of the elevation above sea level to the invert of each pipe. The invert of a pipe is the point on the inside surface of the bottom of the pipe. This elevation is referred to as the invert elevation of the pipe and is given in terms of feet above sea level with two decimal places of precision. A sample invert elevation for a pipe would be: 256.50′. All of this information should also be provided to the structural engineer because the design of the building foundation will have to be coordinated with the utilities below grade.

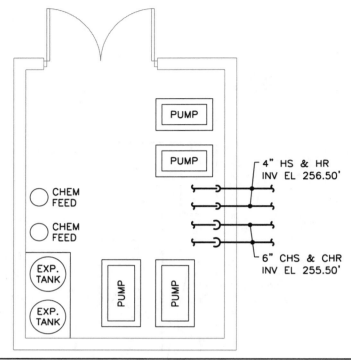

FIGURE 2-11   Coordination with site utilities.

It is necessary to provide this information to the civil and structural engineers at the DD phase in order for the civil engineer to coordinate the locations and elevations of the HVAC utilities with the other utilities on site and for the structural engineer to make any necessary adjustments to the foundation design, such as lowering the elevation of a wall footing, that may be required to accommodate the HVAC utilities. Figure 2-11 is an example of a DD-level floor plan that would be developed and sent to the civil and structural engineers for coordination with the site utilities. The floor plan shows the 4-in. heating water supply and return utilities and the 6-in. chilled water supply and return utilities that are required to penetrate the exterior wall of the building. Also given on the floor plan are the invert elevations of the heating and chilled water utilities at the penetrations of the exterior wall.

Construction Documents

The final phase of the design process is the construction documents (CD) phase. The CDs for a project consist of the final drawings and specifications for all disciplines involved in the project. The CDs are often used to solicit competitive bids from multiple contractors and are then used in the process of obtaining permits. Once all of the necessary permits have been obtained, the CDs are used by the various contractors to construct the building and all of its associated systems.

Although not covered in detail in this book, construction specifications are an integral part of the construction documents. They normally follow the Construction

Specifications Institute (CSI) MasterFormat®, which organizes the work of the various disciplines into 33 divisions. Each division is subdivided into sections, each of which describes different components of that division. Each section is further subdivided into three parts: Part 1—General, Part 2—Products, and Part 3—Execution. Part 1 of each section describes the general conditions associated with that section such as related documents, requirements for product submittals, and quality assurance requirements. Part 2 of each section describes the physical characteristics of the items described in the section, including construction materials, factory testing requirements, and acceptable manufacturers. Part 3 of each section describes the requirements for installation, start-up, and field testing. In the CSI MasterFormat, the HVAC systems are specified in Division 23—Heating, Ventilating, and Air-Conditioning. Each of the HVAC system components is specified in a separate section of Division 23. For example, indoor air handling units are specified in Section 237313—Modular Indoor Central-Station Air Handling Units.

The following is a brief discussion of the final design process and the various parts of the construction drawings. These topics will be further developed in later chapters of the book.

Final HVAC Load Calculations

Space-by-space HVAC calculations are required for the final HVAC system design. The heating and cooling loads of each space (or room), as well as the heating and cooling supply airflows, are required for the final design of the central HVAC equipment, terminal equipment, and ductwork and piping distribution systems.

Final Equipment Selections

Once the final HVAC load calculations are completed, final equipment selections are made by the HVAC system designer. For major pieces of HVAC equipment (such as air handling units, boilers, chillers), the manufacturers' equipment representatives for the equipment that forms the basis of design should be consulted. The manufacturers' representatives have various resources, including equipment selection software and direct contact with factory engineers, that enable them to make equipment selections accurately and efficiently. These representatives are usually very cooperative with consulting engineers because they want their equipment to be specified in the construction documents as the basis of design. For minor pieces of HVAC equipment (such as air devices, VAV terminal units, fan-coil units) the HVAC system designer can normally consult the manufacturers' product catalogs (and the associated equipment selection guidelines contained within them) to make appropriate selections.

Final Coordination With Other Disciplines

Prior to completing the construction documents, it is necessary for the HVAC system designer to communicate the final needs of the HVAC systems to the architect, electrical engineer, structural engineer, and civil engineer on the design team. This information should be communicated as early as possible in its final form so that these other disciplines can incorporate the needs of the HVAC systems into the final design of their systems. Final information should be communicated to the architect, structural engineer, and civil engineer in a similar fashion as the preliminary information was communicated in the DD phase. The structural engineer should review the MEP drawings to confirm that the collateral load allowances are appropriate for the MEP systems.

Company Name
Street Address
City, ST Zip

Revision		
No.	Date	Description

EQUIPMENT LIST

Project: Project
Project No.: Project No.
Date: Date
By: Initials

Status:
___ Concept
___ %
___ %
___ %
X Final

Rev	Item	Desig	Description	Location	hp	kW	FLA	Volt/Ø	Disc. Switch (M/E)	Disc. Cord & Plug (*)	Single-Point Elec (*)	Starter/Contactor Manual (M/E)	Starter/Contactor Automatic (M/E)	Starter/Contactor Reduced Volt. (M/E)	Starter/Contactor VFD (M/E)	Starter/Contactor 2-Speed (M/E)	Starter/Contactor None (*)	Standby (*)	Emergency Power (*)	Explosion Proof (*)	Outdoor (*)	Corrosive (*)	Smoke SA Detector (*)	Smoke RA Detector (*)	Smoke Dampers (Quan)	Emergency Off (*)	Cut Sheet Attached (*)	Notes
	1	AHU-1	Air Handling Unit	Mech Rm	10			208/3	E				E										*					
	2	ACCU-1	Air Cooled Cond. Unit	On-grade			120	208/3	M		*		M								*						*	1
	3	EHC-1	Elec. Heating Coil	Mech Rm		30		208/3	M																			1
	4	ATC	ATC Panel	Mech Rm		.5		120/1	E																			
	5	CUH-1	Cabinet Unit Heater	Lobby		3		208/1	M																			
	6	VAV-1	VAV Terminal Unit	See plans		4.5		208/3	M																			
	7	VAV-2	VAV Terminal Unit	See plans		3		208/3	M																			
	8	VAV-3	VAV Terminal Unit	See plans		7		208/3	M																			
	9	VAV-4	VAV Terminal Unit	See plans		7		208/3	M																			
	10	VAV-5	VAV Terminal Unit	See plans		4		208/3	M																			
	11	VAV-6	VAV Terminal Unit	See plans		2.5		208/3	M																			
	12	EF-1	Roof Exhaust Fan	Toilet Rooms	1			208/3	M				E															
	13	EF-2	Ceiling Exhaust Fan	Unisex Toilet	1/25			120/1	M			E																2

NOTES:
1. ACCU-1 and EHC-1 will not operate simultaneously.
2. Provide wall switch.

FIGURE 2-12 Sample equipment list for electrical coordination.

However, the electrical requirements of the HVAC equipment should be communicated to the electrical engineer through the use of an equipment list, manufacturers' cut sheets (product data sheets), and a highlighted set of HVAC plans that shows all of the HVAC equipment requiring an electrical connection. Figure 2-12 is a sample equipment list that can be used to communicate the electrical characteristics of the HVAC equipment, such as voltage, phase, kilowatts, full load amps, and horsepower, to the electrical engineer. The equipment list also identifies other electrical requirements of the HVAC equipment, such as disconnect switches and starters, and the discipline (mechanical or electrical) that is responsible for including these requirements in the construction documents. In the sample equipment list shown in Fig. 2-12, the disconnect switches for the exhaust fans and electric heating coils for the VAV terminal units will be specified by the HVAC system designer to be furnished with this equipment, whereas the starters for the air handling unit supply fan and roof exhaust fan will be designed by the electrical engineer as part of the electrical power distribution system. The equipment list can also be used for preliminary submissions. If this is done, it is important to keep track of the revisions to the original equipment list by identifying the modified items with the revision number in the revision column and maintaining a list of the revisions (with the associated dates and descriptions) in the revision block at the top of the equipment list.

Construction Drawings

The HVAC construction drawings generally consist of the following parts, listed in the order they will appear in the HVAC drawing set.[12]

Legend, Abbreviations, and General Notes   The legend, abbreviations, and general notes are usually shown on the first drawing in the HVAC drawing set. Or, for smaller projects, the legend, abbreviations, and general notes can be presented on the first HVAC floor plan drawing. The legend is a listing of the symbols used in the HVAC drawings, abbreviations are a listing of the abbreviations used in the HVAC drawings, and general notes describe the general requirements associated with the HVAC systems.

Floor Plans   The HVAC floor plans show the arrangement of the HVAC ductwork and piping distribution systems, as well as the HVAC equipment on each floor of the building. HVAC floor plans will be presented at the same scale as the architectural floor plans. A scale of 1/8 in. = 1 ft is commonly used by architects to present the floor plans for projects exceeding approximately 4,000 ft². Architects use this scale because 16,000 ft² of floor space can easily be presented on a 24-in. × 36-in. drawing at 1/8 in. = 1 ft, whereas only about 4,000 ft² of floor space can be presented on a 24-in. × 36-in. drawing at ¼ in. = 1 ft. Also, a scale of 1/8 in. = 1 ft is large enough to show the detail that is commonly required for the HVAC ductwork and piping distribution systems and the HVAC equipment. Areas requiring close coordination between disciplines, or where there is a significant amount of HVAC work, will be presented on large-scale plans (see below).

It is common for HVAC ductwork and HVAC equipment with a ductwork connection (such as VAV terminal units) to be presented on one set of floor plans and for HVAC piping (such as heating water piping) and HVAC equipment with HVAC piping connection (such as VAV terminal units with hot water heating coils or finned-tube radiators) to be shown on a separate set of floor plans. This is done where the HVAC ductwork, HVAC equipment, and HVAC piping, if presented on the same set of floor plans, would be cluttered and difficult to read.

Sections   Sections are usually presented at a ¼ in. = 1 ft scale. These sections are used to clarify the arrangement of HVAC systems in areas where close coordination between disciplines is required or where there is a significant amount of HVAC work involved that cannot be clearly conveyed through the use of floor plans alone, such as at the branch connections to risers in vertical shafts and selected areas within equipment rooms.

Large-Scale Plans   Similar to sections, large-scale plans are also presented at a ¼ in. = 1 ft scale and are used for areas where close coordination between disciplines is required or where there is a significant amount of HVAC work involved. Equipment rooms are almost always presented as large-scale plans because of the amount of HVAC work that must be shown and because of the close coordination between disciplines that is required. Because a larger scale is used, it is common for the HVAC ductwork, HVAC equipment, and HVAC piping to be presented on the same large-scale plan. However, if it would provide a clearer presentation, the HVAC ductwork and HVAC equipment can be shown on one large-scale plan while the HVAC piping and HVAC equipment is shown on a separate large-scale plan. Also, it is sometimes necessary to present the HVAC ductwork, HVAC equipment, and HVAC piping at different elevations within equipment rooms on separate large-scale plans when the equipment rooms are higher than one story of the building.

Details   Because it is not possible to show all of the details necessary for a proper installation of certain pieces of HVAC equipment on the floor plans or large-scale plans, it is necessary for the HVAC system designer to show this information in equipment connection details. These details will show all of the required ductwork and piping connections, as well as support requirements and miscellaneous appurtenances such as thermometers, pressure gauges, and flexible pipe connectors. Also, it is common for details to be developed that describe miscellaneous items associated with the HVAC systems such as pipe hangers, roof curbs, and penetrations through the building envelope.

Equipment Schedules   Equipment schedules are used to present the pertinent information associated with the HVAC equipment for a project in tabular format. The equipment schedules list the identification numbers, capacities, electrical characteristics, dimensions, weights, and manufacturers and model numbers that form the basis of design for the HVAC equipment. Separate schedules are provided for each type of HVAC equipment. For example, all of the pumps for a project will be presented in a pump schedule and all of the air handling units will be presented in an air handling unit schedule.

Diagrams   Diagrams of the automatic temperature control (ATC) systems associated with the HVAC systems for a project are necessary to show all of the components that are required to ensure that the HVAC systems operate in accordance with the design intent. A sequence of operation is developed for each ATC diagram to describe how the HVAC system is to perform in all of its modes of operation. A direct digital control (DDC) point list is also prepared for DDC ATC systems. Refer to Chap. 9 for a discussion of ATC systems and their associated diagrams and DDC point lists.

For steam, heating water, and chilled water systems, flow diagrams are commonly used to give an overview of the central plant for these systems. These diagrams show all of the HVAC equipment that is required, as well as the arrangement of this equipment within the systems. Pipe sizes, flow rates, and the HVAC equipment designations are shown on the flow diagrams. It is not necessary to show every equipment isolation valve, thermometer, pressure gauge, or other equipment appurtenance on the flow diagrams because this information is shown in the equipment connection details. However, it is helpful to show shutoff valves that are used to isolate major portions of each system. These valves will also be shown on the HVAC floor plans or large-scale plans.

Finally, riser diagrams are often employed for projects having three or more stories. These diagrams can be used for piping or ductwork systems to show the sizes of the piping or ductwork risers, flow rates, and the branch connections at each floor of the building.

Endnotes

1. A thermostat is an electrical device that performs an action (such as closing a contact) in response to a change in temperature. The term *thermostat* is used in a generic sense in the earlier chapters of this book to describe the device that controls the space temperature. However, modern space temperature control systems serving occupied areas commonly utilize electronic control systems, which would consist of a space temperature sensor (senses temperature only) that is connected to a controller. The controller is the device that actually performs the desired action based upon the input it receives from the space temperature sensor.

2. The building envelope consists of the exterior walls, windows, doors, roofs, and skylights, which enclose the building and separate the indoor environment from the outdoors.

3. The types of air systems are usually finalized after the schematic design submission. Prior to the final selection of the air systems, the simplest type of air system—constant air volume—can be used to perform the block load HVAC load calculations. Refer to the Design Submissions section in this chapter for more information on the schematic design submission and block load HVAC calculations.

4. Dew point is the temperature of moist air below which condensation will occur. The dew point temperature of air gives an indication of the total amount of moisture in the air. The higher the dew point, the more moisture is contained within the air and vice versa.

5. A floor plan is a drawing that depicts a view of a floor, or partial floor, of a building looking vertically downward.

6. A section is a drawing that depicts a view looking horizontally in a certain direction. Sections are identified on the floor plans by a number designation, a reference arrow depicting the direction of the view, and the drawing where the section is presented.

7. The finished ceiling is the underside of the ceiling that is exposed to the finished space.

8. Wide flange steel beams are structural members that are mounted horizontally to support the floors or flat roof of a building. Because they have a section that looks like the letter I, they are sometimes referred to as I-beams. The top and bottom (horizontal) portions of the beam are the flanges and the vertical portion of the beam, which joins the flanges, is the web. Roof support beams can also be mounted at an angle for sloped roofs.

9. Bar joists are composite members, also mounted horizontally or at an angle, that have an open web.

10. The free area velocity for a louver is the velocity of the air through the total open area (free area) of the louver.

11. The face velocity for a louver is the velocity of the air across the entire face of the louver, which includes the louver blade area and the louver free area.

12. A further discussion of construction drawings is given in Chap. 10.

CINCINNATI STATE LIBRARY

CHAPTER **3**

Piping, Valves, and Specialties

Piping systems are used to convey the heated and cooled fluids from the central plant to the air systems and terminal equipment where the energy from these fluids is used to heat and cool the building. Because piping systems are such an integral part of HVAC systems, a basic understanding of their fundamental components is necessary in order to properly design these systems.

Piping systems are used in industry for a variety of different purposes. Pipes, which range in diameter (pipe size) from very small to very large, convey different types of fluids, which exist over a wide range of temperatures and pressures. For this reason, we will limit our discussion of piping systems by pipe size, working fluid, fluid temperature, and fluid pressure. This will enable us to focus on the piping system components that are commonly used in HVAC systems for commercial buildings. The following is a breakdown of these characteristics:

1. Pipe size: Pipe sizes used in commercial HVAC systems normally range from ¾ to 12 in. Pipes larger than 12 in. are used for very large commercial buildings, industrial applications, and district heating and cooling systems, which are beyond the scope of this book.

2. Working fluids: Common working fluids are water, steam, brine,[1] and refrigerant.

3. Fluid temperature: Fluid temperatures range from as low as 15°F for ice-making brine systems to as high as 406°F for high-pressure steam systems [250 pounds per square inch gauge (psig)].[2] Common fluid temperature ranges for various HVAC piping systems are listed below:
   a. Cold water systems
      (1) Ice-making brine: 15–32°F
      (2) Chilled water: 35–65°F
      (3) Heat pump water: 50–100°F
      (4) Condenser water: 70–100°F
   b. Hot water systems
      (1) Heating water: 100–200°F
   c. Steam systems
      (1) Low-pressure (up to 15 psig) steam: 212–250°F
      (2) Medium-pressure (above 15 psig, up to 125 psig) steam: above 250°F, up to 352°F

(3) High-pressure (above 125 psig, up to 250 psig) steam: above 352°F, up to 406°F

4. Fluid pressure: Fluid pressures are normally below 125 psig for water, steam, and brine systems. However, pressures as high as 250 psig are possible for high-pressure steam systems, water and brine systems in high-rise buildings, and refrigerant systems. It is not desirable for water, steam, brine, or refrigerant piping systems to operate below atmospheric pressure (negative gauge pressure).

Piping systems comprise the following three basic parts, which will be discussed in this chapter:

- Pipe, fittings, and joints
- Valves
- Specialties such as meters and gauges

Symbols and abbreviations are used to represent the piping systems, valves, and specialties on the HVAC construction drawings. Refer to the sample legend and abbreviations (Figs. 10-1 and 10-2) in Chap. 10 for the piping system designations and symbols for the valves and specialties discussed in this chapter.

Pipe, Fittings, and Joints

The two most common materials for HVAC pipe are carbon steel and copper. Carbon-steel pipe is referred to as black steel pipe or simply steel pipe. Copper pipe is normally referred to as copper tube.

Pipe fittings, such as elbows and tees, are required at all turns and branch connections in piping systems.

Joints refer to the connections between pipe and fittings, unions, flanges, valves, specialties, and similar junctions. Joints in HVAC piping systems are normally threaded[3] or welded for steel pipe and soldered or brazed for copper tube. Joints consisting of grooved pipe and mechanical couplings are also used. However, this method of joining has not gained universal acceptance in the HVAC industry.

It should be noted that steel pipe and copper tube used in HVAC piping systems are not compatible for direct contact with each other.[4] When it is necessary for steel pipe and copper tube to connect in HVAC piping systems or for steel pipe to connect to copper connections on HVAC equipment (or vice versa), dielectric fittings are required to prevent direct contact of these two dissimilar metals. Dielectric fittings are normally constructed of ferrous material (for connection to the steel component) and copper alloy (for connection to the copper component), both of which are separated by a plastic component that prevents metal-to-metal contact.

Steel Pipe

Steel pipe is often used in HVAC piping systems because it is readily available, cost-effective, and durable. Steel pipe, manufactured in accordance with the American Society for Testing and Materials (ASTM International) specification A53, is available in two grades: Grade A and Grade B. Grade B steel pipe is commonly used in HVAC piping systems because it has a higher tensile strength than Grade A steel pipe.

Steel pipe is also available in two types: welded and seamless. Welded pipe is manufactured by rolling a sheet of steel into a cylinder and welding the longitudinal seam. The weld can be either furnace butt-welded (Type F) or electric-resistance-welded (Type E). Type F pipe, which has a lower allowable stress than Type E pipe, is suitable for HVAC applications where the pipe sizes are 2 in. and smaller. Type E pipe is typically used for pipe sizes from 2½ to 12 in. Seamless pipe (Type S) is manufactured by extruding and, therefore, has no longitudinal seam. As a result, Type S pipe has a higher allowable stress than welded pipe, but it is also more costly. Type S pipe is not required for commercial HVAC piping systems unless the project requires a higher degree of quality than normal.

Steel pipe is manufactured with different wall thicknesses, identified by schedule (or weight class). Schedule 40 steel pipe is used for most HVAC applications for pipe sizes up to 12 in. Schedule 80 steel pipe, which has a thicker wall than Schedule 40 steel pipe, is commonly used for steam condensate piping because the thicker wall provides some allowance for corrosion. Schedule 40 steel pipe and Schedule 80 steel pipe have the same outside diameter for corresponding pipe sizes. Steel pipe is sold in straight lengths.

Steel Pipe Fittings and Joints

Fittings for steel pipe used in HVAC piping systems are normally constructed of cast iron,[5] malleable iron,[6] or wrought (worked) steel.

Fittings, as well as unions, flanges, valves, and specialties, used in steel piping systems are rated for different working pressures, which are defined by class. Class, followed by a dimensionless number, refers to the maximum allowable working saturated steam gauge pressure measured in pounds per square inch (psi). Common classes for steel pipe components used in HVAC piping systems are: Class 125, Class 150, Class 250, and Class 300.

Cast iron is commonly used in the construction of Class 125 and Class 250 components. Cast iron components are normally constructed with threaded ends for connections to the adjoining pipe. Threaded joints are typically used in HVAC piping systems for steel pipe components that are 2 in. and smaller, such as the cast iron elbow shown in Fig. 3-1.

Malleable iron is used in the construction of Class 150 and Class 300 components. Malleable iron fittings are normally constructed with threaded ends and are also used for pipe sizes that are 2 in. and smaller. Malleable iron unions with threaded ends are commonly used at connections of steel piping (2 in. and smaller) to HVAC equipment where ease of disassembly is required (Fig. 3-2).

Welded joints are preferred for larger pipe sizes. Since cast iron and malleable iron are not suitable materials for welding, wrought steel fittings of Class 150 and Class 300 with butt welding (beveled) ends are commonly used for pipe sizes from 2½ to 12 in.

Flanges constructed of wrought steel with butt welding ends are commonly used at connections of steel piping (from 2½ to 12 in.) to HVAC equipment where ease of disassembly is required. Two flanges and a gasket[7] are required to make a flanged joint.

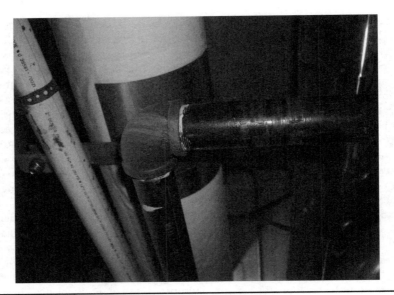

FIGURE 3-1   Photograph of a cast iron elbow with threaded ends.

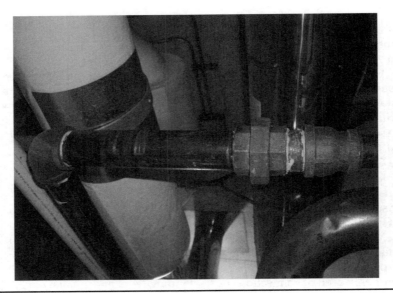

FIGURE 3-2   Photograph of a malleable iron union with threaded ends.

During installation, the gasket is placed between the two flanges, and through-bolts and nuts are tightened to a specified torque thus completing the joint. Pipe fittings and valves are also available with flanged ends for connection to the flanges welded to the ends of pipe. Figure 3-3 is a photograph of a wrought steel flange welded to the end of a steel pipe.

FIGURE 3-3 Photograph of a wrought steel flange with butt welding ends.

Steel pipe fittings, unions, and flanges are manufactured in accordance with specifications defined by the American Society of Mechanical Engineers (ASME) and ASTM. For example, Class 125 and Class 250 cast iron fittings with threaded ends are manufactured in accordance with ASME B16.4, and Class 150 and Class 300 wrought steel fittings with butt welding ends are manufactured in accordance with ASTM A234.

It is common for Class 125 and Class 150 components to be used in the same piping system, and for Class 250 and Class 300 components to be used in the same piping system. The maximum working pressure at any point in a piping system comprising Class 125 and Class 150 components should be 125 psig because the lowest rated component establishes the operating limits of the system. Similarly, 250 psig should be the maximum working pressure at any point in a piping system comprising Class 250 and Class 300 components. It is the responsibility of the HVAC system designer to determine the maximum working pressure in the HVAC piping system and specify the appropriate class of pipe fittings, unions, flanges, valves, and specialties.

Specification for Steel Pipe, Fittings, and Joints

The following are typical specifications for steel pipe, fittings, and joints for steam, water, or brine HVAC piping systems:

1. Piping 2 in. and smaller, above ground
   a. ASTM A53, Grade B, Type F Schedule 40 black steel pipe with Class 125 (or Class 250) cast iron fittings and threaded joints, or
   b. ASTM A53, Grade B, Type F Schedule 40 black steel pipe with Class 150 (or Class 300) malleable iron fittings and threaded joints.

2. Piping 2½ in. through 12 in., above ground
   a. ASTM A53, Grade B, Type E Schedule 40 black steel pipe with Class 150 (or Class 300) wrought steel fittings and welded joints.

Copper Tube

Copper tube is seamless and is manufactured first by extruding[8] a cast copper billet into a tube. The tube is then drawn[9] through a die in several stages to obtain the final tube diameter and wall thickness. Copper tube that is used to convey water or brine for HVAC piping systems is manufactured in accordance with ASTM specification B88. The copper tube is available in different wall thicknesses, identified by Types K, L, M, and DWV (drainage, waste, and vent), which designate descending wall thicknesses. All types of copper tube have the same outside diameter for corresponding pipe sizes.

Copper tube is often used for water or brine HVAC piping systems because of its inherent corrosion resistance and ease of installation. Copper tube can be used for low-pressure steam systems (15 psig or less); however, it is customary to specify steel pipe for steam service. Type L copper tube is usually specified as an option to steel pipe for HVAC piping systems that are 2 in. and smaller for aboveground installations.[10] Type K copper tube is used in HVAC piping systems that are installed below ground or within concrete slabs because it is more durable than Type L copper. Also, copper tube is more corrosion resistant than steel pipe, which is not recommended for direct-buried installations. Type M copper tube is used predominately in residential construction and is not recommended for use in commercial buildings. Type DWV copper tube has the thinnest wall and is suitable for nonpressurized piping systems, such as air-conditioning condensate drainage or plumbing drainage, waste, and vent piping systems.

Type ACR (air-conditioning and refrigeration) copper tube, which is manufactured by extruding and drawing, is also cleaned (to remove any traces of drawing lubricants or contaminants) and capped prior to shipping from the factory. Type ACR copper tube, which is used for refrigerant piping systems, is manufactured in accordance with ASTM B280 and has a different set of wall thicknesses than Types K, L, M, and DWV copper tube.

Copper tube is sold in either the hard-drawn or annealed (soft) state. Hard-drawn copper tube is in the rigid state, which follows the drawing process. Annealed copper tube undergoes a process of annealing[11] at the factory after the drawing process in order to soften the hard-drawn tube. Annealed copper tube differs in appearance from hard-drawn copper tube in that it has a matte surface finish.

For HVAC piping systems in commercial buildings, hard-drawn Type L copper tube is used above ground for water and brine systems due to its rigidity. Annealed Type K copper tube used below ground or within concrete slabs because it requires fewer joints than hard-drawn copper tube. Annealed Type ACR copper tube is generally used for refrigerant pipe sizes that are 1.5 in. and smaller; hard-drawn Type ACR copper tube is used for pipe sizes from 2 to 4 in. Refrigerant pipe sizes larger than 4 in. are rarely used for commercial HVAC applications.

Hard-drawn copper tube is usually sold in straight lengths, and annealed copper tube is usually sold in coils.

Copper Fittings and Joints

Fittings for copper tube used in HVAC piping systems are normally constructed of wrought copper, although cast copper and cast bronze fittings are also used for some applications. Joints for copper tube used in HVAC piping systems are normally soldered or brazed; however, threaded joints are also used for some applications such as equipment connections. Fittings for soldered or brazed joints are constructed with sockets that are sized to fit over the corresponding size of copper tube (Fig. 3-4). Sockets for soldered joints are deeper than the sockets for brazed joints.

Soldered joints use filler metals that have a melting point below 800°F. The filler metal used for soldered joints in copper tube for HVAC piping systems is an alloy of 95% tin and 5% antimony. Copper tube with soldered joints are acceptable for water or brine HVAC piping systems that have a maximum working pressure of 250 psig and are 2 in. and smaller because the joints are rated for a minimum of 250 psig at a service temperature of 200°F. Therefore, it is acceptable to use Type K or Type L copper tube (2 in. and smaller) with soldered joints in the same water or brine HVAC piping system that utilizes steel pipe (2½ through 12 in.) having Class 125, Class 150, or Class 250 fittings, valves, and specialties, provided the maximum temperature for the water or brine is 200°F.

Brazed joints, which use filler metals having a melting point above 800°F, are stronger than soldered joints. The filler metal used for brazed joints in copper tube that is used for water or brine HVAC piping systems is an alloy of copper and phosphorus (Type BCuP). The filler metal used for brazed joints in copper tube that will be used for refrigerant piping systems is a silver alloy (Type BAg1). Brazed joints are recommended for underground water or brine piping and refrigerant piping of all

FIGURE 3-4 Photograph of a wrought copper elbow with socket ends for soldering. (*Photo courtesy of NIBCO INC.*)

sizes because the rated pressure of the joints is equal to the working pressure of the copper tube for temperatures up to 200°F. Therefore, it is acceptable to use Type K or Type L copper tube of any size up through 12 in. with brazed joints in the same water or brine HVAC piping system that utilizes steel pipe having up to Class 300 fittings, valves, and specialties, provided the maximum temperature for the water or brine is 200°F.

Wrought copper unions with soldered, brazed, or threaded ends are commonly used at connections of copper tube to HVAC equipment where ease of disassembly is required.

Specification for Copper Tube, Fittings, and Joints

The following are typical specifications for copper tube, fittings, and joints for water or brine HVAC piping systems:

1. Piping 2 in. and smaller, above ground[12]
    a. ASTM B88, Type L, hard-drawn copper tube with wrought copper fittings and 95-5 tin/antimony soldered joints.
2. Piping 12 in. and smaller, below ground[13]
    a. ASTM B88, Type K, annealed copper tube with wrought copper fittings and BCuP brazed joints

The following are typical specifications for copper tube, fittings, and joints for refrigerant piping systems:

1. Piping 1.5 in. and smaller, above ground
    a. ASTM B280, Type ACR, annealed copper tube with wrought copper fittings and BAg1 brazed joints.
2. Piping 2 in. through 4 in., above ground
    a. ASTM B280, Type ACR, hard-drawn copper tube with wrought copper fittings and BAg1 brazed joints

Valves

Valves are required to perform various functions in HVAC piping systems, including shutoff, throttling (or balancing) of flow, prevention of flow reversal, pressure regulation, and pressure relief. Different types of valves are available to perform these functions, each with its own advantages and disadvantages depending upon the requirements for the valve and the operating conditions in which it will be installed. Furthermore, valves are constructed of various materials and are available with different ends for connection to the piping system. Once again, the proper selection of the construction material and end connections for the valve will depend upon the operating conditions and the type of pipe to which it will be connected. For example, the resilient seating material used to improve the shutoff capabilities of a valve also imposes temperature and fluid limitations on the valve's use. Valves are constructed in accordance with specifications defined by the Manufacturer's Standardization Society (MSS).

Functions

Shutoff

As the name implies, shutoff valves perform the duty of shutting off the flow in a piping system at the point where the valve is installed. Their position will be either fully opened or fully closed. Shutoff valves are commonly installed at HVAC equipment connections and are used to isolate the equipment for maintenance and replacement. Shutoff valves are also installed at major branches in HVAC piping systems, such as a branch serving the floor of a building. They are also installed in the service entrance piping where a building is connected to a district heating and cooling system.

Throttling

HVAC piping systems that circulate water or brine normally consist of main supply and return pipes that serve various branches in the piping system. The branches of the piping system are connected in a parallel arrangement, similar to the connection of components in a parallel electrical circuit. Pipe branches can serve a single piece of equipment, multiple pieces of equipment, or an entire floor of a building.

Each branch in the piping system is designed to receive a certain portion of the overall system flow. Therefore, it is necessary to throttle (or balance) the fluid flow in each pipe branch to the design flow for that branch. Valves that are suitable for balancing the flow of the working fluid are installed in the pipe branches for this purpose. Balancing valves are also installed in the discharge piping of pumps for the purpose of balancing the pumps to their design flow. The flow through a balancing valve is usually measured at a flow meter installed near the balancing valve. Refer to the Specialties section later in this chapter for a description of flow meters. Calibrated balancing valves can also be used to balance the flow through minor branches in the piping system (pipe sizes 2 in. and smaller). Refer to the Calibrated Balancing Valves section later.

Prevention of Flow Reversal

Valves used to prevent the reversal of flow in a piping system include check valves, double check assemblies, and reduced pressure zone backflow preventers (listed in the order of increasing level of reverse flow protection). Check valves are most commonly used in water, brine, and steam condensate systems because their purpose is solely to prevent reverse flow in the HVAC piping system itself—they are not required to prevent contamination of a potable water system by nonpotable water systems. Double check valve assemblies (two check valves connected in series) and reduced pressure zone backflow preventers have a higher level of protection against reverse flow and are typically used to protect potable water systems from contamination by nonpotable water systems, such as in the makeup water connections to hydronic and steam systems. Double check assemblies and reduced pressure zone backflow preventers are not used within steam piping systems; however, check valves are used in steam condensate piping systems and may also be used in the steam supply connections of multiple boilers to a common header.

Pressure Regulation

Pressure regulating valves perform the duty of maintaining a constant pressure of the working fluid at a location upstream or downstream of the valve or of maintaining a constant differential pressure between the inlet and outlet of the valve.

They accomplish this by controlling the position of the valve, thereby throttling the fluid flow to maintain the desired pressure.

Pressure Relief

Pressure relief valves are used as safety devices to prevent overpressurization of the components within HVAC piping systems.

Types

The following sections list some of the more common types of valves used in HVAC piping systems. For each valve type, a general description is provided along with a recommended application. The list is not exhaustive of all types of valves that are available, and the applications for the valves may be expanded based on the judgment of the design professional and the standard practices of a particular design firm.

Multi-turn Valves

Multi-turn valves require multiple rotations of the valve stem to change the position the valve from fully opened to fully closed and vice versa. Two types of multi-turn valves commonly used in HVAC piping systems are gate valves and globe valves.

Gate Valve

Description   Gate valves consist of a solid wedge mounted within the valve body, which is connected to the internal end of the valve stem. A handwheel actuator is connected to the external end of the valve stem for manual operation. When the valve stem is turned completely in the counterclockwise direction, the wedge is raised to the open position. When the valve stem is turned completely in the clockwise direction, the wedge is lowered to the closed position. Gate valves have a straight-through body configuration; that is, the flow exits the valve in the same direction that it enters the valve. Gate valves are intended to be fully opened or fully closed; they are not suitable for throttling (i.e., functioning in a partially open position). When in the closed position, the metal-to-metal contact of the wedge within the seat does not provide absolute (bubble-tight) shutoff. Bubble-tight shutoff can be achieved by adding a resilient coating to the wedge. However, resilient wedge gate valves are normally rated for a maximum of 160°F, so their applications for HVAC piping systems are limited.

Gate valves are available in both rising-stem and nonrising-stem configurations. The rising-stem configuration (Figs. 3-5 and 3-6) is preferred because the stem threads are protected from the fluid flow. However, clearance must be provided external to the valve in order to allow for the raised stem when the valve is in the open position. Rising-stem gate valves provide an indication of the valve's position. If the stem is in the raised position, the valve is open; if the stem is in the lowered position, the valve is closed.

Nonrising-stem gate valves (Figs. 3-7 and 3-8) are used when clearance for the raised stem is not available. However, the threads of the valve stem on which the wedge travels are always exposed to the fluid flow. As a result, these threads can become eroded[14] or corroded[15] over time, which can prevent the valve from functioning properly.

Application   Gate valves are recommended for use as shutoff valves for piping up to 12 in. in steam or steam condensate systems.

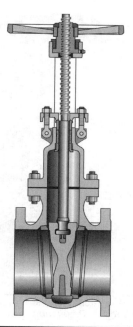

FIGURE 3-5 Drawing of a rising-stem gate valve with flanged ends. (*Reprinted with permission from Crane Co. All Rights Reserved.*)

FIGURE 3-6 Photograph of a rising-stem gate valve with flanged ends.

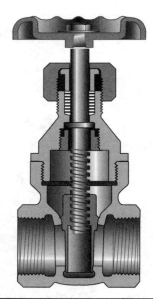

Figure 3-7   Drawing of a nonrising-stem gate valve with threaded ends. (*Reprinted with permission from Crane Co. All Rights Reserved.*)

Figure 3-8   Photograph of a nonrising-stem gate valve with threaded ends. (*Photo courtesy of NIBCO INC.*)

Globe Valve

Description   Globe valves consist of a solid disc mounted within the valve body, which is connected to the internal end of the valve stem. A handwheel actuator is connected to the external end of the valve stem for manual operation. When the valve stem is turned completely in the counterclockwise direction, the disc is raised to the open position. When the valve stem is turned completely in clockwise direction, the disc is lowered to the closed position. Globe valves, which are intended to be used for throttling, can also provide bubble-tight shutoff when equipped with resilient seating. Globe valves are normally installed with the flow and higher pressure under the disc. Therefore, care must be taken during installation to ensure they are installed with the flow in the proper direction.

Globe valves are available in either a straight-through (Figs. 3-9 through 3-11) or angled body configuration (Fig. 3-12). In the angled configuration, the flow exits the valve perpendicular to the direction that it enters the valve.

Globe valves have a higher fluid pressure drop than gate valves because the disc is always in the fluid stream.

Application   Globe valves are recommended for use as throttling valves for piping up to 12 in. in water, brine, or steam systems. Globe valves with resilient seating can also be used as shutoff valves in water or brine systems. However, less expensive valves are more suitable for this purpose.

Quarter-Turn Valves

Quarter-turn valves require only a quarter rotation of the valve stem to change the position the valve from fully opened to fully closed and vice versa. Two types of quarter-turn valves commonly used in HVAC piping systems are butterfly valves and ball valves.

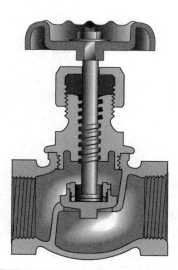

Figure 3-9   Drawing of a straight-through globe valve with threaded ends. (*Reprinted with permission from Crane Co. All Rights Reserved.*)

Figure 3-10 Photograph of a straight-through globe valve with soldered ends. (*Photo courtesy of NIBCO INC.*)

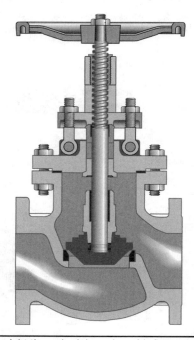

Figure 3-11 Drawing of a straight-through globe valve with flanged ends. (*Reprinted with permission from Crane Co. All Rights Reserved.*)

FIGURE 3-12 Photograph of an angle globe valve with threaded ends. (*Photo courtesy of NIBCO INC.*)

Butterfly Valve

Description   Butterfly valves consist of a solid disc mounted within the valve body, which is connected to the internal end of the valve stem. A hand-lever actuator is connected to the external end of the valve stem for manual operation. When the valve stem is turned such that the disc is parallel to the fluid flow, the valve is in the open position. When the valve stem is turned such that the disc is perpendicular to the fluid flow, the valve is in the closed position. The disc closes against a resilient seat mounted within the valve body, providing bubble-tight shutoff. Because the hand-lever actuator is installed on the valve stem parallel to the disc, the actuator provides an indication of the valve's position. Butterfly valves have a straight-through body configuration and are suitable for throttling and bubble-tight shutoff.

Butterfly valves are available in lug-style (or single-flange) (Figs. 3-13 and 3-14) and wafer-style bodies. Lug-style bodies have tapped lugs that align with the bolt holes in ASTM pipe flanges, allowing them to be bolted to a single flange on one side of the valve. This capability makes lug-style butterfly valves suitable for dead-end service; that is, they can be mounted on a flange at the open end of a pipe and provide bubble-tight shutoff. For this reason, lug-style butterfly valves can be used as shutoff valves for HVAC equipment, enabling the removal of the equipment and piping up to the valve.

Wafer-style butterfly valves (Fig. 3-15) are designed to be installed between two pipe flanges. They do not have tapped lugs that align with the bolt holes in the pipe flanges. Rather, they are held in place by means of compression; that is, the bolts join the two

FIGURE 3-13 Photograph of a lug-style butterfly valve. (*Photo courtesy of NIBCO INC.*)

FIGURE 3-14 Photograph of a lug-style butterfly valve—installed.

FIGURE 3-15   Photograph of a wafer-style butterfly valve. (*Photo courtesy of NIBCO INC.*)

flanges on both sides of the valve. Wafer-style butterfly valves are not suitable for dead-end service and should not be used as shutoff valves for HVAC equipment because the flange on the equipment side of the valve cannot be removed independent of the valve.

Butterfly valves have a slightly higher fluid pressure drop than gate valves but a lower pressure drop than globe valves.

Application   Lug-style butterfly valves are recommended for use as shutoff valves for 2½- through 12-in. piping in water or brine systems.[16] When the hand-levers are equipped with memory stops, butterfly valves can be used as balancing valves for 2½- through 12-in. pipe branches in water or brine systems. Wafer-style butterfly valves are a less expensive option to lug-style butterfly valves, having the same throttling and shutoff characteristics. However, they should not be used where dead-end service is required.

Ball Valve

Description   Ball valves consist of a solid ball held in place within the valve body between two circular resilient seats. The ball is connected to the internal end of the valve stem, and a hand-lever actuator is connected to the external end of the valve stem for manual operation. The ball has a hole (port) in the middle of it. When the valve stem is turned such that the port is parallel to the fluid flow, the valve is in the open position. When the valve stem is turned such that the port is perpendicular to the fluid flow, the valve is in the closed position. Because the hand-lever actuator is installed on the valve stem parallel to the port, the actuator provides an indication of the valve's position. Ball valves have a straight-through body configuration and are suitable for bubble-tight shutoff. Ball valves should not be used for throttling duty.

Ball valves are available in one-piece, two-piece (Figs. 3-16 and 3-17), and three-piece body designs (Fig. 3-18) (listed in the order of increasing cost). One-piece valves cannot be repaired and must be replaced when they fail to function properly. Also, one-piece valves are only available with a reduced-port ball; that is, the port is more than one pipe size smaller than the nominal pipe size. Due to the higher fluid pressure drop associated with the reduced-port, one-piece ball valves are not recommended for use in HVAC piping systems. Repair of two-piece valves is not recommended. However, this type of valve is available with a full- or standard-port ball.[17] Three-piece valves offer in-line reparability and are available with a full- or standard-port ball; however, they are also the most expensive option for ball valves.

Figure 3-16   Photograph of a two-piece ball valve with solder ends. (*Photo courtesy of NIBCO INC.*)

Figure 3-17   Photograph of a two-piece ball valve with solder ends—installed.

FIGURE 3-18   Photograph of a three-piece ball valve with threaded ends. (*Photo courtesy of NIBCO INC.*)

Application   Two-piece or three-piece, full-port ball valves are recommended for use as shutoff valves for piping up to 2 in. in water or brine systems. Extended valve stems are required for installation in insulated piping systems.

Calibrated Balancing Valves

Description   Calibrated balancing valves (Fig. 3-19), also referred to as circuit balancing valves, are available in different configurations. A common configuration consists of a globe valve with pressure taps on the upstream and downstream sides of the valve. During

FIGURE 3-19   Photograph of a calibrated balancing valve with solder ends. (*Photo courtesy of NIBCO INC.*)

the balancing operation, a pressure gauge is connected to the pressure taps to measure the differential pressure across the valve. Flow is balanced through the valve by adjusting the position of the valve until the differential pressure corresponds to the desired flow, as given in the pressure drop/flow data furnished by the valve manufacturer.

Application   Calibrated balancing valves are recommended for balancing the flow through minor branches in the piping system (pipe sizes 2 in. and smaller) and through individual pieces of equipment. A flow meter and associated balancing valve will normally be used for pipe sizes 2½ and larger (refer to the Flow Meters section later).

Check Valves

Check valves are available in various types. The most common types are swing check, lift check, and center-guided check.

Swing Check Valves

Description   Swing check valves (Figs. 3-20 and 3-21) are constructed with a disc that swings by a hinge mounted at the top of the disc. The disc swings open when the fluid flows in the intended direction and swings closed when the fluid flow reverses. Swing check valves have the lowest fluid pressure drop of all check valves. The seating surface of the disc can be equipped with a resilient disc ring to improve the seal against reverse flow. Swing check valves must be mounted horizontally with the hinge pin level to allow gravity to assist in the closing of the disc.

Application   Swing check valves are recommended for use in water, brine, or steam systems where the valve can be mounted horizontally and where silent operation is not required (refer to the description of center-guided check valves later).

Lift Check Valves

Description   Lift check valves (Fig. 3-22) utilize a straight-through globe-style body pattern and consist of a disc that lifts vertically off of a seat when the fluid flows in the intended direction and closes when the fluid flow reverses. Lift check valves have a higher fluid pressure drop than swing check valves because the disc is always in the

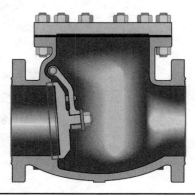

Figure 3-20   Drawing of a swing check valve with flanged ends. (*Reprinted with permission from Crane Co. All Rights Reserved.*)

FIGURE 3-21   Photograph of a swing check valve with flanged ends.

FIGURE 3-22   Photograph of a lift check valve with threaded ends. (*Photo courtesy of NIBCO INC.*)

fluid stream. Lift check valves can also be equipped with resilient seating materials to improve their seal against reverse flow. Lift check valves must be mounted horizontally with the disc stem in the upright position.

Stop check valves, a type of lift check valve, have the added capability of providing positive shutoff. These valves are available in straight-through, Y-pattern, and angled body styles. Stop check valves with a straight-through body style must be mounted horizontally with the disc stem in the upright position. Those with a Y-pattern body style can be mounted horizontally or vertically. Stop check valves with an angled body style are installed in place of a 90° pipe elbow.

Application   Lift check valves have the same applications as swing check valves, although they have no real advantages over swing check valves.

Stop check valves can be used in the discharge piping of pumps or in the piping connections of multiple pieces of equipment to a common header. Stop check valves are commonly used in the steam supply connections of multiple boilers to a common header.

Center-Guided Check Valves

Description   Center-guided check valves (Fig. 3-23) utilize an in-line body style and consist of a center-guided disc that lifts off of a seat when the fluid flows in the intended direction and closes when the fluid flow reverses. Center-guided check valves can be equipped with a spring that actuates the closure of the disc before gravity and fluid reversal slam the valve closed. Spring-actuated, center-guided check valves are referred to as nonslam or silent check valves for this reason. Center-guided check valves have a higher fluid pressure drop than swing check valves because the disc is always in the fluid stream. Spring-actuated, center-guided check valves have an additional fluid

Figure 3-23   Photograph of a center-guided check valve with flanged ends.

pressure drop associated with the closing spring. Center-guided check valves can also be equipped with resilient seating materials to improve their seal against reverse flow. Center-guided check valves can be mounted either horizontally or vertically.

Center-guided check valves are generally used for larger pipe sizes (2½ through 12 in.) and are available with flanged ends or in a wafer-style (similar to wafer-style butterfly valves) for connection between two pipe flanges.

Application   Spring-actuated, center-guided check valves are recommended for use in pump discharge piping (for pipe sizes 2½ through 12 in.) because of their silent operation and also because they can be mounted in either a horizontal or vertical arrangement.

Multi-purpose Valves

Description   Multi-purpose valves (Fig. 3-24), sometimes referred to as triple-duty valves, combine shutoff, check, and balancing duty into a single valve. They consist of a spring-actuated, lift check valve mounted within a straight-through or angled body configuration. The disc stem is equipped with screw threads, which enable adjustment of the spring force on the disc. A calibrated nameplate and stem-position indicator provide a direct reading of the flow through the valve based on the stem position. Pressure taps are provided upstream and downstream of the disc so that the differential pressure across the disc can be measured to verify the stem-position flow measurement against the pressure drop/flow data furnished by the valve manufacturer.

Figure 3-24   Photograph of a multi-purpose valve with flanged ends.

Application   Multi-purpose valves are recommended for use in pump discharge piping, particularly where space for the piping connections to the pump is limited or the length of straight pipe required for proper installation of a flow meter is not available.

Pressure Regulating Valves

Description   Pressure regulating valves are grouped into three categories: direct-acting, pilot-operated, and automatic control valves.

For a direct-acting valve, the desired pressure is set by adjusting a spring that controls the pressure exerted on one side of a main diaphragm located within the valve body. The working fluid exerts a pressure on the other side of the main diaphragm, and the valve is opened to a point where these two pressures equalize. However, direct-acting valves do not have the capability to sense the actual pressure they are controlling and cannot maintain precise pressure control of the working fluid.

Pilot-operated valves also use a spring and main diaphragm to maintain the desired pressure. However, they have an additional pilot diaphragm and pressure control pipes that utilize the actual fluid pressure to maintain more precise pressure control of the working fluid. They are also capable of operating at higher flow rates than direct-acting valves.

Automatic control valves must be controlled by an automatic temperature control (ATC) system and can be either electrically or pneumatically actuated. The ATC system modulates the position of the valve through the valve actuator to maintain the setting of a pressure sensor. The sensor is mounted either upstream or downstream of the valve or is piped to sense the differential pressure between the inlet and outlet of the valve.

The two most common types of pressure regulating valves used in HVAC systems are the pressure reducing valve, which maintains a constant downstream pressure, and the differential pressure valve, which maintains a constant differential pressure between the inlet and outlet of the valve.

Pressure Reducing Valves

Application   For water and brine systems, pressure reducing valves are most commonly used as part of the domestic water make-up assembly. The purpose of the pressure reducing valve is to maintain a constant (downstream) fill pressure for the HVAC piping system regardless of the fluctuating (upstream) pressure in the domestic water system. Figure 3-25 is a photograph of a direct-acting water pressure reducing valve that is used as part of a domestic water make-up assembly.

For steam systems, pressure reducing valves are used to reduce the steam pressure from the (higher) distribution pressure to the (lower) utilization pressure.

Differential Pressure Valves

Application   Differential pressure valves are used in water and brine HVAC piping systems to ensure that the pressure in the system does not exceed a maximum value. For example, an HVAC piping system that consists of a constant speed pump and 2-way automatic control valves,[18] which modulate the fluid flow through the branch piping components of the system, will require a differential pressure valve to be installed in the system to protect the pump from a dead-head (or zero-flow) condition. A dead-head condition could occur if all of the automatic control valves in the system were closed.

Figure 3-25   Photograph of a direct-acting water pressure reducing valve with threaded ends.

Although some types of pumps are able to operate in this condition, the pumps commonly used in HVAC piping systems are not. It is necessary for these pumps to circulate a minimum flow, which is why a bypass pipe with a differential pressure valve would be required.

In this case, a differential pressure valve would be installed in a bypass pipe connecting the main supply and return pipes. The differential pressure valve would have to be selected such that the pressure through the bypass piping and the valve was equal to the maximum head of the pump corresponding to its minimum flow.[19] The differential pressure valve can be installed between the discharge and suction piping at the connection of the system pump or between the main supply and return pipes at the end of the piping system to bypass the minimum pump flow.

Pressure Relief Valves

Description   Pressure relief valves are available in different configurations, the selection of which depends upon the working fluid and the operating conditions (Fig. 3-26).

Application   Generally, pressure relief valves are used to protect pressure vessels (such as boilers and heat exchangers) and steam systems from overpressurization. Under normal operating conditions, pressure relief valves are closed. However, in the case where the pressure at the point of connection to the system exceeds the rating of the valve, the relief valve opens and discharges the working fluid through the discharge pipe to a safe location, thereby relieving the excess pressure in the system. For example, if a burner for a boiler fails to shut off or a steam pressure reducing valve hangs open, the pressure relief valve would open to protect the HVAC piping system components from exposure to pressures exceeding their rated values.

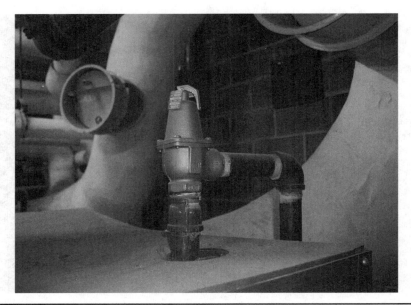

Figure 3-26   Photograph of a hot water pressure relief valve.

Operating Conditions

Operating conditions that affect the selection and installation of valves include the working fluid, fluid pressure, fluid temperature, requirement for bubble-tight shutoff, space available for installation, frequency of operation, and whether the valve will be manually or automatically controlled.

Working Fluid

Proper valve selection depends upon the working fluid in the piping system. The valves used in HVAC piping systems need to be capable of handling fluids in the liquid state for water and brine systems, the gaseous state for steam systems, and both the liquid and gaseous states for steam condensate return systems. The characteristics inherent to the design of some valve types make them more suited to handling liquids than gases, although the inverse is not always true. Valves that are suitable for handling gases are normally suitable for handling liquids as well. Manufacturers' product data should be consulted to determine the suitability of a particular valve for the working fluid.

Fluid Pressure

Valves are constructed to different specifications that determine their pressure rating at a certain temperature. Valves that are suitable for steam service will be rated by saturated steam working pressure (SWP) or pressure class (the two terms are synonymous). Pressure classes such as Class 125, Class 150, Class 250, or Class 300 (which are the same ratings that apply to steel pipe fittings) are common for valves used in HVAC piping systems. For example, a Class 125 valve would be rated for 125 psig steam at the saturated steam temperature of 352°F.

Valves may also have a rating for cold-working pressure (CWP). The CWP rating is the maximum water working pressure of the valve for fluid temperatures up to 150°F for bronze and cast iron body valves and up to 100°F for steel body valves. The CWP rating of a valve can be used for cold water HVAC systems, such as chilled water and condenser water systems. Valves that have only a CWP rating are not suitable for steam service or for HVAC piping systems operating at water (or brine) temperatures in excess of the rated temperature for the CWP.

However, valves that are suitable for steam service will also be suitable for water or brine service. The SWP or pressure class of a valve is used to determine its applicability for use in a water or brine system. For example, a heating water system that has a maximum working pressure of 250 psig would require valves with a minimum rating of Class 250 (or an SWP of 250 psig).

Fluid Temperature

Fluid temperature is also a determining factor in the proper selection of valves because the seating materials used in the valves have both high and low temperature limitations and also have optimum temperature ranges for operation. For example, globe valves with polytetrafluoroethylene (PTFE)[20] seats provide bubble-tight shutoff for heating water systems operating at temperatures between 100 and 200°F. However, softer nitrile-butadiene rubber (NBR) seats are more suitable for chilled water systems operating at temperatures between 35 and 65°F.

Requirement for Bubble-Tight Shutoff

Bubble-tight shutoff is a requirement for shutoff valves but is not a requirement for balancing valves, unless the balancing valves will also be used for shutoff duty.

Space Available for Installation

All valves need to be installed in accessible locations; however, the available space required for proper installation depends upon field conditions. Some valves, such as butterfly valves, require very little pipe length for installation, although they do require space for the lever handle to rotate in a direction that is perpendicular to the pipe. Other valves, such as nonrising-stem gate valves, do not require space above the valve handwheel actuator, which makes them a good choice where space is limited in this area. Pressure regulating valves and calibrated balancing valves will often have requirements for straight piping lengths both upstream and downstream of the valves to ensure proper operation. Also, space must be provided for the discharge piping of pressure relief valves. This piping must be routed to a point, sometimes exterior to the building, where the valves can safely discharge the working fluid when necessary to relieve excess pressure.

Frequency of Operation

Normally, shutoff valves and balancing valves remain in the same position for extended periods of time. For this reason, care should be taken to select valves that do not have moving parts located within the fluid stream because these parts will be subject to erosion and corrosion from the working fluid. On the other hand, automatic control valves may have to change their position multiple times every hour. Therefore, automatic control valves, and other valves that require a high number of opened/closed cycles, should be rated for high-cycle operation.

Manual or Automated Control

Valves used in HVAC piping systems will be either manually or automatically operated. Multi-turn valves requiring manual operation will normally be equipped with handwheel actuators. Chain-wheel actuators are often used in mechanical equipment rooms for valves that are 4 in. and larger and are installed higher than 8 ft above the floor.

Multi-turn valves are typically not used for automatic operation. However, globe valves that have a nonthreaded stem are frequently used for automatic control valves that are 2 in. and smaller. In this case, the automatic actuator opens or closes the valve by stroking the valve stem in and out. Globe valves in both 2-way and 3-way body patterns are used for both two-position and modulating automatic control valves (Figs. 3-27 through 3-29).

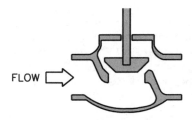

FIGURE 3-27 Drawing of a globe valve with a 2-way body pattern.

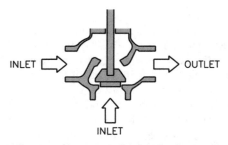

FIGURE 3-28 Drawing of a globe valve with a 3-way body pattern (mixing valve).

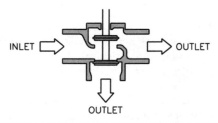

FIGURE 3-29 Drawing of a globe valve with a 3-way body pattern (diverting valve).

FIGURE 3-30 Photograph of linked butterfly valves used as a 3-way valve.

Quarter-turn valves that require manual operation will normally be equipped with hand-lever actuators for valves that are 6 in. and smaller. Gear actuators are recommended for valves that are 8 in. and larger. Chain-wheel actuators are often used in mechanical equipment rooms for valves that are 4 in. and larger and are installed higher than 8 ft above the floor. Quarter-turn valves are well suited for automatic operation and provide bubble-tight shutoff. Butterfly valves are commonly used for automatic control valves that are 2½ in. and larger. Where 3-way automatic control is required, two butterfly valves will be interconnected by a linkage that synchronizes the operation of the two valves (Fig. 3-30); when one valve opens, the other valve closes and vice versa.

Materials

Valves used in HVAC piping systems generally fall into two categories: bronze body and iron body. Bronze body valves are used for connection to copper tube, and iron body valves, constructed of either cast iron or ductile iron,[21] are used for connection to steel pipe. Valve bodies are made of different materials in order to ensure that the valves are compatible with the piping materials to which they are connected.

The internal metal components of valves, including the wedges for gate valves; discs for globe valves, butterfly valves, and check valves; and valve stems are commonly constructed of bronze. The ball for ball valves can be constructed of chrome-plated bronze or solid stainless steel. For higher quality, stainless steel can be used for the valve stems in globe, butterfly, and ball valves and can also be used for the discs in globe and butterfly valves.

Valve seats can be constructed of metal, but resilient (nonmetallic) seats provide a better seal. Resilient materials commonly used for valve seats include:

1. Ethylene-propylene-diene monomer (EPDM) rubber
    a. EPDM seats are commonly used in butterfly valves. EPDM is known for its use in weather seals, such as window and door seals for vehicles.
2. Tetrafluoroethylene (TFE) or polytetrafluoroethylene (PTFE)
    a. TFE and the higher-quality PTFE (or Teflon®) are used as seating materials in globe valves, ball valves, and check valves. PTFE is a hard, slick, chemical-resistant material that has many industrial and consumer applications. Most people are familiar with PTFE's use as a nonstick surface for cookware. Higher-strength, glass-reinforced PTFE (or RPTFE) seats are used in high-performance ball valves and butterfly valves.
3. Nitrile-butadiene rubber (NBR)
    a. NBR, also known as nitrile rubber or Buna-N, is a common seating material for valves used in water or brine HVAC piping systems having a fluid temperature of 100°F or lower. Gate valves can be constructed with NBR-coated wedges. NBR seats can be used in globe, butterfly, and check valves. Nitrile rubber gloves, commonly blue or purple in color, are an application of NBR that many people are familiar with.
4. Viton®
    a. Viton is a high-quality, high-temperature seating material normally used in high-performance valves.

Globe and check valves can be constructed with a metallic (normally bronze) disc and a metallic or resilient seat. The disc itself can also be constructed of a resilient material, such as PTFE, in which case the seat would be metallic.

Ends

Bronze body valves typically have threaded or soldered ends. Iron body valves typically have flanged ends.

Specialties

The term *specialties*, as used in this book, defines a broad category of piping components, which includes anything that is not a pipe, fitting, valve, union, or flange. The specialties we will discuss include meters, pressure gauges, test plugs, strainers, suction diffusers, flexible pipe connectors, air vents, and drains. There are other components that could be classified as specialties, but these are the ones most commonly used.

Meters

The meters most commonly used in HVAC piping systems are thermometers and flow meters.

FIGURE 3-31 Photograph of a liquid-in-glass thermometer.

Thermometers

The thermometers used to measure the fluid temperature within pipes are installed within fluid-filled thermometer wells. For pipes 4 in. and smaller, these wells extend to the center of the pipe; for pipes larger than 4 in., the wells extend a minimum of 2 in. into the fluid. The most commonly used types of thermometers are bimetallic-actuated dial thermometers and liquid-in-glass thermometers (Fig. 3-31).

Flow Meters

Flow meters are used primarily in commercial HVAC water and brine systems in conjunction with throttling valves to enable balancing of the system, that is, to ensure that the proper flow is being circulated by the system pump and that the proper flow is being delivered to the major branches in the piping system. Calibrated balancing valves are normally used to balance the flow through minor branches in the piping system and through individual pieces of equipment (refer to the Calibrated Balancing Valves section earlier).

Balancing of steam systems is not required as long as the steam distribution piping is properly sized. Therefore, flow meters would not be required in steam systems unless it was necessary to continuously measure the steam flow and assess a charge for it. Although it is not common to continuously measure water or steam flow in commercial HVAC systems, a brief discussion of flow measurement is given in Chap. 9.

The two most common types of flow meters used in HVAC water or brine systems are orifice and venturi meters. Both types are differential pressure meters; this means that the flow rate through the meter is determined by measuring the differential pressure across the meter.

An orifice flow meter (Fig. 3-32) consists of a flat plate with a specific-sized hole bored in it (usually about 50% of the cross-sectional area of the pipe). The orifice plate

Figure 3-32   Photograph of an orifice flow meter. (*Photo courtesy of NIBCO INC.*)

is installed between two pipe flanges. Pressure taps are either an integral part of the meter or they are installed in the upstream and downstream pipe flanges. The pressure taps are used to measure the differential pressure across the orifice plate. The orifice meter is furnished with either a permanent indicator (mounted to the flow meter) or a portable indicator and hoses for connection to the pressure taps. The indicators must be calibrated for the particular orifice meter. Because orifice meters have a high pressure drop (in the range of 5 psi), they are normally used to balance the flow in the major branches of the piping system, not the system pump.

A venturi flow meter consists of a steep converging cone on the inlet side of a short reduced-diameter throat section and a shallow diverging cone on the outlet side of the throat section. This configuration allows most of the kinetic (velocity) energy that is developed in the throat section to be converted back to potential (pressure) energy, resulting in a low pressure drop for the meter. The differential pressure is measured between the inlet flange and the throat. The venturi meter is also furnished with either a permanent or portable indicator calibrated for the particular venturi meter. Venturi meters are capable of handling high flow rates with a low pressure drop (in the range of 2 to 3 psi). Thus, they are normally used in the main system piping to balance the flow of the system pump, but can also be used in the major branches of the piping system as well.

The accuracy of both orifice and venturi flow meters is affected by the presence of elbows and/or throttled valves both upstream and downstream of the flow meters. Although the actual requirements vary depending upon the piping configuration, approximately five pipe diameters of straight pipe are required upstream and two pipe diameters of straight pipe are required downstream of orifice and venturi flow meters for accurate measurement of fluid flow. The manufacturer's product data should be consulted for the installation requirements for each flow meter and for the actual pressure drop through the flow meters in the system at the design flow rates.

Pressure Gauges

Pressure gauges measure the difference in pressure between two points. It is common for one of these two points to be the atmosphere,[22] in which case the measured pressure is based on the reference of atmospheric pressure. This is called gauge pressure. Pressure gauges in commercial HVAC systems are most commonly used to measure the gauge pressure of fluids in pipes, such as chilled water, heating water, and steam. However, pressure gauges can also measure the difference in the pressure of the fluid at two points in a piping system, such as the differential pressure between the inlet and outlet of a strainer. In this case, the differential pressure would provide an indication of the extent that the strainer is clogged with debris.

Pressure gauges are connected with ¼-in. brass tubing in threaded openings in large pipes or in pipe tees for smaller pipes. A pressure snubber should be installed between the pipe and the pressure gauge for pipes with liquid flow, and a pressure siphon should be installed between the pipe and the pressure gauge for pipes with steam flow. Pressure snubbers consist of a porous metal disc that dampens the fluctuations in the fluid pressure and reduces wear on the gauge mechanism. Pressure siphons consist of a single coil of brass tubing that is filled with water to prevent live steam from entering the pressure gauge Bourdon tube. A gauge cock, which is typically a needle-type shutoff valve, should also be installed between the pipe and the pressure gauge to enable the pressure gauge to be isolated from the fluid in the pipe. Figure 3-33 shows a typical pressure gauge installation.

FIGURE 3-33 Photograph of a pressure gauge mounted in pipe.

Figure 3-34   Photograph of a test plug (the plug is located just to the right of the pipe hanger).

Test Plugs

Test plugs (Fig. 3-34) are used where fluid temperature and/or pressure readings are required and the cost of permanent thermometers and/or pressure gauges is not warranted. Test plugs consist of a ¼-in. brass or stainless steel body with an EPDM insert. The insert has a small, self-closing opening in the center, which allows for the temporary insertion of a test thermometer or pressure gauge with gauge adapter. The test plug is mounted in a threaded opening in large pipes or in a pipe tee in smaller pipes. A protective cap is placed over the test plug to prevent any leakage when the test plug is not in use. Test plugs are commonly used in the piping connections for minor pieces of HVAC equipment, such as the heating water coils in VAV terminal units and finned-tube radiators.

Strainers

Strainers are used to remove any sediment that may be entrained in the fluid stream of the water, brine, or steam system. Strainers are commonly installed at the main system pumps in water or brine systems and also upstream of sensitive components, such as control valves and pressure regulating valves. Strainers used in commercial HVAC systems typically have a Y-pattern body[23] and are constructed of bronze with soldered or threaded ends for pipe sizes 2 in. and smaller. For pipe sizes 2½ in. and larger, strainers are constructed of cast iron with flanged ends. Sediment is trapped within the strainer by a 20-mesh stainless steel screen.[24] The water pressure drop through a clean strainer depends upon the pipe size and the fluid flow rate and is generally in the range of 1 psi. A water pressure drop of about 2 to 4 psi should be included in the pump head calculation for the main system strainer to account for some loading of the strainer.

FIGURE 3-35   Photograph of a strainer with a Y-pattern body, flanged ends, and a blow-down valve.

Strainers must be flushed periodically to blow-down the trapped sediment. Small strainers (1 in. and smaller) are typically furnished with a plugged blow-down outlet. The blow-down outlet of larger strainers (1¼ in. and larger) is commonly specified with a ½- or ¾-in. blow-down pipe that is equipped with hose-end drain valve.[25] Thus, for small strainers, it is necessary to close a shutoff valve upstream of the strainer before removing the blow-down plug. Then, the shutoff valve can be slowly throttled open to flush the sediment from the strainer. For larger strainers, the blow-down valve enables the strainer to be flushed without shutting off the equipment. Figure 3-35 shows a strainer with flanged ends and a blow-down valve.

Suction Diffusers

Suction diffusers (also called inlet guide fittings) are used to connect the suction piping to end-suction pumps where there is insufficient space for the recommended five pipe diameters of straight pipe upstream of the pump suction connection. A suction diffuser (Fig. 3-36) contains an orifice cylinder and straightening vanes that allow the pipe connection to the pump suction to be made approximately within the distance of a long radius 90° elbow. A pressure drop of about 0.5 psi should be allowed for the suction diffuser in the pump head calculation. The manufacturer's product data should be consulted for the actual pressure drop through the suction diffuser; this value varies depending upon the pipe size and the water flow rate.

Figure 3-36   Photograph of a suction diffuser.

Flexible Pipe Connectors

Flexible pipe connectors, which are used to attenuate the noise and vibration generated by motor-driven equipment such as pumps, chillers, and cooling towers, are constructed of a flexible piping material [either Kevlar®-reinforced EPDM (rubber) or stainless steel hose and braid] that is equal to the pipe size. Flexible pipe connectors are approximately 14 to 28 in. long (depending upon the pipe size) and are available with threaded or flanged ends. They are connected between the shutoff valves and the equipment they serve and are able to compensate for some misalignment between the equipment and piping connections.

Rubber is the preferred material for flexible pipe connectors (Fig. 3-37) because it provides better noise and vibration attenuation than stainless steel hose and braid. However, rubber flexible pipe connectors are limited by their pressure and temperature rating (typically 215 psig at 250°F). As a result, they are acceptable for use in Class 125 and Class 150 water or brine systems for temperatures up to 250°F. For systems with temperatures exceeding 250°F, such as steam systems in excess of 15 psig, stainless steel hose and braid flexible pipe connectors should be used (Fig. 3-38). The manufacturer's product data should be consulted for the actual pressure and temperature ratings of flexible pipe connectors.

Air Vents and Drains

Air vents are required at all high points in water and brine HVAC systems to enable air to be purged from the systems during the initial start-up and subsequent operation and maintenance of the systems. Drains are required at all low points in water, brine, and

Figure 3-37   Photograph of a rubber flexible pipe connector.

Figure 3-38   Photograph of a stainless steel hose and braid flexible pipe connector.

FIGURE 3-39 Photograph of a manual air vent.

steam systems to enable the systems to be drained for repair or replacement of system components.

Manual Air Vents

Manual air vents are used to purge large quantities of air from water and brine systems (Fig. 3-39). This is normally required during the initial start-up and after draining and refilling portions of the systems due to repair or replacement of system components. Manual air vents consist of a needle valve that is manually operated with a thumb screw or screwdriver. Manual air vents typically have a ½-in. inlet connection and a 1/8-in. outlet connection. Manual air vents should be installed at all high points in the distribution piping, as well as at the high points in the equipment piping connections.

Automatic Air Vents

Automatic air vents are used to purge small quantities of air from water and brine systems required during normal operation of the system (Fig. 3-40). An automatic air vent should be installed on the air separator in the system (refer to the Air Separator section in Chap. 4) and anywhere else it is suspected that air may accumulate during normal system operation. Automatic air vents consist of a float-actuated needle valve that opens to discharge any air that enters the valve. Automatic air vents typically have a ½-in. inlet connection and a ½-in. outlet connection. A hose should be connected to the outlet and routed to a point where small quantities of water can be safely discharged during normal operation of the vent.

FIGURE 3-40   Photograph of an automatic air vent.

Drains

Drains in water, brine, and steam systems normally consist of a ½- or ¾-in. pipe connected to a tee or threaded insert in the system piping at the low points in the system. The drain pipe should be equipped with a hose-end drain valve (Figs. 3-41 and 3-42). This valve allows for connection of a hose for discharging the system fluid to a nearby floor drain or other receptacle connected to the building sanitary drainage system.

FIGURE 3-41   Photograph of a hose-end drain valve. (*Photo courtesy of NIBCO INC.*)

FIGURE 3-42 Photograph of a hose-end drain valve—installed.

Endnotes

1. The term *brine* is used to describe a mixture of water and a salt or glycol. Brines, which have a lower freezing point than water, are used in HVAC piping systems to prevent freezing of the solution in outdoor piping and equipment. Solutions of propylene and ethylene glycol are the most common types of brine used in HVAC systems.
2. Gauge pressure is the pressure of a fluid with respect to atmospheric pressure. Thus, atmospheric pressure, which is 14.7 pounds per square inch absolute (psia), is equal to 0 psig.
3. The threaded ends of steel pipe, fittings, valves, unions, etc., are American Standard Taper Pipe Thread [also referred to as National Pipe Thread (NPT)]. The tapered thread causes the threaded joint to tighten as the male threaded end is screwed into the female threaded end. The male threaded end is referred to as male pipe thread (MPT) and the female threaded end is referred to as female pipe thread (FPT).
4. An electrochemical process called galvanic action occurs when two dissimilar metals are coupled together and are immersed in an electrolyte. The more active (less noble) metal will act as an anode and will corrode. The less active (more noble) metal will act as a cathode and will remain unchanged. Such is the case when steel pipe and copper tube are coupled together in a water or brine piping system. The water or brine acts as the electrolyte; the copper tube becomes the cathode; and the steel pipe becomes the anode that corrodes.
5. The cast iron used in the construction of pipe fittings, flanges, and valves is gray cast iron, or "gray iron," which derives its name from the gray color of its fracture.

Gray cast iron is a brittle material, which means, under stress, it yields very little before fracturing.

6. Malleable iron is an alloy of white cast iron that, after casting, is heat treated to increase its tensile strength and ductility (flexibility).

7. The gasket material must be compatible with the working fluid and have the same pressure classification as the flanges.

8. Extruding of copper tube is a hot-working process. Hot-work is performed at temperatures above the thermal critical range of the metal.

9. Drawing of copper tube is a cold-working process. Cold-work is performed at temperatures below the thermal critical range of the metal (usually atmospheric temperature).

10. Type L copper tube can be used for all pipe sizes up to 12 in. for aboveground installations; however, steel pipe is usually more cost-effective for pipe sizes 2½ to 12 in.

11. Annealing is a form of heat treatment in which a cold-worked material is heated to a temperature below the melting point and then slowly cooled. The effects of annealing are the relief of stresses in the material that were imparted during the cold-working process and an increase in the softness of the material.

12. Steel pipe is normally specified for piping 2½ in. and larger.

13. Ductile iron pipe is also suitable for below ground installations due to its corrosion resistance; it is not recommended for below-slab installations because of the joints that are required.

14. Erosion is the physical wearing away of the internal valve surfaces by the fluid stream.

15. Corrosion is the result of the chemical reaction between the fluid and the valve surfaces.

16. Butterfly valves of the high-performance design may be suitable for steam service.

17. A standard-port ball is one in which the port is up to one pipe size smaller than the nominal pipe size.

18. Automatic control valves can be 2-way (having a single inlet and single outlet) or 3-way (having two inlets and one outlet, or one inlet and two outlets). Automatic control valves are either two-position (opened/closed) or modulating (capable of positioning to any point between opened and closed).

19. This information can be obtained from the pump curve for the selected pump.

20. PTFE is also known by the DuPont brand name Teflon.

21. Ductile (or nodular) iron is an iron alloy in which the carbon in the metal occurs in spheroids (or nodules), as opposed to flakes, which is how the carbon occurs in cast iron. The carbon spheroids in ductile iron make it more ductile (flexible) than cast iron and also give it a higher tensile strength. Ductile iron is preferred over cast iron in applications where severe physical impact or hydraulic or thermal shock may occur.

22. Atmospheric pressure is sensed by a port on the pressure gauge itself.

23. Basket strainers are available for increased holding capacity where large amounts of sediment must be removed from the fluid stream.

24. A 20-mesh screen has 0.035-in. square openings with an overall 49% opening area. Other screens having different sized openings are available for strainers, but a 20-mesh screen is the most common size for HVAC piping systems.

25. A hose-end drain valve has a male threaded connection on the outlet of the valve to which a hose can be connected for safely draining the system fluid.

CHAPTER 4

Central Plant

The central plant is the part of the HVAC system that converts energy from fuel sources, such as gas, fuel oil, or electricity, into heating and cooling energy. This energy is transferred from the central plant through the heating and cooling piping systems to the air systems and terminal equipment where it is used to heat and cool the building. In this chapter we will discuss centralized HVAC systems, that is, HVAC systems in which the central plant is remote from the air systems. Decentralized HVAC systems in which the central plant equipment is an integral part of each air system are discussed in Chap. 5.

The main components of a central plant are the heat-generating (heating) equipment and heat-rejecting (cooling) equipment. An example of a piece of central plant heating equipment is a gas-fired or fuel-burning boiler, which produces either hot water (referred to as heating water) or steam. This heating water or steam is distributed from the central plant through the heating piping system to the air systems and terminal equipment where it is used to heat the building. An example of a piece of cooling central plant equipment is an electric water chiller, which produces chilled water. This chilled water is distributed from the central plant through the cooling piping system to the air systems and terminal equipment where it is used to cool the building.

Central plants also consist of equipment that is supplemental to the main heating and cooling equipment, such as a cooling tower for a water-cooled chiller or a feedwater pump for a steam boiler. Pumps and their auxiliary components, which are required for heating and chilled water systems, are also part of central plants. A steam pressure-reducing station and steam-to-hot water heat exchanger are examples of central plant equipment that are commonly used when the building receives its heating energy from a campus steam system or steam utility.

Heating Equipment

The two types of centralized heating equipment we will discuss in this chapter are boilers and heat exchangers.

Boilers

Boilers are commonly used in centralized heating systems to transfer heat from a fuel source to the working fluid of the building heating system. Gas-fired and fuel-burning boilers are the most common types of boilers used in centralized heating systems. Boilers which utilize electricity as the fuel source are rarely used in centralized heating systems because it is usually more practical to utilize (decentralized) electric resistance heaters in the air systems and terminal equipment if electricity is used to heat the building.

Purpose

The purpose of a boiler is to transfer the heat that is released through the combustion of a fuel, such as natural gas, propane,[1] or fuel oil,[2] to the working fluid of the building heating system, which could be water, brine, steam, or a combination of these fluids. This heating water[3] or steam is used by the air systems and terminal equipment to heat the building.

Physical Characteristics

Boilers can be classified in different ways by the following characteristics:

1. Working fluid
   a. Hot water
   b. Steam

2. Fuel
   a. Gas
      (1) Natural gas
      (2) Propane
   b. Liquid
      (1) Fuel oil

3. Draft[4]
   a. Natural draft
   b. Mechanical draft
      (1) Induced draft
      (2) Forced draft

4. Condensing[5] or noncondensing

5. Construction
   a. Condensing
      (1) Stainless steel
      (2) Aluminum
   b. Noncondensing
      (1) Cast iron sectional
      (2) Steel
      (3) Copper

6. Operating pressure: Boilers are normally constructed to meet the requirements of the *American Society of Mechanical Engineers (ASME) Boiler and Pressure Vessel Code*, Section IV, which defines the operating pressures of boilers as follows:
   a. Low pressure: up to 15 psig steam and up to 160 psig hot water[6]
   b. High pressure: above 15 psig steam, or above 160 psig and/or 250°F hot water

Boilers are also available in a range of heating capacities, which are defined for full-load conditions by the input, gross output, and net output ratings. The input rating defines the amount of heat produced by combustion of the fuel. The gross output rating defines the amount of heat that is actually transferred to the working fluid.[7] The net output (or I-B-R[8]) rating is the gross output rating minus an allowance for the operating heat losses and pickup[9] of an average piping system.

Boilers generally consist of the following parts:

1. Burner

2. Control panel

3. Combustion chamber (or firebox)
4. Fluid pressure vessel
5. Ancillary systems
   a. Gas train or fuel oil delivery system
   b. Operating controls
   c. Possible induced draft fan
   d. Vent
      (1) Positive or negative pressure
      (2) Condensing or noncondensing
      (3) Possible power venting equipment and controls
   e. Safeties
      (1) Pressure relief
      (2) Low water cutoff
      (3) High-limit temperature (hot water) or pressure (steam)
      (4) Gas train or fuel oil delivery system safeties
      (5) Burner safeties
         (a) Flame monitor
         (b) Air proving switch
         (c) Others

Connections

Hot water boilers have the following connections:

- Water inlet
- Water outlet
- Gas or fuel oil
- Pressure relief
- Vent
- Combustion air (normally for condensing-type boilers)
- Condensate drain (condensing-type boilers only)
- Automatic temperature controls
- Electrical

Figures 4-1 through 4-3 illustrate a hot water boiler on a floor plan drawing, a hot water boiler, and a connection detail for a hot water boiler, respectively.

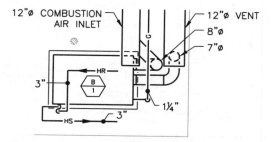

FIGURE 4-1   Floor plan representation of a hot water boiler.

FIGURE 4-2 Photograph of a hot water boiler.

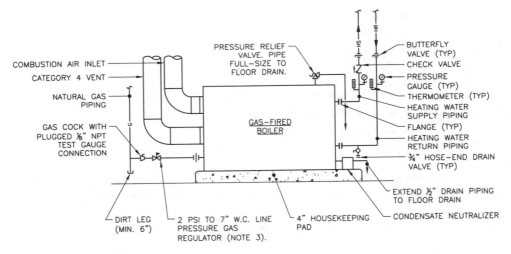

COMBUSTION AIR INLET

CATEGORY 4 VENT

NATURAL GAS PIPING

GAS COCK WITH PLUGGED ⅛" NPT TEST GAUGE CONNECTION

PRESSURE RELIEF VALVE. PIPE FULL-SIZE TO FLOOR DRAIN.

GAS-FIRED BOILER

BUTTERFLY VALVE (TYP)

CHECK VALVE

PRESSURE GAUGE (TYP)

THERMOMETER (TYP)

HEATING WATER SUPPLY PIPING

FLANGE (TYP)

HEATING WATER RETURN PIPING

¾" HOSE-END DRAIN VALVE (TYP)

EXTEND ½" DRAIN PIPING TO FLOOR DRAIN

CONDENSATE NEUTRALIZER

DIRT LEG (MIN. 6")

2 PSI TO 7" W.C. LINE PRESSURE GAS REGULATOR (NOTE 3).

4" HOUSEKEEPING PAD

NOTES:

1. REFER TO PLANS FOR PIPING, INLET, AND VENT SIZES.
2. THERMOMETERS, PRESSURE GAUGES, AND LINE PRESSURE GAS REGULATOR SHALL BE FURNISHED WITH BOILER FOR FIELD INSTALLATION.
3. PROVIDE COPPER VENT PIPING, FULL SIZE OF VENT CONNECTION, TO THE OUTDOORS IN ACCORDANCE WITH NFPA 54.
4. BOILER SHALL BE FURNISHED WITH A COMPLETE GAS TRAIN COMPLYING WITH THE STANDARD FOR CONTROLS AND SAFETY DEVICES FOR AUTOMATICALLY FIRED BOILERS, ANSI/ASME CSD-1.

FIGURE 4-3 Connection detail for a hot water boiler.

Steam boilers have the following connections:

- Water (steam condensate return) inlet
- Steam outlet
- Gas or fuel oil
- Pressure relief
- Vent
- Automatic temperature controls
- Electrical

Design Considerations

Boilers have the same design considerations as all gas-fired and fuel-burning appliances. Therefore, we will discuss these design considerations first and then discuss the design considerations that are specific to boilers.

All Gas-Fired and Fuel-Burning Appliances   The following design considerations are common for all gas-fired and fuel-burning appliances.

Combustion Air   The combustion of fuel by all gas-fired and fuel-burning appliances[10] requires a supply of air (referred to as combustion air). Combustion air can be (1) obtained from the room in which the appliances are installed, (2) transferred from spaces that are adjacent to the room in which the appliances are installed, (3) obtained from the outdoors by openings in the room, (4) obtained from the outdoors by a mechanical forced-air system, or (5) obtained by a direct connection of outdoor combustion air to each appliance.

The amount of combustion air that is required for proper operation of gas-fired and fuel-burning appliances is proportional to the combined input rating of the burners for all appliances installed within the room. The *National Fire Protection Association (NFPA) Standard 54—National Fuel Gas Code* defines the combustion air requirements for gas-fired appliances. *NFPA Standard 31—Standard for the Installation of Oil-Burning Equipment* defines the combustion air requirements for fuel-burning appliances.

Draft   All gas-fired and fuel-burning appliances fall into one of three categories:

- Natural draft: No draft is applied to the flue gas outlet of the appliance. A draft hood (Fig. 4-4) is normally installed on the flue gas outlet to maintain this point at atmospheric pressure and also to isolate the combustion process in the combustion chamber of the appliance from variations in the draft of the vent system.
- Induced draft: Draft is applied at the flue gas outlet of the appliance to induce the flow of flue gas through the combustion chamber. This is accomplished either through the buoyancy of the flue gas in the vent system, through an induced draft fan mounted on the flue gas outlet of the appliance, or both. A barometric draft regulator (Fig. 4-5) is often used in the absence of an induced draft fan to control excessive draft from the vent system. It does this by allowing room air to enter the vent system when the draft exceeds the setting of the weighted damper in the barometric draft regulator.

FIGURE 4-4 Photograph of a draft hood.

FIGURE 4-5 Photograph of a barometric draft regulator.

- Forced draft: The blower on the appliance burner forces the flue gas through the combustion chamber with a positive pressure that results in a positive pressure at the flue gas outlet. This pressure may be sufficient to force the flue gas through the vent system to the outdoor vent terminal. However, it does not need to do this (it depends upon the static pressure capabilities of the burner blower and the vent system design). Forced draft appliances do not have a draft hood or barometric draft regulator on the flue gas outlet; the vent connects directly to the flue gas outlet.

Vent System   A vent system is required for all gas-fired and fuel-burning appliances to safely convey the flue gas from the appliances to the outdoors. A properly designed vent system must address the following factors:

- The type of fuel utilized by the appliances
- The temperature of the flue gas
- The height of the chimney[11]
- The length of each section of the vent system and the combined input rating of the appliances that are connected to each section
- The type of draft of the appliances (natural, induced, or forced)
- The pressures at the appliance flue gas outlets
- The presence of draft regulating devices for the appliances such as draft hoods or barometric draft regulators
- The potential for condensation within the vent system
- The potential need for power venting equipment in the vent system to supplement the draft created by the height of the chimney and the buoyancy of the flue gas

Incorporating these factors into the vent system design is an involved process, the details of which are beyond the scope of this book. The HVAC system designer should contact a representative of a manufacturer of factory-prefabricated vent systems to assist in the design of the vent system for each project. These representatives are commonly skilled in the design of vent systems and also have access to computer programs, which assist in this task. The HVAC system designer should provide the manufacturer's representative with a floor plan showing the locations of the appliances and the desired outdoor vent terminal, as well as the other pertinent information listed above. This will enable the manufacturer's representative to design a proper vent system for the project.

Although it is not important for the HVAC system designer to fully understand all of the details associated with designing the vent system, it is important to have a basic understanding of the various types of vent systems. *NFPA Standard 211—Standard for Chimneys, Fireplaces, Vents, and Solid Fuel-Burning Appliances* defines the requirements for vent systems, which vary depending upon the type of appliance and the fuel utilized; *NFPA Standard 54—National Fuel Gas Code* describes the requirements for the installation of gas-fired appliances; and *NFPA Standard 31—Standard for the Installation of Oil-Burning Equipment* describes the requirements for the installation of fuel-burning appliances.

Gas-Fired Appliances   Vented gas-fired appliances have been grouped by *NFPA Standard 54—National Fuel Gas Code* into the following four categories, depending upon whether the static pressure in the vent system is positive or nonpositive (negative or neutral) and also whether the flue gas temperature is sufficiently high enough to prevent condensation or not:

- Category I: nonpositive, noncondensing
- Category II: nonpositive, condensing
- Category III: positive, noncondensing
- Category IV: positive, condensing

Appliances with dissimilar vent characteristics should be vented separately; that is, each vent system should only serve the same category of appliances. For example, multiple Category I appliances may be combined into a single Category I vent system and multiple Category IV appliances may be combined into a single Category IV vent system provided the vent systems are designed properly. However, Category I appliances should not be combined into the same vent system as Category IV appliances because the positive pressure of the Category IV appliances will affect the pressure at the flue gas outlet of the Category I appliances. This could cause the flue gas from the Category I appliances to spill out into the equipment room and could also affect the operation of the Category I appliance burners. Similarly, condensing appliances should not be combined into the same vent system as noncondensing appliances because the low flue gas temperatures of the condensing appliances could cause undesirable condensation within the vent system and will also affect the buoyancy of the flue gas within the vent system, which may be required for proper operation of the noncondensing appliances.

It is common for all categories of vent systems to be double-walled; that is, they are constructed of an inner conduit, which carries the flue gas, and an outer conduit, which is separated from the inner conduit by an air space or insulation. This double-wall construction lowers the temperature drop of the flue gas in the vent system, which is an important factor in maintaining the buoyancy of the flue gas and in preventing condensation of the flue gas for noncondensing appliances. It also allows the clearance from the outer wall of the vent to combustible building materials to be greatly reduced depending upon the applicable building code.

Category I gas-fired appliances normally utilize a draft hood at the connection of the vent to each appliance. The draft hood, which allows room air to be entrained into the vent system to compensate for variations in the draft of the vent system, ensures that the vent system remains under neutral or negative pressurization. The type of vent utilized for Category I gas-fired appliances is referred to as Type B gas vent.

The seams in the vent system for Category I gas-fired appliances are not required to be pressure-tight because any leakage in the vent system would allow room air to leak into the vent system and not vice versa. The seams in the vent system for Category III and Category IV gas-fired appliances must be pressure-tight in order to prevent the leakage of flue gas out of the vent system. The seams in the vent system for Category II and IV gas-fired appliances must be liquid-tight to prevent the leakage of condensation out of the vent system.

Fuel-Burning Appliances   A Type L vent is utilized for venting oil-burning appliances. Refer to Chap. 6, *NFPA Standard 31—Standard for the Installation of Oil-Burning Equipment* for the venting requirements of oil-burning appliances.

Power Venting   Power venting may be required if there is insufficient draft in the vent system to convey the flue gas from the appliances to the outdoor vent terminal. Power venting may consist of a fan[12] mounted in the vent system between the appliances and the vent terminal or it could consist of a fan mounted on the vent terminal itself. The purpose of the power venting fan is to provide the required boost for the draft in the vent system to convey the flue gas safely to the outdoors. In order to provide the proper amount of boost, the power venting fan must have a means of controlling its capacity based on the need for draft in the vent system. Typically, a static pressure sensor is installed in the vent system and a variable frequency drive is utilized to control the capacity of the power venting fan. A control system is required to vary the speed of the fan in order to maintain the setpoint of the static pressure sensor. The setpoint can be estimated during the design, but it will need to be adjusted to the appropriate setting during start-up of the appliances. Once again, the HVAC system designer should consult with a representative of the vent system manufacturer for assistance in determining the need for power venting and designing the appropriate vent system for the project.

Outdoor Vent Terminals   *NFPA Standard 54—National Fuel Gas Code, NFPA Standard 31—Standard for the Installation of Oil-Burning Equipment*, and applicable local codes describe clearances that must be maintained between vent terminals for gas-fired and fuel-burning appliances and outdoor air intakes for HVAC systems, doors, operable windows, and the like. The purpose of these clearances is to minimize the re-entrainment of flue gas into the building.

Gas Piping and Delivery Pressure   It is common for the burners on gas-fired appliances to require a minimum gas pressure of approximately 4 to 6 in. water column (w.c.) and a maximum gas pressure of 12 to 14 in. w.c. Since the typical delivery pressure for a low-pressure gas system is approximately ¼ psig (7 in. w.c.), direct connection of the low-pressure gas piping to the burner is acceptable. However, if a medium-pressure gas distribution system is utilized in the building, such as a 2-psig (56-in. w.c.) gas system, a 2-psig to ¼-psig regulator will be required at each gas-fired appliance. Also, most 2-psig gas regulators require that the regulator be vented through a pipe to the outdoors. The vent pipe allows the regulator to vent gas to the outdoors should the regulator fail or need to relieve excess pressure. *NFPA Standard 54—National Fuel Gas Code* and applicable local codes should be consulted for all gas piping requirements.

Boilers   The following design considerations are specific to boilers.

Capacity Control   Because the load of any heating system fluctuates over time, it is necessary to vary the heat output of each boiler in the system to match the system load as closely as possible. The heat output of a boiler is controlled by varying the amount of fuel that is burned by the burner (referred to as the firing rate of the burner). There are several ways to control a burner's firing rate, each of which requires a different control system and burner configuration. These methods are as follows:

- On/off control: The simplest form of controlling the firing rate is by cycling the burner on and off. This is referred to as on/off control. When the outlet water temperature (or steam pressure) drops below the setpoint of the outlet water temperature (or steam pressure) sensor, the burner fires at full capacity (high fire). When the outlet water temperature (or steam pressure) rises above the setpoint, the burner is shut off. On/off control is the least energy-efficient means of capacity control because the pre- and postpurge airflows of the boiler required for each cycle remove the heat stored in the boiler and send it to the outdoors.

- Low/high/off control: For somewhat closer capacity control (and increased energy efficiency), the burner can be energized at a low fire setting when the outlet water temperature (or steam pressure) drops below the low fire setpoint. When the outlet water temperature (or steam pressure) drops below the high fire setpoint, the firing rate is increased to the high fire setting. When the outlet water temperature (or steam pressure) rises above the high fire setpoint, the firing rate is reduced to low fire. Upon a further rise in the outlet water temperature (or steam pressure) above the low fire setpoint, the burner is shut off. Low/high/off control reduces the on/off cycles, thereby reducing the heat lost during the pre- and postpurge airflows of the boiler. Keeping the boiler online also improves the response of the boiler to changes in the heating load.

- Modulating control: The most energy-efficient means of capacity control is full modulation of the burner firing rate. The firing rate is modulated by the outlet water temperature (or steam pressure) sensor in proportion to the deviation from the temperature or pressure setpoint. A modulating burner most closely matches the heat output of the boiler to the load of the heating system.

Efficiencies   The efficiency of the boiler(s) in a centralized heating system is an important design consideration because it directly affects the operating cost of the heating system. The type of boiler that is selected, either noncondensing or condensing, is a major factor in the boiler efficiency that can be achieved.

Noncondensing Boilers   Noncondensing boilers (typically cast iron sectional, steel, and copper boilers) must maintain a flue gas temperature that is at least 140°F above the dew point[13] of the water vapor in the flue gas in order to keep this water vapor in the gaseous phase. The typical flue gas temperature for noncondensing boilers is approximately 360°F for typical natural gas combustion. The stack loss associated with this high flue gas temperature results in combustion efficiencies[14] of noncondensing boilers in the range of 70 to 86% and in thermal efficiencies[15] generally in the 80% range. If the flue gas temperature is allowed to drop below 140°F above the dew point of the water vapor in the flue gas, the water vapor may condense in the combustion chamber of the boiler or vent system and cause corrosion of these elements. Noncondensing boilers can produce either hot water or steam.

Condensing Boilers   Condensation of the water vapor in the flue gas within the combustion chamber is not a concern for condensing boilers (typically stainless steel and aluminum boilers) whose flue gas temperatures are less than 140°F above the dew point of the water vapor in the flue gas. In fact, these boilers are designed to condense and drain the water vapor in the flue gas in the combustion chamber of the boiler in order to extract the latent heat of vaporization from the condensed water vapor. This additional heat extraction (and resulting low stack loss) gives condensing boilers combustion and thermal efficiencies in excess of 95%, with higher efficiencies achieved at lower return water temperatures. Condensing boilers must produce hot water—they cannot produce steam.

   Instead of discharging flue gas that is 200°F warmer than the return water temperature (as is the case with noncondensing boilers), the flue gas that is discharged by condensing boilers is in the range of 0 to 100°F warmer than the return water temperature for condensing boilers (depending upon the firing rate and return water temperature).

This reduction in stack loss and lower average water temperature within the boiler is the reason condensing boilers have higher combustion and thermal efficiencies than noncondensing boilers.

Swing Joints on Supply Piping   Swing joints are used to minimize the stresses on the piping connections to boilers. A swing joint is constructed with standard piping elbows. However, its geometry is such that any expansion in the return or supply pipes has the effect of torsionally rotating the pipe within the tapping of the boiler, as opposed to imposing a bending moment on the tapping.

Clearances   Clearances must be maintained around boilers for inspection and maintenance, as required by the manufacturer and also by the applicable codes. Section 1004.3 of the 2009 *International Mechanical Code* requires 18 in. of clearance on all sides, unless otherwise approved. Therefore, if the boiler manufacturer allows less than 18 in. clearance around the boiler, the authority having jurisdiction should be consulted before incorporating this clearance into the design.

Hot Water Boilers   The following design considerations are specific to hot water boilers.

Full-Load (System) Design Temperature Difference   Typical (full-load) design water temperatures for heating water systems are 180 to 200°F heating water supply and 140 to 180°F heating water return. The temperature difference between the heating water supply and heating water return is referred to as the delta t (Δt). A heating water Δt of 20 to 40°F is common for all heating water boilers (noncondensing and condensing). Although a higher Δt can be used for condensing boilers, this is not customary because the terminal heating equipment is usually designed to accommodate a 20 to 40°F Δt.

The higher (full-load) design heating water supply temperature of 200°F may be desirable in colder climates to increase the heat transfer efficiency[16] of the heating equipment, thus reducing the first cost of this equipment. That is, a heating coil with 200°F heating water supply will transfer more heat to the surrounding air than the same heating coil with 180°F heating water supply (assuming equal heating water flows, airflows, and entering air temperatures). This increased heat transfer will result in a greater heating water Δt for the coil with 200°F heating water supply temperature than the coil with the 180°F heating water supply temperature. However, heat losses from the boilers and piping system also increase with an increase in the heating water supply temperature, thus increasing the operating cost.

Operating Pressure   When a heating water system is initially filled, the water is at its coldest temperature and the heating water system pump is not operating, the water pressure in the boiler will be equal to the static water pressure of the piping system on the boiler (the height of the water column of piping installed above the boiler). Once the system has been filled and most of the air has been released from the high points, it is common to adjust the fill pressure of the cold water makeup to ensure that a positive pressure of approximately 10 psig is measured at the highest point in the system when it is cold and not operating.

When the water in the heating water system is heated to the operating temperature and the heating water system pump is operating, the static water pressure in the entire heating water system will increase due to the expansion of the heated water.[17] In addition, the static water pressure everywhere in the system [except at the point where the (closed) expansion tank connects to the system[18]] will increase due to the total dynamic head of the heating water system pump. Because heating water boilers are rated by the maximum water temperature and pressure they are designed to withstand, a high-limit temperature sensor and pressure relief valve are required to be installed on the boiler with settings that do not exceed these ratings of the boiler. Although the purpose of these safeties is to protect the boiler, it is important for the HVAC system designer to ensure that the static water pressure in the boiler does not exceed the pressure setting of the pressure relief valve when the heating water system is hot and operating.

Location of Boiler in Heating Water System   In most cases, it is appropriate for the boiler to be installed on the suction side of the heating water system pump with the point where the (closed) expansion tank connects to the system located between the boiler and the pump. The reason is that the pressure at this point is not affected by the total dynamic head of the heating water system pump. Therefore, the difference between the static pressure rating of the boiler and the (hot) operating static pressure of the system on the boiler can be developed by the system pump to overcome the friction losses of the heating water system. However, if the boiler is installed on the discharge side of the system pump, the static water pressure in the boiler will be equal to the (hot) operating static pressure of the system on the boiler plus the total dynamic head of the pump. This configuration decreases the total dynamic head that the system pump can develop to overcome the friction losses of the heating water system.

Noncondensing Boilers   The following design considerations are specific to noncondensing hot water boilers.

Minimum Return Water Temperature   The cast iron, steel, and copper combustion chambers of noncondensing boilers can be corroded if the water vapor in the flue gas is allowed to condense within the combustion chamber. Condensation will occur within the combustion chamber when the return water temperature is at or below the dew point of the water vapor in the flue gas, which is approximately 130°F for typical natural gas combustion. Furthermore, cast iron and steel boilers are susceptible to thermal shock,[19] which can occur when the return water temperature is too low. As a result, a minimum return water temperature of 140°F is generally recommended for noncondensing boilers not only to prevent condensation in the combustion chamber but also to minimize the risk of thermal shock.

Constant Outlet Water Temperature   The firing rate of the burner on a noncondensing boiler is usually controlled to maintain a constant outlet water temperature, typically either 180 or 200°F. The heating water supply temperature and heating water system Δt need to be selected to ensure that the heating water return temperature does not drop below 140°F.

Heating Water Temperature Reset   Because the maximum heating water supply temperature of 180 to 200°F is only needed to meet the design heating load of the building (when the outdoor temperature is at its lowest), it is common for the heating water supply temperature to be automatically reset[20] (or adjusted) by the heating water control system to a lower temperature during part-load heating conditions. During part-load conditions, the cooler heating water supply temperature enables the control systems for the heating equipment to more closely match the heat output of this equipment to the loads they serve. If the heating water supply temperature is not reset for part-load conditions, the heating equipment control systems may overshoot the controlled variable (such as space temperature), causing uncomfortable conditions in the spaces they serve.

For noncondensing boilers, the heating water supply temperature at the outlet of the boiler must remain constant for the reasons stated earlier. Reset of the heating water supply temperature for the heating water system must be accomplished by mixing a portion of the heating water that is returned from the building with the heating water that is supplied by the boilers. This is accomplished by installing a 3-way mixing valve in the heating water supply pipe, with one inlet port connected to the heating water returned from the building and the other inlet port connected to the heating water supplied by the boiler (Fig. 4-37). The outlet port is connected to the heating water that is supplied to the building. A temperature sensor is installed in the heating water supply to the building downstream of the 3-way mixing valve. This heating water supply temperature is input to the control system, and the 3-way mixing valve is modulated as required to maintain the (varying) setpoint of the heating water supply temperature, which is reset based on the outdoor temperature. A common heating water supply temperature reset schedule supplies heating water to the building at 180°F when the outdoor temperature is 0°F and supplies 110°F heating water to the building when the outdoor temperature is 50°F.

Condensing Boilers   The following design considerations are specific to condensing hot water boilers.

Flue Gas Condensate   Because the acids in the flue gas condensate are corrosive, the combustion chamber (also referred to as the heat exchanger) in condensing boilers must be constructed of corrosion-resistant materials such as stainless steel or aluminum. Condensing boilers have a condensate drain connection that must be piped to the building sanitary system, in most cases (depending on the applicable local code). All condensate that is discharged to the building sanitary system must be below 140°F. Therefore, a condensate cooler may be required. The cooler mixes domestic water with the hot flue gas condensate to maintain a maximum discharge temperature of 140°F. Also, because the flue gas condensate is acidic, an acid neutralizer (which normally consists of a canister filled with limestone) is often utilized to neutralize the pH of the condensate before it is discharged to the building sanitary system. This is of particular concern if the building sanitary piping is constructed of cast iron.

Low Return Water Temperature Desirable   Thermal shock is not an issue with condensing boilers. In fact, return water temperatures below 140°F are desirable to increase the condensation of water vapor in the flue gas.

Constant Water Flow   Because condensing boilers generally contain a smaller volume of water than noncondensing boilers, it is necessary for the water flow through these boilers to remain constant in order to prevent "hot spots" in the heat exchangers. Some condensing boilers are equipped with factory-installed circulating pumps that are integral to the boiler; others require a pump to be field-installed external to the boiler.

One method of ensuring a constant water flow through each boiler is to utilize a primary-secondary pumping system (Fig. 4-38). In this type of system, there are two loops: the primary loop, which includes the boiler, and the secondary loop, which includes the heating water system in the building. The two loops are connected by the "common pipe." Normally, the primary pumps circulate a constant water flow through the boilers, while the secondary pump can circulate either a constant or variable flow through the heating water system. Flow through the common pipe will vary depending upon the number of primary pumps that are operating and the flow through the secondary loop.

Variable Outlet Water Temperature   In contrast to noncondensing boilers, the outlet water temperature for condensing boilers can be varied to reset the heating water temperature based on the outdoor temperature (as discussed in the Noncondensing Boiler section earlier). This capability of varying the outlet water temperature eliminates the need for the 3-way mixing valve in the heating water supply pipe and also results in higher thermal efficiencies at part-load heating conditions because the lower return water temperature is capable of extracting more latent heat from the flue gas than a higher return water temperature can.

Steam Boilers   The following design considerations are specific to steam boilers.

Operating Pressure   Typically, commercial steam boilers in the sizes required to serve a single building are rated for a maximum operating pressure of 15 psig. A common operating pressure for commercial steam boilers serving a single building is 2 psig. This steam pressure is adequate to overcome the pressure losses in the distribution piping for the relatively small heating system contained within a single building with a minimum amount of wasted energy due to flash steam.[21] Also, the 219°F saturation temperature of 2 psig steam is not so hot that certain pieces of heating terminal equipment, such as finned-tube radiators, cannot be safely enclosed and mounted within the occupied areas of the building.

However, district or campus steam heating systems, which may utilize miles of steam piping to serve multiple buildings, require a higher initial steam pressure to overcome the significant pressure losses in the distribution piping. For this reason, the boilers serving district or campus steam heating systems commonly operate at steam pressures in excess of 100 psig.

Condensate Recovery   For most commercial buildings, the steam condensate that is formed at the steam utilization equipment (air system heating coils, terminal heating equipment, heat exchangers, humidifiers, etc.), as well as condensate that is formed in the steam distribution piping itself, is returned to the central plant where it is reused as feedwater for the steam boiler.

The steam condensate is usually returned by steam pressure motivation[22] from the steam utilization equipment to the condensate transfer equipment, which consists of one or more condensate receiver/pump sets located within the building. Each condensate receiver/pump set consists of a receiver (tank) and either a single condensate pump (simplex arrangement) or two condensate pumps (duplex arrangement). A float switch installed in the receiver energizes the condensate pump at the high level and de-energizes the pump at the low level. Steam condensate is pumped by the condensate pump through the pumped condensate return piping to the boiler feedwater receiver located in the central plant. If the central plant is located below all of the steam utilization equipment, the steam condensate may return directly to the boiler feedwater receiver, eliminating the need for the condensate receiver/pump sets.

Makeup water is added to the boiler feedwater receiver as required by the water level controls in the receiver. The boiler feedwater is pumped from the boiler feedwater receiver to the steam boiler(s) by the boiler feed pump(s). Typically, each boiler is fed by a dedicated feedwater pump. These pumps are controlled by the water level controls in the steam boilers to maintain the proper water levels in the boilers.

The advantages of recovering condensate, as opposed to discharging it to the sanitary drainage system, are as follows: the recovery of the heat contained within the hot steam condensate, lower makeup water requirements, and reduced water treatment. If it is anticipated that more than 10% of the feedwater will consist of makeup water, a feedwater heater should be added to the boiler feedwater receiver to keep the water in this receiver at 180 to 200°F. Maintaining the feedwater at this temperature ensures that the dissolved oxygen[23] in the feedwater is removed before it is pumped to the boilers. This also minimizes the risk of thermal shock to the boilers.

Heat Exchangers

Heat exchangers are commonly used in centralized heating systems and water-source heat pump systems.

Purpose

Heat exchangers are used in HVAC systems where it is necessary to transfer heat from one fluid stream to another without mixing the two fluid streams. Heat exchangers can be used to transfer heat from steam to water, water to water, water to brine, or brine to water. A common application of a steam-to-water heat exchanger is in the transfer of heat from a district or campus steam heating system to a heating hot water system within a building. Another application of a heat exchanger—where separation of the two fluid streams is of primary concern—is in the transfer of heat from a closed heat pump water system to an open cooling tower system. Heat exchangers can also be used to isolate fluid systems in tall buildings, thereby preventing overpressurization of these systems due to static head.

Physical Characteristics

There are two types of heat exchangers commonly used in HVAC systems: U-tube shell and tube heat exchangers and plate and frame heat exchangers.

U-tube Shell and Tube Heat Exchangers   U-tube shell and tube heat exchangers consist of a copper U-tube bundle mounted within a steel cylindrical shell. The cold fluid stream normally circulates through the tubes of the tube bundle, and the hot fluid stream normally circulates through the shell (around the tube bundle). Heat is transferred from the hot fluid to the cold fluid through the walls of the tubes. Shell and tube heat exchangers are commonly used to transfer heat from steam to water or brine. However, they can also be used to transfer heat from water to water, water to brine, or brine to water. Baffles are installed in the shell to direct water flow across the tubes if the heat exchanger is used to transfer heat from water to water, water to brine, or brine to water. A minimum of two passes of the fluid through the tube bundle are required for U-tube shell and tube heat exchangers. For most HVAC applications, shell and tube heat exchangers are between 3 and 6 ft long and 6 and 12 in. in diameter, although larger heat exchangers are available. Figure 4-6 illustrates the piping connections to a steam-to-hot water shell and tube heat exchanger.

Plate and Frame Heat Exchangers   Plate and frame heat exchangers consist of two main components:

- Plates: Multiple plates, normally constructed of stainless steel, form the surface through which heat is transferred from the hot fluid to the cold fluid. Channels formed between the stainless heat exchanger plates provide the narrow passages through which the two fluid streams circulate. The hot fluid circulates through the channels on one side of each plate while the cold fluid circulates through the

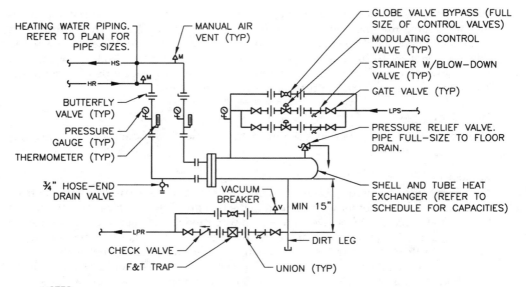

FIGURE 4-6 Connection detail for a steam-to-hot water shell and tube heat exchanger.

NOTES:

1. ARRANGE PIPING CONNECTIONS TO FACILITATE REMOVAL OF TUBE BUNDLE FROM HEAT EXCHANGER WITH A MINIMUM DISTURBANCE TO PIPING.

2. PROVIDE TWO CONTROL VALVES IN PARALLEL AS SHOWN FOR CONTROL VALVE SIZES 2½" AND LARGER.

3. GLOBE VALVE BYPASS AROUND TRAP SHALL BE FULL SIZE OF TRAP.

4. SET PRESSURE RELIEF VALVE FOR 5 PSIG.

channels on the other side of each plate. The heat exchanger plates separate the two fluid streams and conduct heat from the hot fluid to the cold fluid. Elastomer gaskets are used to seal the joints between the plates. Plate and frame heat exchangers are used strictly to transfer heat from water to water, water to brine, or brine to water; they cannot be used to transfer heat from steam to water. A steel shroud surrounds the plate pack to protect the plates from damage.

*   Frame: The frame is normally constructed of steel and holds the plates together. The frame consists of a fixed steel front plate (called the fixed head) and a movable steel rear plate (called the movable head). The stainless steel heat exchanger plates are sandwiched between the two heads with compression bolts. There may also be a top carrying bar and a bottom guide bar; these bars support the plates. The carrying bar and guide bar are connected between the fixed head and a rear vertical support frame. Heat exchanger plates can be added after the initial installation if additional capacity is required as long as the carrying bar, guide bar, and compression bolts are long enough to accommodate the additional plates and the system pump has sufficient capacity for the increased flow rates and pressure losses.

Plate and frame heat exchangers are commonly 5 to 7 ft high, 5 to 8 ft long, and 3 to 5 ft wide, although larger heat exchangers are available.

Refer to Chap. 47 of the 2008 *ASHRAE Handbook—HVAC Systems and Equipment* for more information on the physical characteristics of various types of heat exchangers.

Connections

Connections to heat exchangers are limited to the piping for the entering and leaving hot and cold fluids, and a pressure relief valve (for shell and tube heat exchangers).

U-tube Shell and Tube Heat Exchangers   For shell and tube heat exchangers, there is a connection on the top of the shell for the entering hot fluid and a connection on the bottom of the shell for the leaving hot fluid. If the shell and tube heat exchanger is used to transfer heat from steam to water (or brine), the steam supply connection is on the top of the shell and the steam condensate return connection is on the bottom of the shell. The tube bundle (through which the cold fluid circulates) is mounted to a tube sheet that is sandwiched between the shell and head flanges. The shell flange is mounted to the shell of the heat exchanger, and the head flange contains the entering and leaving cold fluid piping connections. This connection of the tube sheet between the shell and head flanges facilitates the removal of the tube bundle for maintenance or replacement. Also, a tapping is normally provided in the top of the shell for installation of a pressure relief valve, which functions to protect the shell from overpressurization. If there is no pressure relief valve tapping in the shell, the pressure relief valve is connected to the hot entering fluid piping on the equipment side of the isolation valve.

Plate and Frame Heat Exchangers   Connections for plate and frame heat exchangers are limited to the entering and leaving hot and cold fluid connections, which are normally an integral part of the (front) fixed head of the heat exchanger [although connections can also be provided in the (rear) movable head of the heat exchanger]. Depending upon the channel configuration of the heat exchanger plates, the entering and leaving connections for the hot and cold fluids can be located either on the same side of the heat exchanger fixed head or in a diagonal arrangement. It is common for the hot and cold fluids to circulate through plate and frame heat exchangers in a counterflow configuration; that is, the hot and cold fluids flow in opposite directions through the heat exchanger. This arrangement, in which the temperature gradient between the hot and cold fluids remains essentially constant, maximizes the heat transfer efficiency of the heat exchanger and also allows for a crossover temperature between the hot and cold fluids. Crossover temperature will be discussed in the Design Considerations section later. Figures 4-7 through 4-9 are graphical representations of a plate and frame heat exchanger. Note that

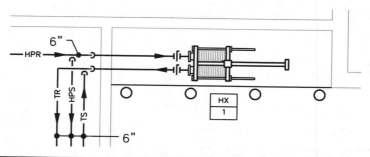

FIGURE 4-7   Floor plan representation of a plate and frame heat exchanger.

FIGURE 4-8 Photograph of a plate and frame heat exchanger.

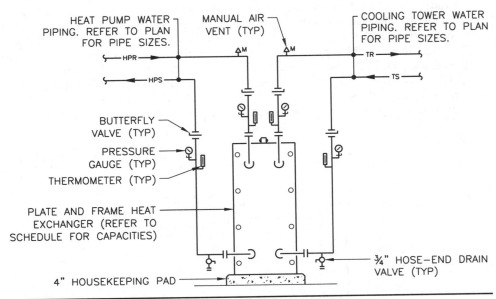

FIGURE 4-9 Connection detail for a plate and frame heat exchanger.

in Fig. 4-7, only two pipes are shown connecting to the heat exchanger on the floor plan because two of the four pipes connected to the heat exchanger are directly over the other two pipes. All four pipes could have been shown on the floor plan slightly staggered, but it is clear that there are actually four pipes connected to the heat exchanger by the pipes shown in the adjacent room and also by the connection detail.

Design Considerations

The following are design considerations for heat exchangers.

Definitions

1. Approach temperature
   a. Heating: The difference between the entering hot-side fluid temperature and the leaving cold-side fluid temperature. For example, the approach temperature for a heat exchanger designed to utilize 200°F (entering) heating water to produce 180°F (leaving) brine is 20°F.[24]
   b. Cooling: The difference between the leaving hot-side fluid temperature and the entering cold-side fluid temperature. For example, the approach temperature for a heat exchanger designed to utilize 38°F (entering) brine to produce 45°F (leaving) chilled water is 7°F.[25]

2. Crossover temperature
   a. Crossover temperature occurs when the fluid leaving the cold-side of a heat exchanger is warmer than the fluid leaving the hot-side of the heat exchanger. Refer to the discussion of crossover temperature in the Plate and Frame Heat Exchanger section later.

3. Fouling factor
   a. Over time, scale and other deposits coat the heat transfer surfaces of heat exchangers. In order to account for this increase in thermal resistance of the heat transfer surfaces and the resultant loss of capacity, a fouling factor[26] is used during the selection process to ensure that the heat exchanger retains sufficient capacity for the life of the equipment. Fouling factors vary for different fluids and depend upon the nature and temperature of the fluid and the fluid velocity through the heat exchanger. Typical fouling factors for fluids flowing at a velocity of 3 feet per second (fps) or higher are as follows:
      (1) Treated water, closed system: 0.0005 ft²·h·°F/Btu (125°F or less), 0.0010 ft²·h·°F/Btu (over 125°F)
      (2) Treated water, open system (125°F or less): 0.0010 ft²·h·°F/Btu
      (3) Untreated water, open system (125°F or less): 0.0030 ft²·h·°F/Btu
   b. The heat exchanger manufacturer's product data and/or the manufacturer's representative should be consulted to determine the appropriate fouling factors that should be used for each fluid.

U-tube Shell and Tube Heat Exchangers

1. Capacity
   a. The heat transfer capacity of shell and tube heat exchangers is proportional to the surface area of the tubes. Therefore, utilizing a larger tube bundle (which requires either a larger diameter or longer shell) increases the heat transfer capacity of the shell and tube heat exchanger.

2. Tube bundle—Number of passes
   a. U-tube heat exchangers utilize either two- or four-pass tube bundles. Typically, a two-pass tube bundle is sufficient for most applications. However, if length is a concern, a four-pass tube bundle can be utilized; this would increase the diameter of the shell. Conversely, if length is not a concern, but the diameter of the shell is, a longer heat exchanger with a two-pass tube bundle can be utilized.
   b. The entering and leaving fluid connections to the U-tube bundle are on the head end of the heat exchanger.

3. Pressure rating
   a. The shell of a shell and tube heat exchanger is normally constructed to meet the requirements of the *ASME Boiler and Pressure Vessel Code,* Section VIII, as an unfired pressure vessel. The operating pressure of the heat exchanger shell cannot exceed the ASME rating of the shell.
   b. Heat exchanger shells and tubes are typically rated for an operating pressure of 150 psig at 375°F.
   c. A pressure relief valve is required to protect the shell from over-pressurization.

4. Vacuum breaker
   a. A vacuum breaker is required to be installed in the shell of a steam-to-hot water shell and tube heat exchanger if the steam flow through the heat exchanger is controlled by a 2-way control valve. The vacuum breaker prevents negative pressurization of the shell, which can occur when there is no steam flow and the steam within the shell cools and contracts. If a negative pressure develops within the shell of the heat exchanger, it can inhibit proper operation of the steam trap (which requires a positive pressure to operate) and cause condensate flooding within the shell of the heat exchanger.

5. Flow rates
   a. It is common for both the hot- and cold-side fluid flow rates and temperature differentials to be equal. However, unequal flow rates and temperature differentials can be designed if one fluid is water and the other fluid is brine or for other reasons that may be necessary.

6. Fluid velocity
   a. The fluid velocity within the tubes should be a minimum of 3.3 fps to reduce the potential for fouling and a maximum of 11 fps to minimize erosion of the tubes.
   b. The manufacturer's product data and/or the manufacturer's representative should be consulted for recommended fluid velocities.

7. Pressure loss
   a. Reasonable fluid pressure losses for both the shell and tube sides of a water-to-water shell and tube heat exchanger are between 1 and 5 ft w.c.

8. Mounting
   a. Shell and tube heat exchangers are usually mounted approximately 4 ft above the floor on a steel frame. This mounting height facilitates maintenance and provides elevation for the steam condensate piping to be pitched to the condensate receiver.

9. Steam trap selection
   a. If a 2-way control valve is used to control the steam flow through the heat exchanger (which is normally the case), the steam pressure within the shell of a shell and tube heat exchanger will fluctuate depending upon the load on the heat exchanger. Therefore, a float and thermostatic (F&T) trap is the appropriate type of steam trap for this application because it adjusts automatically to the condensate load and is not adversely affected by changes in steam pressure. The F&T trap should be mounted a minimum of 15 in. below the bottom of the shell in order for there to be 15 in. of water column (approximately 0.5 psi) of positive pressure (minus the approximately 0.25 psi pressure loss through the vacuum breaker) on the trap to facilitate its operation should the pressure within the heat exchanger drop to 0 psig.

10. Approach temperature
   a. Shell and tube heat exchangers are typically utilized for systems having approach temperatures greater than 10°F.

11. Crossover temperature not possible
   a. Although a crossover temperature is possible for a plate and frame heat exchanger (see discussion below), it is not possible for a shell and tube heat exchanger. The fluid leaving the cold side of a shell and tube heat exchanger must be cooler than the fluid leaving the hot side of the heat exchanger.

12. Automatic temperature control
   a. Typically, the fluid temperature leaving the load side of a heat exchanger is the controlled variable. In order to maintain this temperature, the flow rate of the source fluid through the heat exchanger is modulated.
   b. If the hot-side fluid is steam, a 2-way control valve is used to modulate the steam flow rate through the shell of the heat exchanger. In cases where the required steam flow rate exceeds the capacity of a 2½-in. steam control valve, two steam control valves connected in a parallel arrangement are recommended to control the steam flow rate. This parallel arrangement allows for closer control of the steam flow rate at part-load conditions.

13. Clearances
   a. Clear space equal to the length of the shell should be provided beyond the head end of the heat exchanger to allow the tube bundle to be removed for maintenance or replacement. In order to provide this clear space, the shutoff valves, specialties, and unions or flanges for the tube inlet and outlet connections should be installed in the vertical pipe drops at an elevation above the heat exchanger. Sufficient access should also be provided for piping connections to the shell, which will be on the top and bottom of the shell.

Plate and Frame Heat Exchangers

1. Capacity
   a. The heat transfer capacity of plate and frame heat exchangers is proportional to the surface area of the heat exchanger plates. Therefore, more plates and/or larger plates increase the heat transfer capacity of the heat exchanger.

2. Pressure rating
   a. Plate and frame heat exchangers are normally constructed to meet the requirements of the *ASME Boiler and Pressure Vessel Code*, Section VIII, as an unfired pressure vessel and are typically rated for an operating pressure of 150 psig.

3. Flow rates
   a. Similar to shell and tube heat exchangers, unequal hot- and cold-side fluid flow rates and temperature differentials can be accommodated with plate and frame heat exchangers.

4. Pressure loss
   a. The passages through which fluid flows in a plate and frame heat exchanger are much smaller than those in the shell or tubes of a shell and tube heat exchanger. Consequently, the fluid pressure loss for a plate and frame heat exchanger will be much higher than the fluid pressure loss through a shell and tube heat exchanger.
   b. The fluid pressure loss through a plate and frame heat exchanger is proportional to the fluid velocity through the channels between the plates.
   c. A reasonable fluid pressure loss through a plate and frame heat exchanger is 5 psi.

5. Mounting
   a. Plate and frame heat exchangers are usually mounted on a concrete housekeeping pad that is 4 in. high and 4 in. larger than the equipment on all sides.

6. Temperature rating
   a. The temperature rating of plate and frame heat exchangers is determined based on the elastomer gasket material that is used to seal the joints between the plates. These gaskets are typically constructed of NBR, but EPDM and Viton® gaskets can also be utilized for higher operating temperatures. NBR has a maximum temperature rating of 230°F, EPDM has a maximum temperature rating of 300°F, and Viton has a maximum temperature rating of 365°F.

7. Approach temperature
   a. Plate and frame heat exchangers are typically utilized for systems having approach temperatures less than 10°F.
   b. The approach temperature can be reduced by utilizing more plates and/or larger plates.

8. Crossover temperature possible
   a. A crossover temperature is possible with a plate and frame heat exchanger. The fluid leaving the cold side of a plate and frame heat exchanger can be warmer than the fluid leaving the hot side of the heat exchanger. This is often desirable, particularly when the approach temperature is less than 10°F. A 5°F crossover temperature is common for a water-source heat pump application, as illustrated in Fig. 4-10.

9. Automatic temperature control
   a. Depending upon the pumping arrangement and control of the source- and load-side fluid systems, either the temperature or flow rate through the hot or cold sides of the heat exchanger may be modulated to control the heat that is absorbed or rejected by the heat exchanger. For example, if the heat

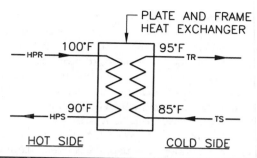

PLATE AND FRAME
HEAT EXCHANGER

100°F   95°F
HPR →   TR →

90°F   85°F
← HPS   TS →

HOT SIDE   COLD SIDE

FIGURE 4-10   Crossover temperature for a typical water-source heat pump application.

exchanger is used in a water-source heat pump system to reject heat to a cooling tower, the temperature of the hot-side fluid (cooling tower loop) entering the heat exchanger will be modulated by varying the flow of air through the cooling tower to maintain a maximum leaving cold-side water temperature. Thus, the hot-side fluid flow rate will remain constant and the cold-side fluid flow rate may be modulated or remain constant.

  b. Modulation of the hot- or cold-side fluid flow rates through the heat exchanger can be accomplished by using a 2-way or 3-way control valve, the choice of which depends upon the pumping arrangement and control of the fluid systems.

10. Clearances

  a. Clear space should be provided behind the heat exchanger for removal of the movable head and heat exchanger plates for maintenance or replacement.

  b. Approximately 3 to 6 ft of space (depending upon the pipe sizes) should be allowed in front of the heat exchanger for the elbows, valves, and specialties that are associated with the hot- and cold-side fluid pipe connections to the fixed head. Space should also be provided in the rear of the heat exchanger for piping connections to the movable head, if required.

Cooling Equipment

The two types of centralized cooling equipment we will discuss in this chapter are chillers and cooling towers.

Chillers

Chillers are commonly used in centralized cooling systems to extract heat from the working fluid of the building cooling system and reject this heat to the outdoors. Electric chillers are the most common types of chillers used in centralized cooling systems. Chillers which utilize steam or natural gas as the fuel source are also used, but are less common.

Purpose

Most people are familiar with the air-conditioning system in their home, which uses a cooling coil containing refrigerant to cool the air that is blown across it by the supply fan. This type of cooling coil, which is referred to as a direct expansion (DX) coil,[27] is also used in HVAC systems for commercial buildings. However, DX coils, and their associated refrigeration systems, have certain limitations (particularly limits on the distance between the cooling coil and the heat rejection equipment), which make them unsuitable for certain HVAC systems, particularly centralized HVAC systems. Chilled water cooling systems, on the other hand, do not have the limitations of DX cooling systems. These systems are utilized extensively in commercial HVAC systems because water can be cooled in a central location and distributed at great lengths to the various air systems and terminal equipment, which utilize this chilled water to cool the building.

At the heart of a chilled water system is the chiller. A chiller is a central piece of equipment whose sole purpose is to cool (or chill) the water returned from the building cooling equipment to a temperature usually between 42 and 45°F. This water is then pumped from the central chilled water plant to the building cooling equipment where it is used to cool the building.

If the chiller is used as part of a thermal storage system, it will produce approximately 18°F brine during off-peak (nighttime) hours. The brine is circulated through the thermal storage system to store thermal (cooling) energy in the thermal storage media, which is typically either ice or brine. During on-peak (afternoon) hours, the cooling energy stored in the thermal storage media is used to cool the building, which reduces the electrical demand of the facility and the associated energy cost during the cooling season.

Physical Characteristics

There are many different types of chillers. However, all chillers essentially consist of equipment that utilizes energy to extract heat from the chilled water loop and reject this heat to the outdoors. The energy source can be electricity, steam, or even natural gas. In this book, we will focus on chillers that utilize electricity as the source of energy to compress a refrigerant in a closed vapor compression system to produce chilled water.

Chillers that utilize steam or natural gas as the energy source are called absorption chillers. Absorption chillers use an absorption process that occurs between a refrigerant (typically water) and an absorbent (typically lithium bromide) to produce chilled water. Absorption chillers are suitable for applications where there is an abundant source of steam or natural gas to drive the absorption process or where it is not desirable to utilize electricity for cooling. Although these types of chillers will not be discussed in this book, their application within a central chilled water plant is similar to that of an electric chiller.

Electric chillers contain the four main components that are a necessary part of a vapor compression system: the evaporator, the compressor, the condenser, and the expansion device.

The evaporator is contained within the cooler, which is one part of a chiller. The cooler is a shell and tube heat exchanger in which the refrigerant circulates through either the shell or the tubes of the cooler[28] and is boiled (or evaporated) by the warmer chilled water that is circulated through the other component of the cooler. The chilled water is cooled by the heat that is absorbed by the evaporation of the refrigerant.

The compressor draws the low-pressure gaseous refrigerant from the evaporator, compresses it, and discharges it at high pressure to the condenser. The most common types of compressors used for electric chillers are reciprocating, scroll, screw, and centrifugal compressors. Each of these compressors utilizes a different approach for compressing the refrigerant and each has characteristics that make it suitable for certain applications and size ranges of chillers.

The condenser rejects all of the heat from the vapor compression system, including the heat that is absorbed by the evaporator in the cooler and the heat that is added to the system by the compressor. Condensers can be water-cooled, air-cooled, or evaporative (air-cooled with water spray). A water-cooled condenser is usually part of a packaged indoor chiller and normally consists of a shell and tube heat exchanger in which the (warmer) refrigerant circulates through the shell of the heat exchanger and is condensed by the (cooler) condenser water that is circulated through the tubes of the heat exchanger. The condenser water is warmed by the heat that is rejected by the condensation of the refrigerant. The heat that is absorbed by the condenser water is commonly rejected to the outdoors through a cooling tower.

An air-cooled condenser is usually part of a packaged outdoor chiller, although it may also be the outdoor component of an indoor chiller. It consists of a finned-tube coil through which the refrigerant is circulated and condensed by outdoor air that is drawn across the coil by one or more condenser fans. An evaporative condenser is an air-cooled condenser which has an additional recirculating water spray system. The evaporative cooling effect of the water sprayed over the condenser coil increases the heat rejection capacity of the condenser.

A thermostatic expansion valve is installed in the refrigerant liquid line between the condenser and the evaporator, which throttles (or regulates) the liquid refrigerant flow to the evaporator. As a result, the pressure in the evaporator is below the saturation pressure of the liquid refrigerant, which causes the refrigerant to boil. The heat required to boil the refrigerant is absorbed from the chilled water circulating through the waterside of the cooler, thereby cooling the chilled water. The thermostatic expansion valve regulates the refrigerant flow through the evaporator to maintain a constant refrigerant suction temperature, ensuring that only refrigerant gas leaves the evaporator.

Connections

Connections for electric chillers consist of the following:

1. Water-cooled chiller (Figs. 4-11 and 4-12)
    a. Chilled water supply and return piping connections to the cooler.
    b. Condenser water supply and return piping connections to the condenser.
    c. Refrigerant pressure relief piping connection.
        (1) The refrigerant pressure relief piping will be connected to the refrigerant pressure relief device (a relief valve, rupture disk, or both) on the chiller and will be routed to the outdoors in accordance with the requirements of *ANSI/ASHRAE Standard 15-2010—Safety Standard for Refrigeration Systems*.
    d. Electrical power connection: The electrical power connection can be either single-point or dual-point.
        (1) Single-point electrical power connection
            (a) In order to simplify the electrical power connection, a single-point electrical connection and unit-mounted disconnect switch are often available as optional accessories. In this case, the three-phase electrical power from the building electrical power distribution system would connect to the chiller at a single point: the unit-mounted disconnect switch. All wiring, unit controls, safeties, starters for the compressor(s) and oil pump, and other electrical components necessary for proper operation of the chiller would be factory-installed so the only field electrical connection required would be made at the unit-mounted disconnect switch. Also, a 120 volt (V)/1 Phase (Ø) control transformer is often available as an

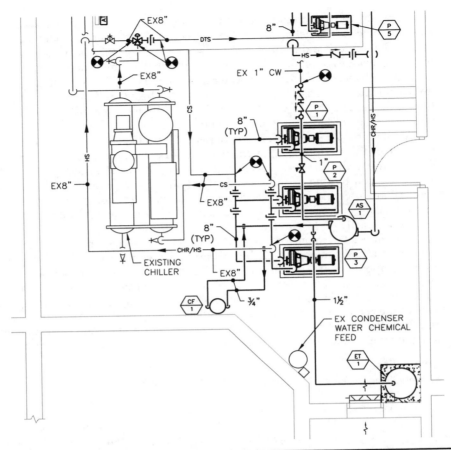

FIGURE 4-11   Floor plan representation of a water-cooled chiller and auxiliary hydronic equipment.

optional accessory so that the 120V/1Ø power required for the 120V/1Ø components on the chiller is derived through this transformer from the three-phase power supply. This option avoids the requirement for a separate 120V/1Ø electrical power connection from the building electrical power distribution system.

(2) Dual-point electrical power connection

(a) If the chiller has two compressors, the standard offering is normally a dual-point electrical connection. In this case, two three-phase electrical power connections are required for the chiller, one for each compressor.

(b) Unit-mounted disconnect switches are normally available for this configuration as an optional accessory and so is the 120V/1Ø control transformer.

(3) Whether the electrical power connection to the chiller is single point or dual point should be coordinated with the project electrical engineer because the design of the building electrical power distribution system may favor one configuration over the other.

Figure 4-12   Photograph of a water-cooled chiller.

   e. Automatic temperature controls (ATC) connection.
      (1) The ATC connection will be made at the chiller control panel and will integrate the operation of the chiller into the overall chilled water system.
2. Air-cooled chiller (Figs. 4-13 through 4-15)
   a. Chilled water supply and return piping connections to the cooler.
   b. Electrical power connection.
      (1) The electrical power connection is the same as is required for a water-cooled chiller, except additional factory-installed wiring and starters will be provided for the condenser fans.
   c. ATC connection.
      (1) The ATC connection is the same as is required for a water-cooled chiller.

Design Considerations
The following are design considerations for chillers.

   1. Chiller selection
      a. Type of compressor: Although it varies from one manufacturer to another, the type of compressors used and the associated range of chiller capacities are generally as follows:
         (1) Reciprocating compressors: 25- to 75-ton chillers
         (2) Scroll compressors: 25- to 500-ton chillers

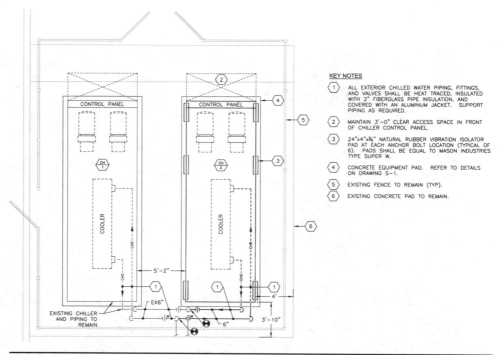

KEY NOTES

1. ALL EXTERIOR CHILLED WATER PIPING, FITTINGS, AND VALVES SHALL BE HEAT TRACED, INSULATED WITH 2" FIBERGLASS PIPE INSULATION, AND COVERED WITH AN ALUMINUM JACKET. SUPPORT PIPING AS REQUIRED.

2. MAINTAIN 3'-0" CLEAR ACCESS SPACE IN FRONT OF CHILLER CONTROL PANEL.

3. 24"x4"x¾" NATURAL RUBBER VIBRATION ISOLATOR PAD AT EACH ANCHOR BOLT LOCATION (TYPICAL OF 6). PADS SHALL BE EQUAL TO MASON INDUSTRIES TYPE SUPER W.

4. CONCRETE EQUIPMENT PAD. REFER TO DETAILS ON DRAWING S-1.

5. EXISTING FENCE TO REMAIN (TYP).

6. EXISTING CONCRETE PAD TO REMAIN.

FIGURE 4-13   Floor plan representation of two air-cooled chillers.

FIGURE 4-14   Photograph of an air-cooled chiller.

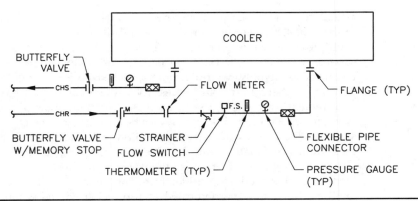

FIGURE 4-15 Connection detail for an air-cooled chiller.

(3) Screw compressors: 75- to 800-ton chillers
(4) Centrifugal compressors: 100- to 4,000-ton chillers
A chiller manufacturer's representative should be consulted to determine the appropriate type of chiller for each application.

b. Refrigerant: Refrigerants most commonly used in HVAC equipment fall into one of the following three categories: chlorofluorocarbon- (CFC), hydrochlorofluorocarbon- (HCFC), and hydrofluorocarbon- (HFC) based refrigerants.

In accordance with the Montreal Protocol, an international treaty administered by the United Nations Environment Programme (UNEP), which controls the consumption and production of ozone-depleting substances, CFC-based refrigerants, such as R-11 and R-12, are no longer produced due to their ozone-depletion potential (ODP). Today, CFC-based refrigerants are available only through stockpiled resources and are used to replenish refrigerant in existing equipment. HCFC-based refrigerants, such as R-22 and R-123, have a lesser ODP and are therefore not scheduled to cease production until the year 2030. HFC-based refrigerants, such as R-134a and R-410A, do not have any ODP and are therefore not regulated by the Montreal Protocol.

Chillers utilizing HFC-based refrigerants should be selected for use because they are not regulated by the Montreal Protocol. Chillers utilizing HCFC-based refrigerants should not be selected for use because, according to ASHRAE, the expected service life of chillers is 20 years or more, which would mean this equipment may still be operational after the year 2030 when HCFC-based refrigerants have ceased production.

Cost, availability in competing manufacturers' equipment, equipment efficiency, and the owner's requirements should also be given careful consideration when selecting the refrigerant for a chiller.

c. Capacity: Chiller capacity should equal the estimated peak cooling load of the areas served by the chiller. Some excess capacity can be allowed for future loads if these loads will be added within 5 years or so of the chiller installation. However, it is not advisable to oversize a chiller because of the added cost of the larger equipment and also because chillers typically operate most efficiently when they are fully loaded.

d. Multiple chillers: It is not common for there to be any redundancy in chiller capacity when the chilled water system serves cooling equipment providing comfort cooling only. Multiple chillers, each sized for the full load or some percentage thereof, are only necessary if the areas served have a critical requirement for cooling. Examples include certain areas of hospitals and data centers. Multiple chillers may also be desired for large-capacity cooling systems where the owner may want to limit each individual chiller to some maximum size. Finally, multiple chillers may be justified for a chilled water system that utilizes a thermal storage system. In this case, one chiller may be dedicated to producing chilled brine for the thermal storage system and another chiller may be dedicated to producing chilled water for the air systems and terminal equipment.

e. Capacity control: Because a chiller is seldom fully loaded, it is necessary for the chiller to be able to reduce its cooling capacity during part-load operation. The simplest way to reduce chiller capacity is to cycle the compressor. However, if this occurs too frequently, the life of the compressor will be shortened. If the only means of reducing the chiller capacity is through compressor cycling, multiple compressors for each chiller are recommended. This will reduce the frequency of cycling for each compressor and provide closer control of the chilled water supply temperature. Other means of controlling chiller capacity are available and depend upon the type of compressor used. Individual cylinders can be unloaded for reciprocating compressors, a slide valve can be used for screw compressors, and prerotation vanes can be used for centrifugal compressors. Although the HVAC system designer should know how the capacity is controlled, it is more important to know *that* it can be controlled and to what degree. Capacity control to 25% of a chiller's full-load capacity is normally adequate for HVAC applications. For example, a 100-ton chiller should be capable of reducing its capacity down to 25 tons.

f. Efficiency: A chiller's energy performance is rated in accordance with the guidelines established by the *Air-Conditioning, Heating and Refrigeration Institute (AHRI) Standard 550/590* (formerly *ARI Standard 550/590*). The two most common measures of a chiller's full-load energy performance are:

(1) kW/ton: Electrical power input expressed in terms of kilowatts (kW) per cooling output expressed in terms of tons.[29] A lower kW/ton indicates better energy performance.

(2) Energy efficiency ratio (EER): Cooling output expressed in terms of British thermal units per hour (Btuh) per electrical power input expressed in terms of watts (W). A higher EER indicates better energy performance.

For example, a 100-ton air-cooled chiller that requires 110 kW input has a full-load energy performance of 1.10 kW/ton or 10.9 EER.[30]

Generally, water-cooled chillers have better energy performance than air-cooled chillers. However, one must be careful when comparing the energy performance ratings of water-cooled chillers to air-cooled chillers because the full-load electrical power input for water-cooled chillers is limited to the compressor power only. For air-cooled chillers, the full-load electrical power input includes the compressor power plus the condenser fan power. For example, a water-cooled chiller (unit) may be rated at 0.65 kW/ton compared to the same size air-cooled chiller (package) rated at 1.10 kW/ton. This is not a valid comparison because other equipment is associated with the water-cooled chiller system in addition to the chiller itself. When comparing the overall energy performance (kW/ton) of a water-cooled chiller (system) to an air-cooled chiller (package), the electrical power input (kW) for all of the equipment associated with the water-cooled chiller (system) must be included in the comparison. This includes the water-cooled chiller compressor(s), condenser water pump, and cooling tower fan(s). The combined electrical input (kW) of all of this equipment should be divided by the water-cooled chiller's full-load cooling output in tons. This value can be used to compare the energy performance (kW/ton) of the water-cooled chiller (system) to the air-cooled chiller (package).

g. Voltage: Chillers used for HVAC applications will utilize three-phase electrical power. The highest secondary three-phase voltage available in the building will normally be used to serve the chiller in order to keep the wire sizes of the feeder serving the chiller as small as possible. For example, if a building has both 480V/3Ø and 208V/3Ø power available, the chiller will normally be selected to utilize 480V/3Ø power. The project electrical engineer should be consulted to determine the appropriate voltage for the chiller. The HVAC system designer should also ensure that the chiller can be furnished by the manufacturer at the desired voltage.

2. Freeze protection for outdoor chiller installations: For areas subject to freezing temperatures, chillers installed outdoors must have some means of protecting the water in the outdoor pipes and cooler from freezing. There are three ways to accomplish this:

a. Drain the outdoor piping and cooler if the chiller will not be used during freezing conditions.

(1) If the outdoor piping and cooler are empty, freezing is not a concern. Isolation valves for the outdoor components must be provided indoors and drains must be provided in the low points of the outdoor components. Draining of the outdoor components becomes a maintenance responsibility for the owner and must be performed every year prior to the onset of freezing temperatures. Refilling of the outdoor components must also be performed prior to the onset of cooling operation in the spring of each year.

b. Provide electric heating for all components subject to freezing if water is used in the chilled water loop.

(1) Thermostatically controlled heat tape is required on all outdoor piping, valves, and specialties. The heat output of the heat tape, expressed in terms of watts/foot (W/ft), depends upon the size of the pipe, thickness of insulation, and the design winter outdoor temperature. The HVAC system designer should consult with the heat tape manufacturer's representative or catalog data and specify the level of protection (W/ft) required for all exterior chilled water piping. Heat output of 5 W/ft is

typical for heat tape applications. The manufacturer's installation instructions must also be followed by the installing contractor because it is necessary to wrap the heat tape around the piping, valves, and specialties in an appropriate fashion in order to provide effective freeze protection. Furthermore, the chiller should be specified with an electric heater in the cooler in order to provide freeze protection down to 0°F. Freeze protection can also be provided down to −20°F if the chiller controls are able to start the chilled water pump and circulate chilled water through the cooler. The HVAC system designer should be aware that electric heat tape and an electric heater in the cooler do not provide any freeze protection in the event of a power outage unless these electric freeze protection components are connected to a backup electric power system.

   c. Utilize brine (typically a solution of propylene or ethylene glycol) in the chilled water loop.

     (1) The freezing point of the brine should be 15°F below the lowest expected ambient temperature. A solution of 40% propylene glycol is common because it provides freeze protection down to about −8°F.

     (2) Although glycol solutions provide effective freeze protection, the brine has a higher specific gravity, higher viscosity, and lower specific heat than water. Therefore, glycol solutions are less-efficient fluids than water for the following reasons:

       (*a*) The higher specific gravity means glycol solutions are more dense than water and therefore require more pumping power to circulate the same flow rate of water.

       (*b*) The higher viscosity of the glycol solution increases the friction losses in the system, which cause the overall system head to be higher than the same flow rate of water. Also, the higher viscosity reduces the heat transfer efficiency of the heat exchangers within the system, such as the cooler in the chiller(s) and cooling coils in the air systems and terminal equipment. This requires the chilled water system flow rate for glycol solutions to be greater than that of water. These two factors—higher system head and greater chilled water flow rate—require more chilled water pumping power than the same system circulating water.

       (*c*) The lower specific heat of the glycol solution means less heat can be transferred by 1 gal of glycol solution than by 1 gal of water. Thus, a greater flow rate of glycol solution is required to transfer the same amount of heat as a system utilizing water as the heat transfer fluid.

     (3) Some further considerations for glycol systems are:

       (*a*) A glycol solution must be used to make up for leaks in the system; domestic water alone cannot be used. This requires a glycol feed system consisting of a glycol solution storage tank and feed pump.

       (*b*) The glycol should be diluted with soft water[31] and have a low concentration of chloride and sulfate ions. This may require the addition of a water softener[32] to pretreat the (domestic) fill water used to dilute the glycol. Inappropriate fill water chemistry can lead to a number of problems, including the creation of sludge in the glycol solution (which can reduce heat transfer efficiency) and the depletion of the corrosion inhibitor in the glycol solution.

3. Primary-secondary pumping: All chillers have a certain minimum chilled water flow rate that must be maintained through the cooler. The easiest way to meet this requirement is to maintain a constant chilled water flow rate through each chiller cooler regardless of the building cooling load. In order to ensure constant chilled water flow rate through the coolers of all chillers in a system, a primary-secondary pumping system is normally employed. A primary pump is dedicated to each chiller, and the secondary pump (which can be a variable speed pump) circulates flow through the chilled water system in the building. The primary and secondary pumping loops share a common pipe that joins the primary and secondary pumping loops.

   Variable speed pumps should not be used to circulate chilled water through the chiller coolers unless the chilled water system is specially designed as a variable primary flow system, as described below.

4. Variable primary flow: If the chillers are suitable for variable chilled water flow rates through the coolers, a variable speed pumping system, consisting of a single, variable speed primary (system) pump, can be utilized for greater energy efficiency. In this arrangement, the chilled water flow rate is modulated to meet the fluctuating cooling load of the building. Pumping energy is saved because there is only one pump in the system and also because the total system flow rate, including the flow rates through the chillers, is reduced during part-load conditions. Again, it is very important that the chillers be suitable for variable chilled water flow rates through the coolers, otherwise the chillers may shut down due to a low refrigerant pressure safety.

5. Space requirements: The space required for proper installation of indoor and outdoor chillers must be carefully coordinated with the chillers that form the basis of the system design. The HVAC system designer must realize that the actual chillers installed for the project may be from a different manufacturer than the basis of design and, as a result, will have somewhat different physical dimensions and characteristics. Some allowance should be made in the layout of the mechanical room for these differences between equipment manufacturers. However, if a two-pass cooler (having piping connections on the same end of the cooler) is used as the basis of design, it is not reasonable to assume that a one-pass or three-pass cooler (having piping connections on opposite ends of the cooler) will be acceptable. The owner and installing contractor must understand that the HVAC system design cannot be capable of accommodating drastic departures from the design intent.

6. Clearances
   a. Maintenance: Refer to the manufacturer's product data for the maintenance clearances required on all sides of the chiller. Typically, 36 in. of clearance is recommended on all sides of the chiller. If two chillers are installed side by side, the space between the chillers can usually be shared for maintenance clearance so that a double-wide clearance is not required between the chillers.
   b. Tube pull: Sufficient clearance, usually the full length of the cooler or condenser, is required for the tubes in either of these heat exchangers to be cleaned or removed and replaced. The manufacturer's product data will identify the appropriate end of each heat exchanger for this clearance to be provided. It is common for this clearance requirement to be accommodated

through the equipment room doorway so that all of the clear space is not required within the equipment room.

c. Electrical: Article 110.26 of the *National Electric Code* (NEC) defines working space requirements around electrical equipment that is 600 V nominal, or less. The depth of the working space required in front of the point(s) where the electrical power wiring connects to the chiller will be 36 in., 42 in., or 48 in., depending upon the voltage of the live parts and whether the parts on the other side of the working space are: (1) guarded by insulating materials, (2) grounded, or (3) live. The project electrical engineer should be consulted to determine the appropriate working space required for the electrical connections to the chiller.

7. Installation

a. Indoors: An indoor chiller is normally installed on a concrete housekeeping pad that is 4 in. high and 4 to 6 in. larger than the chiller base frame on all sides. The concrete pad is reinforced with wire mesh, and anchor bolts are cast into the pad in the locations of the bolt holes in the chiller base frame. Neoprene vibration isolation pads, typically ¾ in. thick and of the quantity required to support the load, should be used to provide vibration isolation between the chiller base frame and the housekeeping pad.

b. Outdoors: An outdoor chiller installed on grade will normally be installed on a reinforced concrete pad that is thick enough to support the weight of the chiller. The project structural engineer normally designs the concrete pad, including the appropriate reinforcement. Similar to an indoor housekeeping pad, anchor bolts are cast into the pad and neoprene vibration isolation pads are used to isolate the chiller base frame from the concrete pad. If vandalism is a concern, the chiller may be surrounded by a fence in order to maintain the appropriate maintenance and electrical clearances around the chiller and also to allow for tube pull clearance (usually through a gate in the fence). The manufacturer's product data should be consulted to determine the minimum clearance from solid walls, fences, or other heat-rejecting equipment to ensure that adequate space for airflow and heat rejection is provided around the outdoor chiller.

An outdoor chiller that is installed on a building roof will normally be installed on a horizontal structural steel frame that is supported 12 to 18 in. above the roof by vertical members that are rigidly attached to the roof structure. This height above the roof allows sufficient clearance for the roofing system to be replaced under the steel frame without disturbing the chiller. Holes are provided in the horizontal structural steel frame to coordinate with the locations of the bolt holes in the chiller base frame. The chiller is connected to the steel frame with anchor bolts, and neoprene pads are used to isolate the chiller base frame from the steel frame.

c. Coordination: The project structural engineer will design the indoor housekeeping pads, outdoor concrete pads on grade, and structural steel equipment supports for roof-mounted equipment. Therefore, the HVAC system designer must communicate the physical dimensions, weight, and anchor bolt locations of the chiller(s) that form the basis of design with the project structural engineer during the design phase. During construction, the project structural engineer should also review and approve the chiller equipment submittal and modify the design of the equipment bases as required to coordinate with the physical dimensions, weight, and anchor bolt holes of the actual chillers that will be installed for the project.

8. Compressor starter: Normally, a full-voltage, across-the-line starter is provided for each compressor as a standard feature by the chiller manufacturer. However, the full-voltage starting (or inrush) current of compressors can be quite high, making the use of this type of starter undesirable from an electrical power distribution standpoint. Therefore, it is common for chiller manufacturers to offer a wye-start, delta-run reduced-voltage starter for each compressor as an option to accomplish what is referred to as a "soft" start of the chiller compressors. This type of starter, commonly referred to as a wye-delta starter, can only be utilized if the compressors are allowed to start unloaded. That is because the reduced voltage of the wye-start configuration (58% of full voltage) applies only about one-third the starting torque of a full-voltage starter. Once the compressor attains full speed, the windings are reconfigured to the delta-run configuration, which applies full voltage to the motor windings and allows the motor full torque capability.

A variable frequency drive (VFD) can also be used to accomplish a soft start for certain types of compressor; however, VFDs are rarely used for their soft-start characteristics only. The benefit of varying the speed of a compressor through a VFD is the reduction in energy use that can be realized during part-load operation of the chiller. The chiller manufacturer should be consulted to determine whether VFDs for the compressors are an option and, if so, how to design the chilled water system to take full advantage of their energy-saving potential.

9. Water temperatures
   a. Chilled water: A chilled water supply temperature of 45°F is common, although a lower chilled water supply temperature can be used if required by the chilled water system.[33] A chilled water temperature rise of 10°F is common, although a chilled water temperature rise as high as 15 to 17°F can be used to reduce chilled water pumping energy as long as the chiller and the chilled water coils in the air systems and terminal equipment are selected based on these operating parameters.[34]
   b. Condenser water: A maximum condenser water supply temperature of 85°F with a 10°F temperature rise through the chiller condenser is a common criterion for selecting water-cooled chillers and their associated cooling towers.[35] In addition to ensuring the chiller and cooling tower have sufficient capacity at the maximum condenser water supply temperature, it is also necessary to determine the minimum condenser water supply temperature allowed by the chillers selected for the project. Generally, chillers cannot receive condenser water with a temperature below approximately 70°F without shutting down through a low refrigerant pressure safety. Constant-speed condenser water pumps are normally used to circulate condenser water through the chiller condenser and cooling tower because it is not desirable to vary the condenser water flow rate through cooling towers. For

this reason, the cooling tower fan(s) must be cycled on and off or the speed of the fans must be modulated through two-speed motors or variable frequency drives to maintain the temperature of the condenser water supplied to the chiller condenser between its maximum and minimum levels.

10. Chemical treatment: Proper chemical treatment of the water in both the chilled water and condenser water loops is necessary to reduce corrosion and scaling of the internal pipe walls and heat exchange surfaces in these systems.
    a. Chilled water: Refer to the Chemical Treatment subsection of the Auxiliary Hydronic Equipment section of this chapter for a discussion of the chemical treatment required for the closed chilled water loop.
    b. Condenser water: Refer to the Design Considerations subsection of the Cooling Tower section of this chapter for a discussion of the chemical treatment required for the open condenser water loop.

11. Chilled water loop volume: The volume of the chilled water loop should exceed 3 gal/ton of chiller capacity in order to achieve stability and accuracy of the chilled water supply temperature. If the chilled water loop volume is less than 3 gal/ton, the chiller may have to cycle frequently at part-load conditions and the chilled water supply temperature may fluctuate unacceptably as a result. In order to compensate for a low chilled water system volume, it may be necessary to design a chilled water storage tank to add the necessary water volume to the chilled water system.

12. Fouling factor: A fouling factor should be utilized in the selection of both the cooler and water-cooled condenser for chillers. *AHRI Standard 550/590* recommends a fouling factor of 0.0001 $ft^2 \cdot h \cdot °F/Btu$ for the cooler selection and 0.00025 $ft^2 \cdot h \cdot °F/Btu$ for the water-cooled condenser selection.

13. *ANSI/ASHRAE Standard 15-2010—Safety Standard for Refrigeration Systems* requirements:
    a. Because refrigerants have varying levels of toxicity and flammability, ASHRAE has developed the guidelines contained within *ANSI/ASHRAE Standard 15-2010* for their safe use indoors.
    b. *ANSI/ASHRAE Standard 15-2010*, Section 8.11, describes the refrigerant detection, exhaust, and alarm requirements if a machinery room is required by Section 7.4 to house all of the components containing refrigerant. The formula in Section 8.11.5 is used to calculate the exhaust airflow.
    c. *ANSI/ASHRAE Standard 15-2010*, Section 8.11.6, prohibits combustion equipment from being installed in the same room as refrigerant-containing equipment unless the combustion equipment has a direct combustion air connection to the outdoors or the room is equipped with a refrigerant detection system that will automatically shut down the combustion process if a refrigerant leak is detected.
    d. *ANSI/ASHRAE Standard 15-2010*, Section 9.7.8, requires refrigerant pressure relief piping to be routed from the refrigerant pressure relief device on the chiller to the outdoors and "discharge to the atmosphere at a location not less than 15 ft above the adjoining ground level and not less than 20 ft from any window, ventilation opening, or exit in any building." This piping must be sized for a maximum back pressure in the refrigerant pressure

relief piping, which depends upon the refrigerant flow and the total equivalent length of the piping. The manufacturer's product data should be consulted for pipe sizing guidelines.

14. Noise: Refer to Chap. 8 for recommendations to control noise from chillers.

15. Controls: Chillers are equipped with on-board systems that control the internal functions of the chiller. These functions include control of the refrigeration system to maintain the setpoint of the controlled variable (normally leaving chilled water temperature). In addition, the on-board control system regulates other necessary operating parameters that are not of particular interest to the HVAC system designer, such as oil temperature and refrigerant suction pressure. Many of the on-board control system parameters can be accessed and adjusted through the chiller control panel or they can be communicated to the building automation system (BAS) (if there is one) for central monitoring and control.[36] Of particular concern for the HVAC system designer is how the chiller operates within the overall chilled water system. This is summarized as follows:

   a. First, the chiller must be placed into the "ready" mode. This can be performed automatically by sending a signal to the chiller control panel from the BAS or can be performed manually at the chiller control panel. If the chiller operates only during the cooling season, it will be placed into the ready mode at the beginning of the cooling season and deactivated at the end of the cooling season of each year. While in the ready mode, the chiller will operate under its on-board control system.[37]

   b. Prior to starting the chiller refrigeration system, the chiller control panel will start the chilled water pump (and condenser water pump if the chiller is water-cooled) and verify that there is water flowing through the cooler (and condenser if the chiller is water-cooled). Chilled water (and condenser water) flow is proven either by a flow switch installed in the inlet or outlet connections to the cooler (and condenser) or by a differential pressure switch that is piped between the inlet and outlet connections on the cooler (and condenser).[38] After receiving the start signal, starting the associated pumps, and proving water flow, the chiller refrigeration system will be started.

   c. During operation, the chiller varies its cooling capacity (as discussed previously) through its on-board control system to maintain the setpoint of the controlled variable (normally leaving chilled water temperature). It is also necessary to control the entering condenser water temperature for water-cooled chillers. This is done by cycling the cooling tower fan(s) on and off, by varying the speed of the cooling tower fan(s), or by diverting condenser water from the cooling tower distribution nozzles to the cooling tower sump.[39] Simple on/off control of a single cooling tower fan and control of a bypass valve can be accomplished by the chiller control panel. However, on/off or variable speed control of multiple cooling tower fans is normally accomplished by the cooling tower control panel.

   d. When one of the operating parameters of the chiller deviates beyond established limits, an alarm will signal at the chiller control panel and may also signal at the BAS if designed to do so. The fault can be either critical, requiring a shutdown of the chiller, or noncritical, requiring notification only. Upon experiencing a critical fault, the redundant chiller (if there is one) will be placed into the ready mode through a communication link between the two chiller control panels or through the BAS and a start signal will be given. If there is no redundant chiller, chilled water will not be produced until the problem that caused the critical fault is resolved and it is cleared in the chiller control panel. Alarms are typically logged electronically at the chiller control panel and can also be logged at the BAS.

Cooling Towers

Cooling towers are common components of centralized cooling systems. Cooling towers are used either in conjunction with water-cooled chillers for chilled water systems or with plate and frame heat exchangers for water source heat pump systems.

Purpose

Cooling towers are used for HVAC applications in conjunction with water-cooled HVAC equipment to cool water by exposing it either directly or indirectly to the ambient air.

Physical Characteristics

Cooling towers that expose the cooling water directly to the ambient air are called open cooling towers. Those that expose the cooling water indirectly to the ambient air are called closed-circuit cooling towers.

For open cooling towers, the water circulating through the water-cooled equipment is the same water that is circulated through the cooling tower; this water is in direct contact with the ambient air. Thus, for open cooling towers, the cooling water system is an open system.[40] An open cooling tower is illustrated in Chap. 39, Fig. 3 of the 2008 *ASHRAE Handbook—HVAC Systems and Equipment.*

For closed-circuit cooling towers, the water circulating through the water-cooled equipment is circulated through a separate heat exchanger coil mounted within the cooling tower. This water is not in direct contact with the ambient air. Water circulated within a separate open system within the cooling tower is in direct contact with the ambient air and is sprayed over the heat exchanger coil. Thus, for closed-circuit cooling towers, there are two separate cooling water systems: the closed system,[41] which circulates water through the water-cooled equipment, and the open system, which circulates water through the cooling tower and is sprayed over the heat exchanger coil. A closed-circuit cooling tower is illustrated in Chap. 39, Fig. 4 of the 2008 *ASHRAE Handbook—HVAC Systems and Equipment.*

Cooling towers do not have any refrigeration components. For open cooling towers, water is distributed through nozzles over the fill, which is typically corrugated polyvinyl chloride (PVC). Air is drawn or blown through the fill by a fan. As some of the water in contact with the air evaporates (normally about 1.1% of the total water flow at design conditions), the water remaining in the liquid state is cooled through the latent heat of vaporization (the heat absorbed by the evaporated water during the evaporation process) and flows through the fill to the sump located at the bottom of the cooling tower. This water is piped to the circulating pump, which circulates it through the water-cooled equipment. The water that is lost through evaporation is replaced by domestic water through the makeup water connection. Closed-circuit cooling towers do not have any fill. Instead, the open-loop water is sprayed through the distribution nozzles across the closed-loop heat exchanger coil. The open-loop water also flows to the cooling tower sump and is circulated from the sump to the distribution nozzles by a pump normally located below the cooling tower.

There are various types of mechanical-draft[42] cooling towers, which are defined by the orientation of the water flow with respect to the air flow and by the arrangement of the fan with respect to the fill. The types of mechanical-draft cooling towers are described as follows:

Crossflow: Air and water flow in opposite directions.

Counterflow: Air and water flow at right angles to each other (Figs. 4-16 and 4-17).

Forced air: Air is blown through the fill.

Induced air: Air is drawn through the fill.

The various types of mechanical-draft cooling towers are illustrated in Chap. 39, Fig. 10 of the 2008 *ASHRAE Handbook—HVAC Systems and Equipment*.

Connections

Because cooling towers are frequently used to reject heat from water-cooled refrigeration equipment, the pipe conveying water from the cooling tower is often referred to as the condenser water supply pipe and the pipe conveying water to the cooling tower is often referred to as the condenser water return pipe. This is the naming convention that will be used in this section. However, cooling towers may also be used to reject heat from nonrefrigeration equipment, such as a heat exchanger. In this case, a naming convention for the piping (such as cooling tower supply and cooling tower return) must be developed by the HVAC system designer and used consistently for the project.

Connections for cooling towers consist of the following:

1. Condenser water return piping connection.

2. Condenser water supply piping connection.

3. Equalizer piping connection (where multiple cooling towers will be installed).

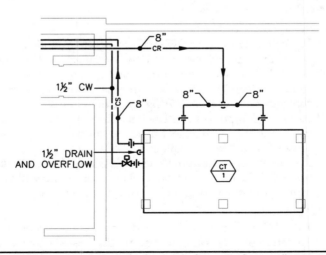

Figure 4-16 Floor plan representation of a forced draft counterflow cooling tower.

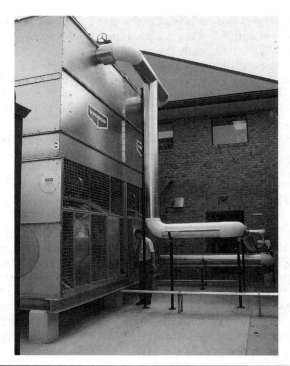

FIGURE 4-17   Photograph of a forced draft counterflow cooling tower.

4. Makeup water piping connection.

5. Drain piping connection.

6. Overflow piping connection.

7. Electrical power connection.
   a. Typically, the electric power connection to a cooling tower will be a single-point, three-phase electrical power connection to a unit-mounted control panel that contains the starters or VFDs for the fans, contactors[43] for the sump heaters (if winter operation is required), water level controls, fan controls, and safeties. Wiring from the control panel to the various unit-mounted components will be installed at the factory. Other wiring, such as the connection to the field-installed condenser water supply temperature sensor, will be field-installed.
   b. A separate electrical power connection, typically single-phase, for freeze-protection heat tape on insulated exterior piping will also be required if the cooling tower will be operated during the winter.

8. ATC connection.
   a. The ATC connection will be made at the control panel and will integrate the operation of the cooling tower into the overall cooling system.

Design Considerations

For all cooling tower systems (open and closed-circuit), cooling water is pumped through the water-cooled equipment where it absorbs the heat rejected by the equipment. This water is then circulated to the cooling tower where it is cooled (directly or indirectly) mainly by the evaporation of a portion of the water that is in direct contact with the ambient air. For closed-circuit cooling towers, the water contained within the closed-loop heat exchanger coil is cooled by the evaporative effect of the open-loop cooling tower water. The greater the rate of evaporation, the closer the leaving water temperature will approach[44] the wet bulb temperature of the ambient air.[45] The rate at which the water evaporates in the cooling tower is proportional to the surface area of the water in the cooling tower and the airspeed, and is inversely proportional to the wet bulb temperature of the ambient air. Thus, an increase in the surface area of the water and/or airspeed in the cooling tower and/or a decrease in the wet bulb temperature of the ambient air results in a decreased approach temperature. A typical approach temperature for HVAC applications is between 5 and 10°F.

Range is defined as the difference between the water temperatures entering and leaving the cooling tower and is determined by the heat load on the cooling tower and the water flow rate.

Typical design temperatures for a cooling tower system are as follows:

* Cooling tower entering water temperature: 95°F
* Cooling tower leaving water temperature: 85°F
* Ambient wet bulb (wb) temperature: 78°F

From these design temperatures, it can be seen that the approach is 7°F[46] and the range is 10°F.[47]

Other design considerations are as follows:

1. The entering and leaving water temperatures and the water flow rate through a cooling tower must be consistent with the water-cooled equipment that it serves.

2. A typical flow rate for a cooling tower serving water-cooled refrigeration equipment is 3 gpm per ton of refrigeration capacity.

3. The water flow rate through a cooling tower should be full flow to the distribution nozzles or full bypass to the cooling tower sump.

4. Cooling tower capacity is modulated by varying the airflow through the cooling tower in order to maintain a constant leaving water (condenser water supply) temperature. This is accomplished by cycling the cooling tower fan(s) on and off, by varying the speed of the cooling tower fan(s) or by diverting condenser water from the cooling tower distribution nozzles to the cooling tower sump.

5. During start-up of a water-cooled chiller, condenser water should be diverted from the distribution nozzles in the cooling tower to the cooling tower sump until the condenser water supply temperature rises to the desired temperature, which must be above the minimum condenser water supply temperature required by the chiller.

6. The condenser water supply temperature is typically maintained at 85°F. However, because refrigeration equipment typically operates more efficiently at a lower condenser water supply temperature, one way to improve energy performance is to allow the condenser water temperature to float down to the minimum condenser water supply temperature (approximately 70°F) whenever possible.

7. If winter operation of the cooling tower is required, the cooling tower and all exterior piping will need to be winterized. Winterization includes the installation of a heater in the cooling tower sump, which can be electric, steam, or hot water. The heater should be sized to keep the water in the sump at 40°F at the design winter outdoor temperature. Also, the exterior piping, including the condenser water piping, drain piping, and overflow piping, should be heat-taped and insulated to prevent it from freezing. Typically, heat tape is self-regulating and does not require any external controls, only an electrical power connection.

8. Some types of cooling towers are more suitable for winter operation than others. The manufacturer's representative should be contacted to determine which type is most suitable for each application.

9. Cooling towers can be used for waterside economizer operation in the winter. During waterside economizer operation, both condenser water and chilled water are diverted from the chiller to a plate and frame heat exchanger by 3-way diverting valves installed in the condenser water and chilled water loops. In this configuration, the hot-side chilled water loop is cooled through the plate and frame heat exchanger by the cold-side cooling tower loop. Typical temperatures for the plate and frame heat exchanger during waterside economizer operation are:
    a. Hot side (chilled water):
        (1) 55°F entering water temperature (EWT)
        (2) 45°F leaving water temperature (LWT)
    b. Cold side (cooling tower water):
        (1) 38°F EWT
        (2) 48°F LWT
    Further details regarding the design of waterside economizer operation are beyond the scope of this book.

10. If the cooling tower is used for waterside economizer, the exterior cooling tower piping should be heat taped and insulated for freeze protection. The interior cooling tower piping should be insulated to prevent condensation on the surface of the piping.

11. Condenser water piping does not need to be insulated (either indoors or outdoors) unless the cooling tower is used for winter operation utilizing standard condenser water temperatures (outdoor insulation required) or waterside economizer using depressed condenser water temperatures (indoor and outdoor insulation required). (Uninsulated condenser water piping assumes the cooling tower sump and the outdoor piping will be drained during the winter to prevent freezing.)

12. Because cooling towers are open systems, the following should be observed:
    a. The cooling tower sump must be higher in elevation than the condenser water pump. This will ensure that the pump suction is always flooded.

b. Piping and fittings between the cooling tower and the condenser water pump should be minimized to reduce friction losses. This will minimize the reductions to the net positive suction head that is available for the pump.

c. The net positive suction head available (NPSHA) for the condenser water pump must exceed the net positive suction head required (NPSHR) by the condenser water pump.[48]

d. The height of the distribution nozzles above the surface of the water in the sump (i.e., the elevation head of the system) must be included in the condenser water pump head calculation because the condenser water pump will have to lift the condenser water this height. The pump head calculation for the rest of the condenser water piping system is the same as for a closed system.

13. If makeup water piping is not connected to the makeup water connection on the cooling tower, it may be connected to the piping between the cooling tower and the condenser water pump, preferably indoors to avoid the need for freeze protection. A slow-acting solenoid valve should be specified to reduce the potential for water hammer in the makeup water piping. An electric water level sensor in the cooling tower would be required for this configuration.

14. A backflow preventer is required in the connection of the makeup water piping to the domestic water system.

15. The water in the sump of the cooling tower must be maintained at a certain level. In order to maintain an appropriate water level, the cooling tower must be equipped with a water level sensor, which is either the (standard) mechanical float with integral makeup water valve or an (optional) electric water level sensor. For the electric water level sensor, the water level controller (normally located in the cooling tower control panel) receives an input signal from the electric water level sensor and sends an output signal to the solenoid valve on the makeup water line to open or close the valve as required to maintain the proper water level in the cooling tower sump.

16. The following should be observed for multiple cooling tower (or multiple-cell cooling tower) installations with common supply and return piping:

a. The piping between the cooling towers and condenser water pumps should be as symmetrical as possible to obtain balanced flow through the cooling towers (or cooling tower cells).

b. Balancing valves should be designed for the return piping connection to each cooling tower (or cooling tower cell) to balance the water flow to the distribution nozzles.

c. An equalizing pipe with shutoff valve should be designed to connect multiple cooling towers. The equalizing pipe will correct any flow imbalances that may arise due to clogged distribution nozzles or strainers. The equalizing pipe should be designed in accordance with the cooling tower manufacturer's instructions. The general rule is that the equalizing pipe should accommodate approximately 15% of the largest cooling tower's flow at a pressure differential, including valves and fittings, not to exceed 1 in. water column. This will ensure that the water level in the cooling tower sumps will differ by no more than 1 in.

   d. Automatic shutoff valves should be designed for the supply, return, and equalizer piping connections to isolate idle cooling towers (or cooling tower cells).

   e. Separate condenser water pumps should be designed for each cooling tower (or cooling tower cell) to be energized in conjunction with each cooling tower (or cooling tower cell).

   f. The cooling tower overflow pipes should be installed at the same elevation.

17. Flexible pipe connectors are required for all pipe connections to cooling towers to isolate the vibration generated by the cooling tower fan(s) from the piping systems.

18. An immersion-type temperature sensor is required in the condenser water supply pipe which is common to all cooling towers. The condenser water supply temperature is used as input for the condenser water supply temperature controller.

19. A strainer in the outlet connection is standard equipment for cooling towers to prevent debris from entering the condenser water system. Therefore, it is not necessary to design a strainer in the suction piping connection for the condenser water pump.

20. If additional filtration of the condenser water is desired, the following options are available:

   a. A side-stream filtration system that circulates a portion of the condenser water flow through a sand filter using the head developed by the condenser water pump. The sand filter must be backwashed periodically when it becomes clogged.

   b. A sump water filtration system that consists of discharge nozzles installed in the cooling tower sump. These nozzles force high-velocity water jets across the bottom of the sump to inlet openings also mounted at the bottom elevation of the sump. This water is circulated by a separate filtration system pump through a sand filter, which traps the debris. The sand filter must be backwashed periodically when it becomes clogged.

21. Typically, cooling towers are mounted on posts at an elevation of at least 18 in. above grade or the building roof to allow access under the cooling towers for inspection and maintenance.

22. Noise generated by cooling towers is a major concern. Induced draft cooling towers with propeller fans are typically the noisiest type of cooling tower. Forced draft cooling towers with centrifugal fans are typically the quietest mechanical-draft cooling towers.[49] A sound attenuator may be available to reduce the noise generated by a cooling tower. The manufacturer's representative should be contacted to discuss the availability and suitability of sound attenuators for each application.

23. Drift, which is the entrainment of water droplets in the air passing through a cooling tower, is largely eliminated by the drift eliminator in the cooling tower. However, some drift escapes from cooling towers (known as drift loss). Therefore, cooling towers should not be placed too close to large windows or building components, which are sensitive to staining or scale deposits.

24. The warm, saturated air leaving cooling towers may produce fog (cooling tower plume) during conditions when the ambient air is unable to absorb all of

the moisture in the discharge air. Consideration should be given to the placement of cooling towers so that the cooling tower plume will not be objectionable.

25. Section 908.3 of the 2009 *International Mechanical Code* requires the plume discharge from cooling towers to "... be not less than 5 ft above or 20 ft away from any ventilation inlet to a building."

26. It is necessary to chemically treat cooling tower water to control the alkalinity (pH) of the water, corrosion, scale formation, and biological growth (such as bacteria and algae). Typically, the makeup water flow is measured by a water meter and chemicals are injected into the system (in proportion to the makeup water flow) by a pump that is mounted on the top of a 50-gal polyethylene chemical storage tank. The conductivity of the cooling tower water is also measured to determine the level of dissolved solids in the water (the greater the conductivity, the greater the level of dissolved solids). When the water conductivity exceeds a certain level, a portion of the cooling tower water is bled to a sanitary drain so that fresh makeup water can be added to the system, thus controlling the level of dissolved solids in the water. Water bleed is performed by opening a solenoid valve in the bleed line, which discharges to a funnel-type floor drain connected to the building sanitary system.[50] The amount of makeup water required is equal to the sum of the bleed water and the water that is evaporated by the cooling tower. Makeup water flow is approximately 2.2% of the total condenser water flow for electric chillers (1.1% bleed and 1.1% evaporation).

27. Galvanized steel is the standard construction material for cooling towers. One manufacturer offers epoxy-coated galvanized steel to provide improved corrosion resistance. Stainless steel is the best, though most expensive, material for cooling towers. A good compromise is to specify a cooling tower constructed of epoxy-coated galvanized steel with a stainless steel sump.

28. Some oxidation of the galvanized surfaces of cooling towers is normal. In fact, if allowed to form properly, the normal zinc oxide layer provides a protective barrier against the effects of the environment. However, if proper water treatment is not maintained, this zinc oxide layer is not allowed to form and a harmful type of corrosion, called white rust, may occur.[51] The risk of white rust formation can be reduced through the use of epoxy-coated galvanized steel and/or an ongoing program of proper water treatment. The problem of white rust can be avoided altogether by specifying cooling towers that are constructed of stainless steel.

29. A remote sump, an option that is available with many cooling towers, allows the sump of the cooling tower to be located indoors. This is particularly desirable for cooling towers that must operate during the winter because the sump would not have to be protected from freezing. Obviously, the cooling tower would have to be located at a higher elevation than the remote sump. This is quite feasible for a cooling tower that is mounted on the roof of a building, but may be a problem for a cooling tower that is mounted on grade.

30. Freeze protection for closed-circuit cooling towers includes not only an immersion heater in the sump and heat tape on the exterior piping but also freeze protection for the heat exchanger coil. This typically requires the use of a glycol solution within the closed-loop system.

31. Coordination with other disciplines includes the following:
   a. Architectural
      (1) The location of the cooling tower should be coordinated, whether it is mounted outside on grade or on the roof. If the cooling tower will be mounted on grade, the project architect will most likely want to locate it in the rear of the building or near a service entrance. If the cooling tower will be mounted on the roof, the architect will most likely want to locate it near the center of the roof.
   b. Structural
      (1) The structural engineer will need the weight and footprint of the cooling tower and will also need to coordinate the mounting points of the cooling tower for design of the structural steel support. The cooling tower should be mounted at least 18 in. above grade or above the roof to allow access under the cooling tower for inspection and maintenance.
      (2) Normally, the fans for cooling towers are furnished with internal vibration isolation; that is, the fans are mounted on spring vibration isolators within the cooling towers. If so, external spring vibration isolators between the cooling tower and the support structure are not required or desired. Neoprene pads are all that are required to isolate any minor vibrations of the cooling tower from the support structure.
   c. Electrical
      (1) Cooling towers will require three-phase electrical power for all electrical components, except electric heat tape, which requires single-phase electrical power.
      (2) The electrical engineer will need the following information:
         (a) The size (hp) of the cooling tower fan motor(s).
         (b) The voltage (V) and phase (∅) of the fan motor(s).
         (c) The wattage, V, and ∅ of electric sump heaters and heat tape, if any.
      (3) The electrical engineer will also need to know if a control panel will be furnished with the cooling tower, requiring a single-point electrical connection, and whether it would include the necessary disconnect switches, starters, or VFDs for the fan motors, and contactor for the electric sump heater. If a control panel is not furnished with the cooling tower, the electrical engineer will have to design these components to be located either indoors or outdoors, depending upon the requirements. The electrical engineer will also need to know if a separate 120V/1∅ power connection is required for the automatic temperature controls or if this 120V/1∅ power will be derived from the cooling tower's three-phase power source through a step-down control transformer.
      (4) Any exterior electrical components will have to be enclosed in a minimum National Electric Manufacturers Association (NEMA) 3R (rainproof) enclosure. For a more durable enclosure, a NEMA 4 (watertight) enclosure should be specified.

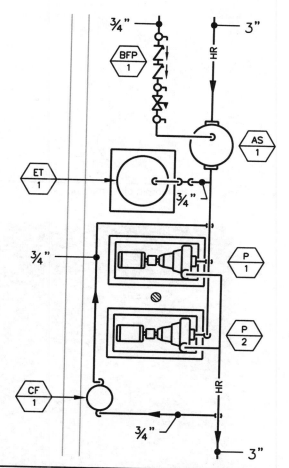

FIGURE 4-18   Floor plan representation of auxiliary hydronic equipment.

Auxiliary Hydronic Equipment

In this section we will look specifically at the auxiliary equipment used in heating, chilled, and heat pump water systems (commonly referred to as hydronic systems). Some examples of auxiliary hydronic equipment are air separators, chemical treatment (chemical feed), expansion tanks, makeup water assemblies, and pumps. Figure 4-18 shows how this equipment would appear on a mechanical room floor plan.

Air Separator

Purpose

Air separators are used to remove the air that is released when the hydronic system water is heated (Figs. 4-19 and 4-20).

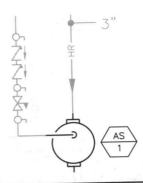

FIGURE 4-19 Floor plan representation of an air separator.

FIGURE 4-20 Photograph of an air separator.

Physical Characteristics

Air separators range in diameter from about 12 to 24 in., with the diameter increasing in proportion to the flow. The height of air separators ranges from 24 to 48 in., again with the larger sizes required for increased flow.

Connections

Connections to air separators are piping connections only; there are no electrical or ATC connections.

The air separator is usually located on the suction side of the system pump because air should be eliminated from the water prior to being circulated by the system pump. This ensures that the pump impeller is not handling water that is entrained with air. Also, the makeup water connection is usually made at the air separator. This location is desirable because makeup water should be added to the system near the expansion tank connection to the system, which is normally located between the air separator and the pump.

The inlet and outlet sizes of the air separator are usually the same size as the system piping. Connections are made to the air separator with unions for sizes 2 in. and smaller and with flanges for connection sizes 2½ in. and larger.

Makeup water piping, typically ¾ or 1 in. in size, is connected to a threaded fitting in the top of the air separator (Fig. 4-21). The makeup water assembly, which consists of a backflow preventer, pressure-reducing valve, and shutoff valves, is shown on the mechanical room floor plan.

An automatic air vent, typically ½ in. in size, is connected to a threaded fitting in the top of the air separator for elimination of the air that is removed from the system water. The automatic air vent is not shown on the mechanical room floor plan. It is shown in the detail for the air separator/makeup water assembly.

Design Considerations

Selection of the air separator is based on the water pressure drop through the unit. A general design criterion is to keep the pressure drop through the air separator to 2 psi or less. This usually results in the inlet and outlet connections being the same size as the system piping.

It is generally not recommended to select the air separator with an integral strainer because a strainer should be located at the suction connection to the system pump. Therefore, an additional strainer in the air separator would be redundant and would require additional maintenance.

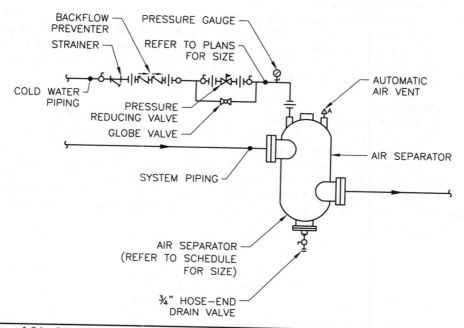

FIGURE 4-21 Detail of air separator/makeup water assembly.

The operating weight of air separators (the weight of the air separator plus the weight of the system water contained within it) ranges from 100 to 400 lb. Typically, there is no special structural support required. Pipe hangers are used to suspend the air separator from the structure above, which should be designed by the structural engineer for the collateral load of the mechanical equipment in the equipment room.

Chemical Treatment

Purpose

The purpose of providing chemical treatment (chemical feed) for water in closed hydronic systems is to reduce corrosion of the system piping and equipment that occurs as a result of the air within the system. Proper chemical treatment also reduces the deposition of minerals contained within the system water onto the inner walls of the system piping and equipment, which is known as scaling. Failure to provide adequate chemical treatment of closed hydronic systems will hasten the deterioration of the system piping and equipment and will also reduce the heat transfer capabilities of the equipment through the corrosion and scale. Reduced heat transfer equates to reduced system efficiency and increased energy costs.

Chemicals are usually added manually to closed hydronic systems through a chemical shot feeder (Figs. 4-22 and 4-23). These types of chemical feed systems require the manual addition of chemicals by either the on-site maintenance personnel or a contracted water treatment company. The chemical composition of the system water should be checked semi-annually and chemicals added, as required, to maintain proper levels of corrosion inhibitors and alkalinity (pH).

Physical Characteristics

The 5-gal shot feeder (typical for hydronic systems) is approximately 12 in. in diameter and 36 in. high. The chemicals are introduced to the system through the funnel at the top. There are no special structural considerations because the equipment is floor-mounted and the weight of the shot feeder is minimal. Access needs to be provided to the shot feeder for routine monitoring and chemical addition.

Connections

There are no electrical or ATC connections to shot feeders. Shot feeders are connected to the system piping in a bypass arrangement. The inlet piping of the shot feeder is connected to the system pump discharge piping and the outlet piping of the shot feeder is connected to the system pump suction piping. The system pump pressure is used to circulate flow through the shot feeder, which flow is only required during routine monitoring and chemical addition. Connections are made to the shot feeder with unions and shutoff valves. The location of the inlet and outlet connections on a shot feeder will

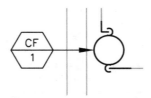

FIGURE 4-22   Floor plan representation of a 5-gal shot feeder.

FIGURE 4-23   Photograph of a 5-gal shot feeder.

differ from one manufacturer to another. A drain pipe with a hose-end drain valve should be provided at the bottom of the shot feeder to enable draining of the shot feeder and bleed-off of the system water, if necessary. The inlet and outlet connections for a shot feeder are typically ¾ in. since it is not necessary for a high flow rate of system water to be circulated through the shot feeder. The drain connection is also typically ¾ in., a common size for an equipment drain.

Expansion Tank

Purpose
An expansion tank is a necessary component in any closed hydronic system. Its purpose is to provide a means for the system water to expand when it is heated without creating excess pressure in the system. It also serves as the reference point for the system pressure, similar to the ground in an electrical circuit.

Physical Characteristics
Within a closed expansion tank is a closed volume of air that may be separated from the system water by a rubber bladder. This volume of (compressible) air provides a means by which the system water can safely expand when it is heated without creating excess pressure in the system.

Expansion tanks range in size from about 8 to more than 100 gal. The diameter of expansion tanks ranges from 12 to more than 48 in. and the height can be from 12 to more than 96 in.

Connections

The expansion tank is connected to the system piping through a piping connection only; there are no electrical or ATC connections (Figs. 4-24 through 4-26). The connection to the system piping should be the same size as the connection on the expansion tank. A 12-in. anti-thermosiphon trap is recommended in the connection to the system piping to prevent gravity heating of the water in the tank. The expansion tank is normally connected between the air separator and the system pump.

Design Considerations

As the hydronic system water is heated, it expands. As the system water expands, the pressure exerted on the system by a closed expansion tank will increase. The pressure exerted by the expansion tank increases because the air trapped within the expansion

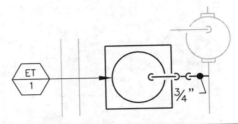

FIGURE 4-24   Floor plan representation of an expansion tank.

FIGURE 4-25   Photograph of an expansion tank.

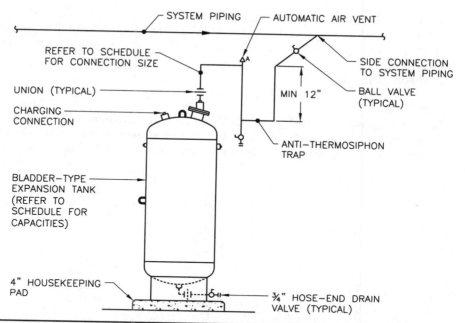

FIGURE 4-26   Detail of expansion tank connection to hydronic system piping.

tank is being compressed. According to the ideal gas law, if the volume of air trapped within the expansion tank is halved, the pressure exerted by the expansion tank on the hydronic system will be doubled. However, the pressure where the expansion tank connects to the hydronic system does not depend upon whether the system pump is operating or not, it only depends upon the temperature of the system water. For this reason, the expansion tank should be connected on the suction side of the system pump so that the pressure on the discharge side of the pump is equal to the pump pressure added to the pressure exerted by the expansion tank when the system pump is operating.

If the expansion tank were connected to the discharge side of the system pump and, as we have previously stated the pressure exerted by the expansion tank at that point does not change due to the system pump pressure, the pressure on the suction side of the system pump when the system pump is operating would be equal to the pump pressure subtracted from the pressure exerted by the expansion tank. This configuration would create the possibility for negative gauge pressure within the hydronic system, particularly at the point where the system pressure would be the lowest—at the suction connection to the pump. This is highly undesirable because it could cause pump cavitation, and could cause air to be drawn into the system water through leaks in the piping or equipment connections wherever a negative gauge pressure exists.

Also, in order to avoid serious consequences, there should only be one expansion tank connection to any closed system. Typically, if a large volume of expansion is anticipated, multiple expansion tanks will be connected together by a common header, which will connect to the system piping at one point.

Expansion tanks can be either floor-mounted or suspended from the building structure. The weight of expansion tanks is a serious consideration since it is possible for the full tank volume to be filled with the system water. Although this is not the design condition, it is necessary to consider the weight of the completely full expansion tank (known as the water-logged weight) when coordinating with the structural engineer.

Makeup Water Assembly

Purpose

The makeup water assembly (Fig. 4-27) for hydronic systems is the source of water supply for the initial system fill and ongoing replenishment of the system water to make up for leaks. The makeup water assembly consists of a backflow preventer (Fig. 4-28),

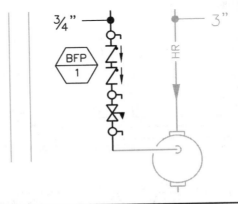

FIGURE 4-27   Floor plan representation of a makeup water assembly.

FIGURE 4-28   Photograph of a backflow preventer.

Figure 4-29   Photograph of a pressure-reducing valve.

pressure-reducing valve (Fig. 4-29), and shutoff valves. A backflow preventer and pressure-reducing valve are necessary components of the makeup water assembly because the makeup water supply pressure, whether originating from a municipality or private source, fluctuates.

During periods when the makeup water pressure is lower than the hydronic system pressure, it is necessary to protect the potable water supply from the contamination that would occur from reverse flow (back siphonage) of the hydronic system water into the potable water system. The backflow preventer is specially designed to prevent the reverse flow of nonpotable hydronic system water from entering the potable water system due to back siphonage or back pressure. The most common type of backflow preventer used for hydronic systems is a reduced pressure zone assembly, which consists of two check valves separated by a relief valve assembly.

During periods when the makeup water pressure is higher than desired for the hydronic system, it is necessary to protect the hydronic system from overpressurization. The pressure-reducing valve accomplishes this and maintains a constant (adjustable) makeup water pressure for the hydronic system.

The purpose of the shutoff valves is simply to isolate each of the components for maintenance or replacement.

Physical Characteristics
Makeup water assemblies range in pipe size from ¾ to 2 in., depending upon the water flow that is desired for the initial hydronic system fill. A ¾-in. makeup water piping connection is common because the flow through a ¾-in. pipe is usually adequate to fill the system within a reasonable amount of time. Larger hydronic systems require larger makeup water assemblies in order to reduce the time required to fill the system. A ¾-in. backflow preventer is approximately 18 in. long and 9 in. high.

Connections

Connection of the makeup water assembly to the hydronic system piping is through a piping connection only; there are no electrical or ATC connections. The makeup water assembly is usually connected to the hydronic system at the system air separator. The connection to the system piping should be the same size as the backflow preventer. Backflow preventers are typically fitted with shutoff valves on the inlet and outlet connections and have a relief valve drain connection. The backflow preventer is also equipped with multiple pressure test connections to facilitate the testing of each of the two check valves and the relief valve assembly within the backflow preventer. Most jurisdictions require testing of backflow preventers on an annual basis. Testing of the backflow preventer is performed with a portable differential pressure gauge.

The direct-acting pressure-reducing valve, located downstream of the backflow preventer, also requires shutoff valves to isolate it for maintenance or replacement. The desired outlet pressure is set through a manual adjustment screw equipped with either a hex head or handwheel.

Design Considerations

Once the hydronic system has been filled, the flow through the makeup water assembly is equal to the loss of system water due to leaks, which should be practically zero. Whenever the backflow preventer operates to prevent back siphonage, some water will be discharged through the relief valve drain connection. Therefore, it is necessary to provide a drain pipe (the same size as the drain connection) that should be routed to a suitable location, such as a floor drain. Sufficient access should be provided for the annual testing of the backflow preventer and maintenance or replacement of the makeup water assembly components, as required. The makeup water assembly should not be mounted so high as to prevent the connection of the test gauge or to make adjustment of the system pressure difficult for the maintenance personnel.

Although it is not necessary, a globe valve bypass may be designed around the pressure-reducing valve, allowing the makeup water assembly to remain in operation while the pressure-reducing valve is repaired or replaced. However, this is not necessary because the hydronic system will operate satisfactorily for a limited amount of time while the makeup water assembly is shut off. A bypass should never be designed around the backflow preventer because this would defeat the purpose of the backflow preventer and would make back siphonage possible if the bypass valve were opened.

Pumps

Purpose

The purpose of a pump in a hydronic system is to circulate the system fluid.

Physical Characteristics

There are many types of pumps including end-suction (Figs. 4-30 through 4-32), close-coupled, in-line (Figs. 4-33 and 4-34), horizontal split-case, vertical split-case, and positive displacement pumps. The most common types of pumps used for hydronic systems are end-suction and in-line pumps, which are both centrifugal pumps.

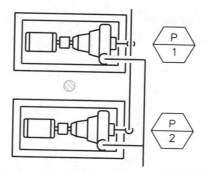

FIGURE 4-30 Floor plan representation of end-suction pumps.

FIGURE 4-31 Photograph of an end-suction pump.

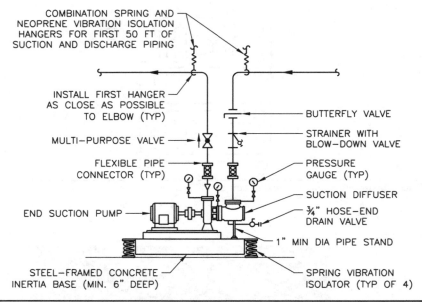

COMBINATION SPRING AND
NEOPRENE VIBRATION ISOLATION
HANGERS FOR FIRST 50 FT OF
SUCTION AND DISCHARGE PIPING

INSTALL FIRST HANGER
AS CLOSE AS POSSIBLE
TO ELBOW (TYP)

MULTI-PURPOSE VALVE

FLEXIBLE PIPE
CONNECTOR (TYP)

END SUCTION PUMP

STEEL-FRAMED CONCRETE
INERTIA BASE (MIN. 6" DEEP)

BUTTERFLY VALVE

STRAINER WITH
BLOW-DOWN VALVE

PRESSURE
GAUGE (TYP)

SUCTION DIFFUSER

¾" HOSE-END
DRAIN VALVE

1" MIN DIA PIPE STAND

SPRING VIBRATION
ISOLATOR (TYP OF 4)

FIGURE 4-32 End-suction pump piping connections.

FIGURE 4-33 Photograph of an in-line pump.

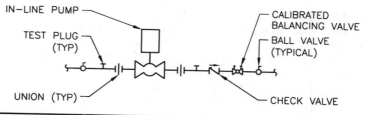

FIGURE 4-34 In-line pump piping connections.

End-suction pumps range in size from 3 to 6 ft long and from 1 to 3 ft wide. The motor shaft is connected to the impeller shaft through a coupling.

In-line pumps are either vertical or horizontal, which describes the orientation of the motor/impeller shaft. The motor shaft is connected directly to the impeller shaft. In-line pumps range in size from 1 to more than 3 ft high (dimension from the impeller to the end of the motor) and 1 to 3 ft between the suction and discharge connections.

Connections

End-suction pumps are attached to an integral steel base frame that is field-mounted to a concrete base. The concrete base can be a 4-in.-high housekeeping pad to which the pump base frame is mounted with spring isolators. However, the preferred mounting is a concrete inertia base to which the pump base frame is bolted. A concrete inertia base is a steel-framed concrete block that is approximately 6 in. high and 6 in. larger than the pump base on all sides which is supported off of the floor by spring isolators. The concrete inertia base provides a rigid base to maintain alignment of the pump shaft and reduce the vibratory motion caused by the rotating pump motor.

The pump suction pipe connection is parallel to the impeller shaft and the discharge pipe connection is perpendicular to the impeller shaft. Flexible pipe connectors are used on the suction and discharge pipe connections for end-suction pumps to isolate the vibration that is generated by the pump from the piping system.

The suction and discharge connections for in-line pumps are in line with each other and are perpendicular to the pump/impeller shaft. Small in-line pumps are supported by the piping system. Large in-line pumps require pipe hangers to be installed near the suction and discharge connections. Very large in-line pumps will be supported from the building floor, usually on a 4-in.-high concrete housekeeping pad.

The piping connections required for pumps include shutoff valves on the pump suction and discharge, balancing valve on the pump discharge, check valve and flow meter on the pump discharge, and pressure gauges. As an option, a multipurpose valve, which performs the duties of a shutoff valve, balancing valve, and check valve, may be installed on the pump discharge. It is common for a suction diffuser, which is similar in size to that of a long radius 90° pipe elbow, to be used on the suction pipe connection for end-suction pumps. This allows the suction pipe to drop vertically into the suction diffuser. Otherwise, it is necessary to provide five pipe diameters of straight pipe upstream of the pump suction connection. If a suction diffuser or the necessary length of straight pipe upstream of the pump suction connection is not provided, undesirable turbulence in the fluid flow will occur at the pump suction connection, which will compromise the performance of the pump and may also damage the pump.

In addition to the piping connections, the motors for pumps require an electrical connection. Typically, motors smaller than ½ hp require single-phase electrical power; motors ½ hp and larger require three-phase electrical power if it is available within the building. All motors require a starter, which can be a motor-rated manual switch for small motors or an across-the-line magnetic motor starter for larger motors. Some applications may also require that the pump motor be controlled by a VFD, which modulates the speed of the motor and functions as the motor starter.

If automatic control of the pump is required through the ATC system, the magnetic motor starter or VFD must be equipped with the capability to receive a signal, such as a start/stop or motor speed signal, from the ATC system. This is discussed in more detail in Chap. 9.

Design Considerations

The pump must be designed to overcome the pressure losses within the system, which include losses associated with the supply and return piping, fittings, valves (shutoff, balancing, and control), strainers, heat transfer equipment (coils, heat exchangers), flow meters, and any other component within the system. The actual pressure drop from the manufacturer's product data should be used for each component. However, 50% of the pressure drop through the piping system is normally used as an allowance for the pressure drop through the pipe fittings, and the shutoff and balancing valves. The elevation head must only be taken into account for an open system (refer to the Cooling Towers section earlier); the elevation of the piping system is not a consideration for a closed system. The following is a sample pump head calculation for a primary-only, closed, chilled water system:

Component	Pressure Drop (ft w.c.)
1. Air separator (2 psi)	4.6
2. Strainer (4 psi)	9.2
3. Suction diffuser (0.5 psi)	1.2
4. Flow meter (5 psi)	11.6
5. Chiller cooler	10.0
6. Piping (400 ft × 4 ft w.c./100 ft)	16.0
7. Valves and fittings (50% of piping)	8.0
8. Control valve (5 psi)	11.6
9. Chilled water coil	10.0
Subtotal	82.2
Safety factor (5%)	4.1
Total head	86.3 (round to 86 ft w.c.)

The preferred selection range for a centrifugal pump is between 85 and 105% of the flow at the best efficiency point (BEP) on the pump curve. Refer to Chap. 43 of the 2008 *ASHRAE Handbook—HVAC Systems and Equipment* for more information on the physical characteristics and selection procedures of centrifugal pumps.

Sufficient access must be provided around the pump for proper maintenance and testing. Typically, 12 to 18 in. of clearance on all sides of an end-suction pump provides sufficient access for maintenance.

4-Pipe and 2-Pipe Heating and Cooling Plants

A 4-pipe heating and cooling plant contains both central heating and cooling equipment and is capable of delivering heating water and chilled water to the building simultaneously through four pipes (one heating water supply, one heating water return, one chilled water supply, and one chilled water return). Heating and cooling equipment within the building that is connected to a 4-pipe system will have four pipe connections, unless the equipment provides either heating only or cooling only. In this case, the equipment would have only two pipe connections.

Figure 4-35 is a schematic diagram of the piping for a 4-pipe heating and cooling plant that utilizes two condensing hot water boilers and two water-cooled chillers. The pumping arrangement is primary-secondary for both the heating water and chilled water systems. Both the heating and chilled water systems are variable flow systems with variable frequency drives controlling the speed of the (secondary) heating and chilled water system pumps. One of the two pumps shown for the heating and chilled water system pumps and one of the condenser water pumps is a standby pump. A separate condenser water pump and cooling tower is dedicated to each chiller. Automatic shutoff valves are designed for the condenser water supply, return, and equalizer piping connections to isolate the idle cooling tower when only one chiller is operating.

A 2-pipe heating and cooling plant contains both central heating and cooling equipment but is not capable of delivering heating water and chilled water to the building simultaneously. It operates either in the heating mode or cooling mode and delivers either heating water or chilled water through two pipes (one dual-temperature water supply and one dual-temperature water return) to the building. Heating and cooling equipment within the building that is connected to a 2-pipe system will have two pipe connections.

Figure 4-36 is a schematic diagram of the piping for a 2-pipe heating and cooling plant that utilizes two condensing hot water boilers and one water-cooled chiller. The pumping arrangement when the plant is operating in the heating mode is a primary-secondary pumping system with a primary pump dedicated to each boiler to ensure a constant flow of water through each condensing boiler. The dual-temperature water system pumps are constant speed and function as the secondary pumps. One of the two pumps shown for the dual-temperature water system pumps and the condenser water pumps is a standby pump.

In the cooling mode, the plant operates in a primary-only pumping arrangement. In this arrangement, the dual-temperature water system has to be a constant-flow system in order to maintain a constant flow of water through the chiller during cooling operation. If a primary pump were designed for the chiller, the dual-temperature water system could be a variable flow system with variable frequency drives controlling the speed of the (secondary) dual-temperature water pumps.

Design Considerations

Design considerations for 4-pipe and 2-pipe heating and cooling plants are as follows:

1. It is common to design redundancy for the equipment in heating systems (such as boilers and pumps) because freezing of the building could occur if the

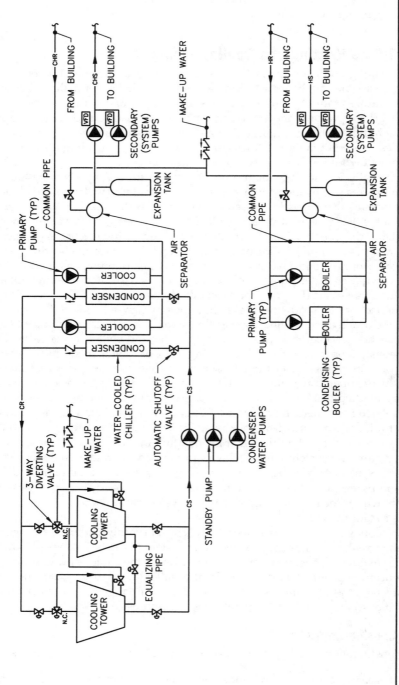

FIGURE 4-35 4-pipe heating and cooling plant schematic piping diagram.

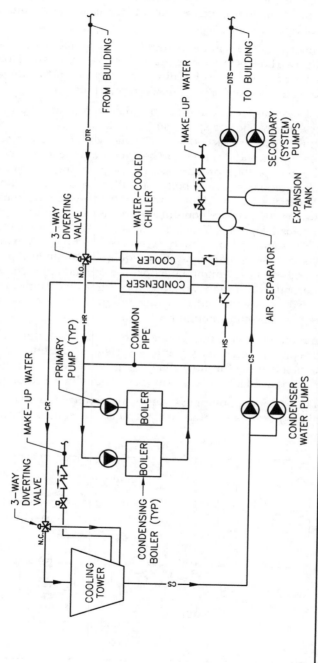

Figure 4-36 2-pipe heating and cooling plant schematic piping diagram.

139

heating system is lost. On the other hand, it is not common to design redundancy for the equipment in cooling systems (such as chillers and pumps) because comfort cooling is generally not considered critical. However, cooling systems serving critical functions, such as computer or health care facilities, may require redundant cooling equipment.

2. Since some redundancy in the boilers is normally required, it is common for each of the two boilers in a 4-pipe or 2-pipe system to be sized for two-thirds of the peak heating load of the building. This provides 67% redundancy to keep the building temperature above freezing if one boiler fails.

3. For small systems, it is common to utilize a constant-speed, primary-only pumping system. However, for larger systems (where pumping energy is significant), a primary-secondary pumping system is recommended because the system (or secondary) flow can be varied to reduce the energy use of the secondary pump. In a primary-secondary pumping system, each piece of primary equipment, such as a boiler or chiller, has a dedicated primary pump. Energy savings are also achieved with primary-secondary pumping systems by staging on the primary equipment (and associated pumps) in response to the system load. Figures 4-37 and 4-38 illustrate a constant-speed, primary-only pumping system and a primary-secondary pumping system. Note that a primary-secondary pumping system requires a common pipe that joins the primary and secondary pumping loops. The common pipe should be sized for the full secondary flow and should be approximately 10 pipe diameters long in order to reduce any unwanted mixing and to keep the pressure loss through this pipe to an absolute minimum.

4. It is common to provide full redundancy for the system pump (or secondary pump in a primary-secondary pumping system) by designing two pumps, each sized to circulate the full flow. One pump will always be running while the other pump is available on a standby basis should the lead pump fail.

5. A primary-secondary pumping system is almost always used for high-efficiency (condensing) boilers because of their need for constant water flow. Some

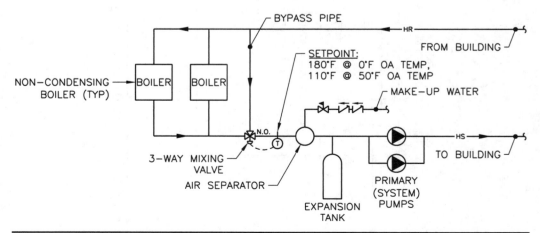

FIGURE 4-37   Constant-speed, primary-only pumping system piping diagram.

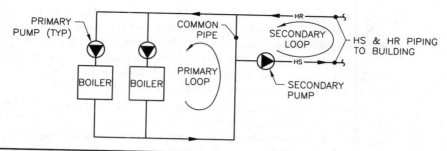

Figure 4-38   Primary-secondary pumping system piping diagram.

high-efficiency boilers are equipped with primary pumps installed within the boilers themselves to ensure that the heat exchangers receive the minimum required water flow.

6. A common control strategy for heating water systems is to reset the temperature of the heating water supplied to the heating equipment in the building based on outdoor temperature. This strategy allows for better control of space temperature and also reduces the heat loss from the heating water piping system during part-load operation.

   a. A common heating water reset schedule is as follows:
      (1) 180°F heating water supply temperature when the outdoor temperature is 0°F.
      (2) 110°F heating water supply temperature when the outdoor temperature is 50°F.

   The heating water supply temperature would vary proportionally between 180 and 110°F as the outdoor temperature varies between 0 and 50°F.

7. Heating water temperature reset is accomplished with condensing boilers simply by resetting the heating water supply temperature from the boilers based on the outdoor temperature. As mentioned previously, the efficiency of condensing boilers increases as the returning water temperature decreases.

8. However, as mentioned earlier, noncondensing boilers must maintain a minimum of 140°F returning water temperature; thus it would not be possible to achieve the reset schedule listed above by resetting the heating water supply temperature from the boilers. Therefore, the addition of a 3-way mixing valve to blend heating water return with heating water supply is required to reset the heating water supply temperature based on outdoor temperature (Fig. 4-37).

9. It is best to utilize the same pipe sizing criteria for the central plant that is used for the distribution system. Refer to Chap. 6 for a discussion of piping distribution systems and associated pipe sizing criteria.

10. The makeup water assembly for all closed systems consists of a backflow preventer, pressure-reducing valve, and shutoff valves.

11. The boiler should be installed at the point of lowest pressure developed by the heating water system pump (suction side of the pump) for the reasons discussed earlier.

12. For cooling plants consisting of multiple water-cooled chillers, it is common for each chiller to have a dedicated cooling tower (or cooling tower cell within a multiple-cell cooling tower) and a dedicated condenser water pump. An additional condenser water pump can serve as a standby pump for every two condenser water systems, provided the systems require the same water flow rate and appropriate valves are installed to isolate the pumps.

13. For central cooling plants having only one chiller and one cooling tower, it is possible for a third pump to function as a standby pump for both the chilled water and condenser water systems, provided the pump has a suitable operating point for both systems.

14. One major disadvantage of 2-pipe heating and cooling systems is the time that it takes to accomplish the changeover from heating operation to cooling operation in the spring of each year because chillers generally cannot tolerate an entering water temperature to the cooler that is greater than 70°F. Therefore, the dual-temperature water loop must cool down from a heating water temperature that is at least 110°F to 70°F before dual-temperature water can be circulated through the chiller cooler and chilled water can be produced. The problem with this is that when the building is calling for cooling, there is no demand for heat. Thus, there is no way for the warm water in the dual-temperature water system to reject its heat. The dual temperature water loop must cool down as the result of heat losses from the insulated dual-temperature water piping, which can take up to 2 or 3 days, depending upon the size of the system. A solution to this problem is available if the chillers are water-cooled. The changeover time can be greatly reduced through the incorporation of a dual-temperature water cool-down system. This system would utilize the cooling tower as a source of heat rejection for the dual-temperature water system. The addition of a plate and frame heat exchanger, 3-way diverting valves, and controls are necessary to accomplish this mode of operation, the details of which are beyond the scope of this book.

Water-Source Heat Pump Plant

A water-source heat pump plant adds heat to, or rejects heat from, the closed-loop heat pump water that is circulated through water-source heat pumps located within the building. Water-source heat pumps, also known as water-to-air heat pumps, are water-cooled, self-contained units with a refrigeration system capable of reversing its cycle to provide either cooling or heating to the spaces in the building that they serve. A water-source heat pump differs from a heating, chilled, or dual-temperature water unit in that a water-source heat pump contains a refrigeration system that heats and cools the conditioned air through a direct expansion refrigerant coil instead of a heating water, chilled water, or dual-temperature water coil. During cooling operation, a water-source heat pump rejects heat to the heat pump water; during heating operation, a water-source heat pump absorbs heat from the heat pump water.

The temperature of the heat pump water must be maintained within certain limits, usually between 60 and 90°F, in order to enable the heat pump units to either reject or absorb heat from the heat pump water, depending upon their mode of operation. The water flow rate for a water-source heat pump system is typically 3 gpm per ton of installed cooling capacity. The range (the design temperature difference between the water leaving the individual water-source heat pump heat exchangers and the water entering the heat exchangers) is typically 10°F. Typical temperatures for the heat pump water are as follows:

- Summer: 90°F heat pump water supply and 100°F heat pump water return
- Winter: 60°F heat pump water supply and 50°F heat pump water return

The advantage of utilizing a water-source heat pump system to condition a building is that some units can be operating in the cooling mode while other units are operating in the heating mode. In fact, a water-source heat pump system is quite energy efficient when simultaneous heating and cooling within the building are required. For example, during the winter, the heat that is rejected to the heat pump water by the units operating in the cooling mode contributes to the warming of the heat pump water, thereby reducing the heat that must be added to the heat pump water by the heat source (which is typically a hot water boiler). The converse is true for units that operate in the heating mode during the summer. This exchange of heat from some areas of the building to others through the heat pump water system reduces the use of purchased energy for the building.

However, when more heat is rejected by the units in the cooling mode than is absorbed by the units in the heating mode, it is necessary for the heat pump water system to have a means to reject the excess heat to the outdoors (referred to as a heat sink). Similarly, when more heat is absorbed by the units in the heating mode than is rejected by the units in the cooling mode, it is necessary for the heat pump water system to have a means to add heat to the system (referred to as a heat source).

Excess heat is normally rejected from the heat pump water either through a plate and frame heat exchanger to an open cooling tower, or through a closed-circuit cooling tower. Heat is added to the heat pump water typically through a hot water boiler. The ground can also be used as a heat sink and heat source for a water-source heat pump system. This type of system is called a ground-source (or geothermal) heat pump system.

Figure 4-39 is a schematic diagram for a typical water-source heat pump plant utilizing a plate and frame heat exchanger, an open cooling tower as the heat sink, and two high-efficiency hot water boilers as the heat source. In the heat addition mode, the 3-way diverting valve diverts the heat pump water to the boilers when the return water temperature drops below 65°F. The lead boiler and its associated primary pump start when the return water temperature drops below 60°F. The lag boiler and its associated primary pump start when the return water temperature drops below 50°F. The firing rate of each boiler burner is modulated to maintain a constant (boiler) outlet water temperature of 90°F. The 3-way mixing valve maintains the heat pump water supply temperature at 60°F. Upon a rise in return water temperature above 65°F, the reverse occurs. In the heat rejection mode, the 3-way diverting valve diverts the heat pump water to the plate and frame heat exchanger when the return water temperature rises above 65°F. When the heat pump return water temperature rises above 90°F, the tower water pump is started and the speed of the cooling tower fan is controlled to maintain the heat pump water supply temperature at 90°F. Upon a drop in return water temperature below 90°F, the reverse occurs.

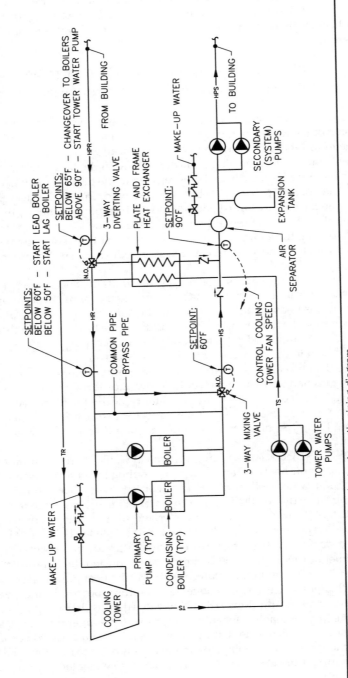

FIGURE 4-39 Water-source heat pump plant schematic piping diagram.

Design Considerations

Design considerations for water-source heat pump plants are as follows:

1. Full redundancy should be designed for the hot water boilers and closed-loop heat pump water system pumps because these boilers and pumps serve the heating needs of the building.

2. Each of the two boilers in a water-source heat pump system should be sized for the sum of the heat absorbed by all the heat pump units operating at full heating capacity. This is done because all of the heat pump units could be operating at full heating capacity during morning warm-up. If a single boiler is not able to maintain the minimum loop temperature, the heat pump units could shut off due to a low refrigerant pressure safety. However, the heat absorbed by the heat pump units from the heat pump water system represents only about 75% of the total heat output of the heat pump units, assuming an average coefficient of performance (COP)[52] for the heat pump units of 4.0. Therefore, the output of each boiler serving a water-source heat pump system does not need to be sized for 100% of the peak heating load of the building, but only for approximately 75% of the peak heating load of the building.

3. The same design considerations regarding constant-speed, primary-only pumping, primary-secondary pumping, pipe sizing criteria, makeup water assembly, and the location of the boiler in the system are the same for heat pump water plants as they are for 4-pipe and 2-pipe heating and cooling plants.

4. Redundancy is not required in the heat rejection equipment if the system provides comfort cooling only (see the discussion of redundancy in the cooling equipment for 4-pipe and 2-pipe heating and cooling plants above).

5. A supply water temperature reset schedule based on outdoor air is not used for heat pump water plants. Rather, the heat pump water supply temperature is allowed to float between 60 and 90°F. For this reason, it is not necessary to add heat to, or reject heat from, the heat pump water system during certain periods of operation when both heating and cooling are required in the building and the loop temperature remains between 60 and 90°F.

Equipment Room Design

The major considerations for the design of equipment rooms are as follows:

1. The overriding principle in equipment room design is to ensure that proper clearance is allowed within and around the room for installation, maintenance, and removal of equipment. This includes allowing access space around each piece of equipment for routine maintenance, allowing space within the equipment room or through doorways for significant maintenance work such as pulling tubes from boilers or chillers, and designing aisles within the equipment room to provide unobstructed clearance for equipment to be removed and reinstalled.

2. Equipment should be located within equipment rooms so that piping and ductwork are routed within the room and between interconnecting pieces of equipment in the most straightforward manner possible. Headroom within the equipment room is reduced wherever piping and ductwork need to cross each other or where they need to cross over equipment.

3. All HVAC, plumbing, and electrical equipment required in each equipment room needs to be identified early in the design phase in order to avoid "surprises" at the end of the project. It is necessary for the HVAC system designer to coordinate with all other design disciplines to ensure that the equipment and associated clearance requirements of all disciplines are coordinated. For example, the HVAC equipment may be fed electrically from multiple panelboards located within the equipment room, or the HVAC equipment may be fed from a single motor control center located within the equipment room or in a different room of the building.

4. Working spaces around electrical equipment, such as the area required in front of electrical equipment connections, and dedicated equipment spaces, such as the clear space above electrical panelboards, must be respected in the HVAC system design. The HVAC system designer should never design piping or ductwork to be routed above electrical panelboards, and HVAC equipment should be located within the equipment room to ensure that the electrical working spaces are maintained.

5. The minimum height from the equipment room floor to the building structure above needs to be coordinated with the project architect and structural engineer. Normally a minimum height of approximately 10 ft is required if there will be piping only within the equipment room. A minimum height of approximately 13 ft is required if there will be ductwork in the equipment room as well as piping. These are only guidelines; it is necessary to coordinate the actual room height requirements with the equipment to be installed within the equipment room. Drawing a section at the most congested location in the equipment room is the best way to determine the minimum clear height required from the equipment room floor to the structure above.

6. Clear space should be allowed to remove plates from plate and frame heat exchangers for maintenance and to add additional plates if additional capacity is required in the future.

7. Chillers cannot be located in the same room as any combustion equipment unless the combustion equipment has a direct combustion air connection to the outdoors or the room is equipped with a refrigerant detection system that will automatically shut down the combustion process if a refrigerant leak is detected.

8. It is best to have at least one exterior building wall in the equipment room to facilitate the installation and removal of equipment. If the mechanical room is in the interior of a building, the corridors and doorways leading from the exterior doorway to the equipment room must be wide enough and high enough to install and remove equipment within the equipment room.

9. It is common to design double doors in equipment rooms for equipment installation and removal. The doors may need to be 8 ft high or higher to provide sufficient space for removal and installation of large equipment.

10. Floor drains connected to the building sanitary system are required near all floor-mounted pumps, backflow preventers, system drains, cooling tower bleed pipe, and cooling tower overflow/drain pipe. It is common to design a floor drain between two adjacent floor-mounted pumps. Floor drains connected

to the building storm water system are required to receive the condensate drainage from cooling coils in air handling units.

11. A makeup water line is required for all closed- and open-loop water systems. Typically a ¾- or 1-in. makeup water line is adequate.

12. ATC panels need to be located within equipment rooms in accessible locations. The size and number of ATC panels need to be coordinated with the ATC representative during the design phase.

13. The locations of the lighting fixtures within equipment rooms need to be coordinated with the equipment locations. The HVAC system designer should communicate the equipment room layout to the project electrical engineer as early in the design phase as possible so the lighting design can be coordinated.

14. The routing of boiler venting and pressure relief piping needs to be coordinated within the equipment room and also outside the building.

15. Unit heaters are generally used to heat equipment rooms. Refer to Chap. 7 for more detailed information.

16. Air-conditioning systems are rarely used to maintain the space temperature within equipment rooms. Instead, ventilation systems are used to draw air from the outdoors in order to keep the room at a maximum of 10°F higher than the outdoor summer design temperature. The heat gains associated with the equipment motors, lighting, and exterior building envelope need to be included in the HVAC load calculation when determining the ventilation airflow requirement. Equipment rooms containing gas-fired or fuel-burning equipment are typically ventilated through a supply fan and relief air system in order to keep the room under positive air pressurization when ventilation is required. The supply fan can be wall-mounted, roof-mounted, or suspended within the room with a connection to the outdoors through a wall louver or rooftop intake air hood. Relief air is typically accommodated through a wall louver or rooftop relief air hood that is protected from backdrafts by either a backdraft damper or a motor-operated damper. If there is no gas-fired or fuel-burning equipment located within the room (or if this equipment has direct connections of outdoor combustion air), an exhaust fan and intake air system with motor-operated damper is preferred for ventilation in order to keep the room under negative pressurization, which reduces the transfer of odors from the equipment room to adjacent occupied spaces. Mechanical room ventilation systems are typically controlled by a space thermostat that is designed to operate the ventilation system whenever the space temperature exceeds 85°F.

17. Combustion air must be provided for all gas-fired and oil-burning appliances located within the equipment room. Refer to *NFPA Standard 54—National Fuel Gas Code* or *NFPA Standard 31—Standard for the Installation of Oil-Burning Equipment* for combustion air requirements.

18. Certain pieces of refrigeration equipment will require refrigerant pressure relief piping designed in accordance with the requirements of *ANSI/ASHRAE Standard 15-2010*.

19. Equipment rooms housing refrigeration equipment may require refrigerant detection, exhaust, and alarm systems designed in accordance with *ANSI/ASHRAE Standard 15-2010*.

Endnotes

1. Natural gas and propane are gaseous fuels.
2. Fuel oil, which can refer to distillate fuel oils (lighter oils) or residual fuel oils (heavier oils), is a liquid fuel. The most commonly used fuel oil for fuel-burning appliances is Grade No. 2 distillate fuel oil (referred to as No. 2 fuel oil).
3. The term *heating water* is used in this chapter to refer to either heated water or heated brine.
4. Draft is the pressure difference that causes the products of combustion to flow through a gas-fired or fuel-burning appliance and vent system.
5. Condensing refers to the water vapor in the flue gas, whether it condenses within the combustion chamber of the boiler or not.
6. Low-pressure hot water boilers are limited to a maximum of 250°F water temperature.
7. The gross output rating is equal to the input rating minus the heat that remains within the flue gas and is discharged through the vent system to the outdoors (referred to as the stack loss) minus the heat lost by the boiler surface to its surroundings.
8. I-B-R refers to the former Institute of Boiler and Radiator Manufacturers.
9. Pickup is an estimate of the load required to initially heat the working fluid and piping of an average system.
10. An appliance refers to any gas-fired or fuel-burning piece of equipment, including the burner.
11. The vertical portion of a vent system is referred to as the chimney.
12. The power venting fan must have a high temperature rating and be approved for installation in a vent system for gas-fired or fuel-burning equipment, whichever is applicable.
13. The dew point of water vapor in flue gas for typical natural gas combustion is approximately 130°F.
14. Combustion efficiency is equal to the input rating minus the stack loss divided by the input rating.
15. Thermal efficiency is the gross output rating divided by the input rating.
16. The heat transfer efficiency of a coil increases as the difference between the average temperature of the coil and the entering air temperature increases.
17. This assumes the heating water system is a closed system; that is, an open expansion tank is not used.
18. Although the water pressure at the point where the (closed) expansion tank connects to the system varies in proportion to the expansion of the water in the system, the water pressure at this point is not affected by the operation of the system pump. For this reason, the point where the (closed) expansion tank connects to the system is sometimes referred to as "the point of no pressure change" in the system.
19. Thermal shock occurs when the cast iron or steel components comprising the combustion chamber and fluid pressure vessel of a boiler are subjected to a large temperature difference between the flue gas and return water. Thermal shock causes accelerated deterioration of these components and ultimately results in cracks and leaks in the combustion chamber and/or fluid pressure vessel.
20. The heating water supply temperature is normally reset based on outdoor temperature. As the outdoor temperature rises, the heating water supply temperature is lowered.

21. When the pressure of steam condensate at pressures greater than atmospheric pressure is reduced to atmospheric pressure (or any pressure lower than the saturation pressure of the condensate), a portion of the condensate re-evaporates into what is called flash steam. For steam systems operating at 2 psig, the percentage of flash steam is minimal (0.7%) and need not be utilized. However, for steam systems operating at pressures exceeding 2 psig, it is recommended that the flash steam be utilized by steam equipment operating at a reduced steam pressure. Otherwise, this flash steam must be vented to the outdoors, which wastes its heating potential.

22. Although the steam condensate is returned by steam pressure motivation, condensate return piping should be pitched ½ in./10 ft in the direction of condensate flow.

23. Excessive dissolved oxygen in the boiler feedwater accelerates the corrosion of the fluid pressure vessel in the boiler.

24. Approach temperature = 200°F − 180°F = 20°F.

25. Approach temperature = 45°F − 38°F = 7°F.

26. Fouling factors have units of thermal resistance (ft²·h·°F/Btu).

27. This type of coil is called a direct expansion coil because the refrigerant within the coil expands, or evaporates, directly within the coil and, in the process, the refrigerant absorbs heat from the air that is blown across the coil.

28. The cooler is called a direct expansion cooler when the refrigerant circulates through the tubes. It is called a flooded cooler when the refrigerant circulates through the shell.

29. One ton of refrigeration equals 12,000 British thermal units per hour (Btuh).

30. EER = [(100 tons) × (12,000 Btuh/ton)] / [(110 kW) × (1,000 W/kW)] = 10.9 EER.

31. Soft water has a low concentration of calcium and magnesium ions.

32. A water softener exchanges sodium ions for the calcium and magnesium ions in the water.

33. Increased space dehumidification can be accomplished and a decreased chilled water flow rate can be utilized with a lower chilled water supply temperature. However, these benefits come at the expense of decreased chiller energy efficiency.

34. A greater chilled water temperature rise requires that the cooling coils in the air systems and terminal equipment transfer heat from the air more efficiently. This increase in heat transfer efficiency is only achieved through a greater cooling coil fin surface, which translates to an increased first cost of the cooling coils and increased fan energy use.

35. Allowing the condenser water supply temperature to drop below 85°F when the outdoor conditions are appropriate is a strategy that can reduce a chiller's energy use.

36. Monitoring refers to receiving an input signal (either analog or digital) from a piece of equipment. Control refers to sending an output signal (either analog or digital) to a piece of equipment. Refer to Chap. 9 for a more detailed discussion of automatic temperature controls and building automation systems.

37. Chiller operation can be controlled remotely through the BAS but is normally performed by the chiller control panel. Common points for monitoring through the BAS include chiller status, water temperatures, and alarms.

38. The flow or differential pressure switches are normally shipped with the chiller for field-installation in the piping.

39. Water flow through a cooling tower should be full flow to the distribution nozzles or full bypass to the sump. Full bypass to the sump is normally performed only during start-up to prevent unacceptably cold condenser water from being circulated

through the chiller condenser. Varying the flow of water through the distribution nozzles while the cooling tower fan(s) are running is not recommended because the percentage of the total water flow through the cooling tower that is evaporated increases, thereby increasing the rate of scale formation on the cooling tower fill.

40. An open system is one that is open to the atmosphere and operates under atmospheric pressure.

41. A closed system is one that is not open to the atmosphere and is not subject to the limitations of atmospheric pressure.

42. Nonmechanical-draft cooling towers that do not contain fill or utilize a fan are available but are not frequently used for HVAC applications.

43. A contactor is a heavy-duty relay. A relay is an electromagnetic switch whose contacts are opened and closed by the presence or absence of current flow through a solenoid coil. Refer to Chap. 9 for a more detailed discussion of relays and their use in automatic temperature control systems.

44. Approach temperature, or simply approach, is equal to the leaving water temperature minus the wet bulb temperature of the ambient air.

45. Wet bulb temperature is a measure of the moisture content of the ambient air and also gives an indication of the rate at which water evaporates in the ambient air. The evaporation rate increases as the wet bulb temperature of the ambient air decreases.

46. Approach = 85°F – 78°F = 7°F approach.

47. Range = 95°F – 85°F = 10°F range.

48. NPSHR is given in the manufacturer's product data for the pump.

49. Nonmechanical-draft cooling towers, which induce air through the aspirating effect of water spray, are typically quieter than mechanical-draft cooling towers. However, this type of cooling tower is not a subject of this book because their application for HVAC systems is relatively limited.

50. Chemically treated cooling tower water is normally not discharged to the building storm water system unless the chemicals are biodegradable and it is acceptable to the authority having jurisdiction.

51. High water alkalinity (high pH) and low water hardness (soft water) are two factors that aggravate the formation of white rust.

52. The coefficient of performance (COP) for a compressor is the ratio of the refrigeration output to the heat equivalent of the electrical input power. For example, a compressor that requires 0.88 kW (3,033 Btuh) of electrical input to produce 1.0 ton (12,000 Btuh) of refrigeration output would have a COP of 4.0 (12,000 Btuh output/ 3,033 Btuh input). A COP of 4.0 is typical for water-source heat pumps.

CHAPTER 5

Air Systems

The term *air system* refers to the equipment that circulates air within a building. Air systems can introduce ventilation air to the building from the outdoors, recirculate air within the building, exhaust air from the building, or perform a combination of any or all of these functions.

The air circulated by air systems can be conditioned (heated, cooled, humidified, or dehumidified) or it can be unconditioned. If heating and/or cooling is required, air systems utilize the heating and/or cooling energy that is supplied by the central plant and impart this energy to the circulated air by means of heating and/or cooling coils. Humidification, if required, is added at the air system either through steam generated at a remote central plant or by a self-contained (or packaged) humidifier located at the air system. Dehumidification, if required, is performed by components such as a reheat coil and automatic temperature controls, which are also associated with the air system.

Common components of air systems that provide conditioned air to the spaces they serve include the following:

- Mixing box: The mixing box is a section of an air system that contains openings for the return air duct and outdoor air duct. The purpose of the mixing box is to provide a point in the system where these two airstreams can be combined. The mixing box may also contain manual or motor-operated dampers, which are used to balance the return and outdoor airflows. If these dampers are not integral to the mixing box, they may be located in the ductwork outside of the mixing box.

- Air filters: Air filters remove the particulates entrained within the air that is circulated through the air system. The efficiency of the filters is determined by the filtration requirements of the spaces served by the air system. A common measure of filtration efficiency is the minimum efficiency reporting value (MERV), as defined by *ANSI/ASHRAE Standard 52.2-2007—Method of Testing General Ventilation Air-Cleaning Devices for Removal Efficiency by Particle Size.* Air filtration efficiency for commercial HVAC applications commonly has a MERV of 8, which can be achieved through the use of 2-in. pleated, disposable air filters. For superior air filtration, such as is required to achieve the Leadership in Energy and Environmental Design (LEED) Indoor Environmental Quality (EQ) Credit 5, an air filtration efficiency of MERV 13 is required. MERV 13 filters are normally 6- or 12-in. rigid-style cartridge filters. In order to increase the service life of these high-efficiency filters, MERV 8 prefilters are normally installed in a separate filter track upstream of the MERV 13 (final) filters.

- Heating coil: The heating coil imparts the heating energy supplied by the central plant to the air that is circulated through the air system.
- Cooling coil: The cooling coil imparts the cooling energy supplied by the central plant to the air that is circulated through the air system.
- Reheat coil: A reheat coil may be a part of an air system when dehumidification is required.
- Supply fan: Air systems that supply air to the spaces served will contain a supply fan.
- Return fan: Air systems that recirculate air from the spaces served may contain a return fan, particularly if the return air ductwork system is extensive.
- Humidifier: A humidifier typically utilizes steam to increase the moisture content of the air that is circulated through the air system to meet the humidification requirements of the spaces served by the air system. Other types of humidifiers that utilize ultrasonic sound waves to atomize water droplets for entrainment into the airstream are also available.

Air systems that do not condition the air before it is introduced to the spaces served have limited use in commercial buildings since most spaces within a building are conditioned to some degree.

Air systems that exhaust air from the building normally consist of exhaust fans only. However, if the exhaust airstream contains hazardous contaminants that cannot be released to the atmosphere, the exhaust air system may also contain air filters. Also, if heat recovery is utilized, air filters and a heat recovery coil will be incorporated into the exhaust air system (as well as the outdoor air system).

Air System Types

For the purpose of distinguishing between air systems that primarily heat and cool from those that primarily ventilate, we will categorize air systems as follows:

1. Heating, ventilating, and air-conditioning (HVAC) air systems
    a. HVAC air systems circulate air primarily to meet the heating and/or cooling needs of the spaces they serve.[1]
    b. HVAC air systems may also meet the ventilation needs of the spaces they serve, but this capability is normally supplemental to their primary function of providing heating and/or cooling. For HVAC air systems, outdoor air is normally a relatively low percentage (less than 30%) of the total airflow supplied by the HVAC air system.[2]
    c. Outdoor air ventilation provided by HVAC air systems meets the outdoor air ventilation requirements of the building occupants and/or provides outdoor air makeup for exhaust air systems.

2. Ventilation air systems
    a. Ventilation air systems circulate air primarily to meet the outdoor air or exhaust air needs of the spaces or equipment they serve.
    b. Ventilation air systems may have heating, cooling, humidification, or dehumidification capabilities. However, these capabilities are for the sole purpose of conditioning outdoor air to a neutral temperature and relative

humidity so that the ventilation air does not become a load on the HVAC air systems.

c. Exhaust air systems also fall into the category of ventilation air systems. Exhaust air systems may serve the exhaust requirements of spaces served and/or may serve the exhaust requirements of certain types of equipment, such as kitchen exhaust hoods.

HVAC Air Systems

As mentioned in Chap. 2, HVAC air systems can be either centralized or decentralized. Centralized air systems receive their cooling and heating energy from a remote central plant. Decentralized HVAC air systems contain the central heating and cooling plant equipment within the air system itself.

HVAC air systems can also be constant air volume (CAV) or variable air volume (VAV). CAV systems deliver constant supply airflow at a variable temperature. VAV systems deliver variable supply airflow at a constant temperature. CAV and VAV air systems can be further subdivided into systems that condition a single temperature zone and systems that condition multiple temperature zones.

Whether the HVAC air systems are centralized or decentralized, CAV or VAV, serve a single zone or multiple zones, the choices for heating and cooling coils within the units are the same. Generally, the heating coil in centralized HVAC air systems will be hot water, steam, or electric, and the cooling coil will be chilled water. The heating coil in decentralized HVAC systems will be electric, direct or indirect gas-fired,[3] or (reverse cycle) direct expansion (DX) refrigerant (for heat pumps only). The cooling coil will be DX refrigerant.

Constant Air Volume Systems

Single-Zone   A single-zone CAV system consists of an air handling unit that delivers constant supply airflow (Fig. 5-1). The heating and/or cooling capacity of the unit is modulated to meet the needs of a single thermostat mounted in one of the spaces served by the unit. Single-zone CAV systems may serve multiple spaces. Therefore, it is important to ensure that all of the spaces served by a single-zone CAV system have similar HVAC load characteristics.

Heating and Ventilating   Heating and ventilating (H&V) systems are used to serve spaces that require heating only and also require outdoor air ventilation for occupant ventilation or exhaust air makeup. H&V units are often configured to deliver 100% outdoor air. However, they may have return air capabilities if the outdoor airflow is less than the supply airflow that is required to provide effective heating of the spaces served during the winter. If the units heat the spaces they serve during unoccupied periods when the exhaust systems are off and the outdoor air dampers are shut, H&V units may also be equipped with return air capabilities. H&V units normally recirculate 100% of the air during the unoccupied mode of operation.

During the summer it is common for H&V units to position their outdoor and return air dampers to deliver 100% outdoor air to the spaces served in order to maintain the space temperature at 5 to 10°F higher than the outdoor temperature. If the H&V units have return air capabilities, provisions must be made within the spaces served to either relieve or exhaust the excess outdoor air that is introduced during the summertime mode of operation.

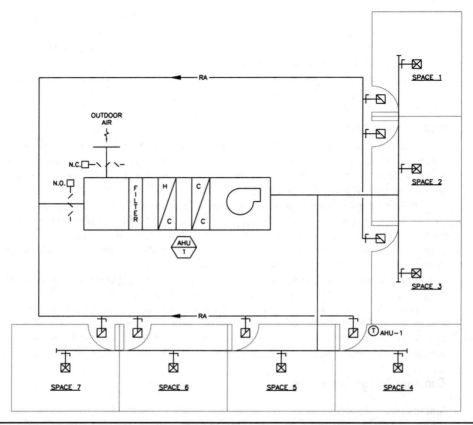

Figure 5-1 Schematic diagram of a single-zone, constant air volume system.

Temperature control for H&V systems is the same as for the single-zone CAV system described earlier, except the space thermostat will control the heating capacity of the H&V unit only. Some common examples of spaces requiring H&V systems are gymnasiums, locker rooms, and kitchens.[4]

Multiple-Zone   CAV systems are not well suited to provide multiple zones of temperature control because serving multiple zones with CAV systems incurs a higher energy cost and, in some cases, a higher first cost than the VAV options that are normally available. However, through the use of reheat, dual-duct, and multizone configurations, CAV systems can serve multiple temperature zones for certain applications, particularly renovations of existing systems.

Reheat   A CAV reheat system consists of a CAV unit whose supply air ductwork branches out to serve multiple zones within the area served by the unit (Fig. 5-2). A reheat coil is mounted within the branch duct to each zone. Typically, the discharge air temperature from the CAV unit is maintained at approximately 55°F so that it can meet the cooling needs of the temperature zones, if required. A heating-only thermostat located in each zone controls the heating output of its associated zone reheat coil. This

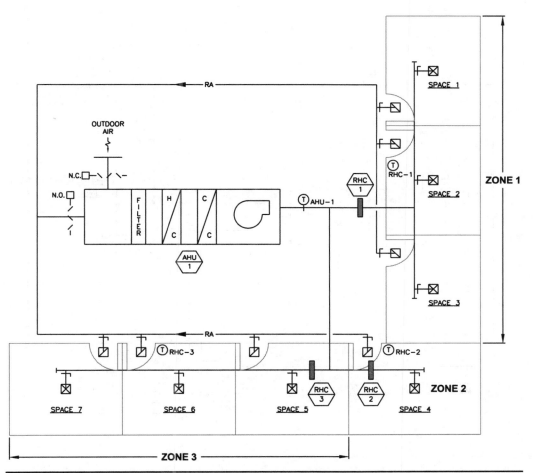

Figure 5-2 Schematic diagram of a constant air volume reheat system.

coil reheats the supply air from the CAV unit as required to maintain the setpoint of the zone thermostat. If no heating is required in the zone, the heating coil will be shut off. The zone reheat coil needs to be sized not only to meet the building envelope heat losses for the zone but also to raise the supply airflow from the discharge air temperature (typically 55°F) to the zone cooling setpoint (typically 75°F).

CAV reheat systems are not energy efficient because simultaneous cooling (at the CAV unit) and heating (at the reheat coil) of the supply airflow occurs. Care must be taken when designing CAV reheat systems because the 2009 *International Energy Conservation Code*, Section 503.4.5, requires the supply airflow to be reduced to at least 30% of the maximum supply airflow to each zone before reheating can occur. One exception to this rule is for zones that have special humidity control requirements.

Dual-Duct   A dual-duct CAV system includes two separate supply air ducts, one hot duct and one cold duct, connected to a specially configured CAV air handling unit (Fig. 5-3). The supply airflow is divided within the unit downstream of the supply fan

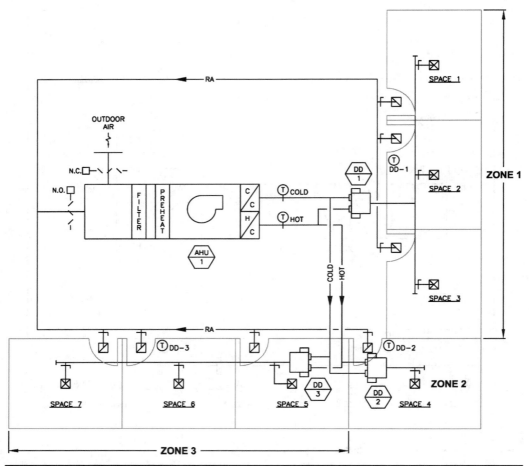

Figure 5-3   Schematic diagram of a dual-duct constant air volume system.

into what are called the hot deck and cold deck. A heating coil is installed within the unit in the hot deck and a cooling coil is installed within the unit in the cold deck. A portion of the supply airflow is blown through the hot deck and discharged through the main hot duct connected to the unit. The remaining supply airflow is blown through the cold deck and discharged through the main cold duct connected to the unit. Both the hot and cold supply ducts are routed parallel to each other through the building and branch out to serve multiple zones within the area served by the unit. Each zone is equipped with a dual-duct mixing box, which has both hot and cold duct inlet connections, each of which is equipped with a motor-operated damper and an inlet airflow sensor. The dual-duct mixing box has a single duct outlet through which air is supplied to the zone. The temperature of the air supplied to the zone is controlled by modulating the hot and cold airflows at the dual-duct mixing box as required to maintain the setpoint of the zone thermostat. For CAV dual-duct mixing boxes, the total supply airflow to the zone remains constant.

Typically, the cold deck air temperature from the CAV air handling unit is maintained at approximately 55°F so that cooling is available if required by the temperature zones. The hot deck temperature from the CAV unit is maintained at approximately 85°F so that heating is available if required by the temperature zones. The heating coil in a dual-duct air handling unit can be hot water, steam, or electric. The cooling coil will normally be chilled water. Dual-duct air handling units rarely, if ever, utilize a DX refrigerant cooling coil.

Dual-duct CAV systems generally do not have an application for commercial buildings because of the high first cost. Thus, we will not discuss dual-duct systems in any more detail. Dual-duct CAV systems were designed years ago mainly to serve laboratory areas within buildings. Therefore, it is necessary for the HVAC system designer to understand this type of system should a project involve the renovation of a building with a dual-duct CAV system. However, today there are more modern HVAC systems available to serve laboratory areas that have a lower first cost and lower operating cost than dual-duct CAV systems. Therefore, it is unlikely for a new building to require a dual-duct CAV system.

Multizone   Multizone CAV systems are similar to dual-duct CAV systems in that there is a hot deck and a cold deck within the air handling unit (Fig. 5-4). The difference is that the hot and cold airstreams for each zone are mixed at the air handling unit. There is a hot and cold air motor-operated damper mounted on the discharge of the air handling unit for each zone; that is, if the unit is a five-zone unit, there will be five hot deck motor-operated dampers and five cold deck motor-operated dampers mounted on the discharge of the unit. There is a single duct connection on the combined outlet of each hot and cold damper serving each zone through which air is supplied to the zone. The temperature of the air supplied to each zone is controlled by modulating the hot and cold airflows at the air handling unit as required to maintain the setpoint of the zone thermostat. For CAV multizone systems, the total supply airflow to each zone remains constant. The hot and cold deck temperatures for multizone CAV systems are similar to the deck temperatures for dual-duct CAV systems.

Multizone CAV systems are not commonly used in commercial buildings because the VAV options that are available typically have a lower first cost and lower operating cost. Thus, we will not discuss multizone systems in any more detail. Also, the number of zones that can be accommodated by a multizone unit is limited by the physical space available to install the hot and cold motor-operated dampers for each zone on the discharge of the air handling unit. Therefore, multizone units can serve no more than about eight zones. The number of zones that can be served by a VAV system is not limited by the equipment but is determined by the needs of the areas served.

Variable Air Volume Systems

VAV systems are the most widely used types of HVAC air systems for medium- and large-sized commercial building projects (projects larger than 10,000 ft²) because VAV systems are flexible, energy efficient, and provide a comfortable indoor environment.

As mentioned previously, VAV systems deliver variable supply airflow at a constant temperature (typically 55°F) through the primary air duct[5] to multiple VAV terminal units, each of which serves a separate temperature zone. Each VAV terminal unit contains a motor-operated damper that modulates the primary airflow to the zone, an inlet airflow sensor, and, in some instances, a heating coil and a small recirculating air fan.[6]

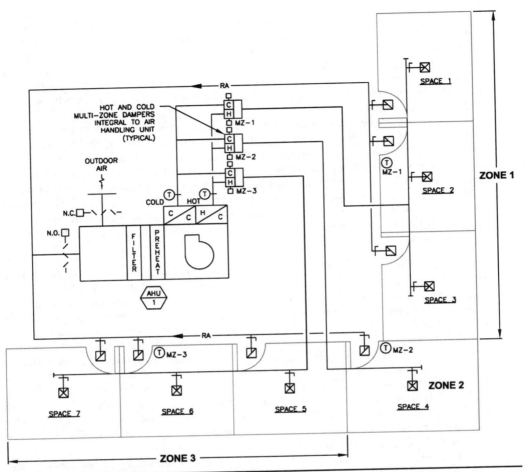

Figure 5-4 Schematic diagram of a multizone CAV system.

The supply airflow from the VAV air handling unit is normally modulated to maintain a constant static pressure within the primary air duct system. This is measured by a duct static pressure sensor, which is typically located two-thirds the way down the primary air duct system. The modulation of the supply airflow tracks with the needs of the VAV terminal units; that is, as more primary air dampers in the VAV terminal units open to supply more air to the zones, the static pressure in the primary air duct system decreases and more supply airflow is delivered by the VAV air handling unit to restore the static pressure in the duct system. Conversely, as the primary air dampers close, the primary air duct static pressure increases and less supply airflow is delivered by the VAV air handling unit to compensate.

The air handling unit for a VAV system is the same as would be required for a CAV system with the exception that there is a means of modulating the supply airflow delivered by the unit. The most common way of modulating the supply airflow of the unit is by controlling the frequency of the signal sent to the supply fan motor through a variable frequency drive (VFD).[7] The speed of an alternating current (ac) motor is directly proportional to the frequency of the input signal to the motor. Therefore, as the frequency of the VFD output signal to the motor is reduced, motor speed is reduced, and supply airflow is also reduced. The converse is true as the frequency of the VFD output signal to the motor is increased. The maximum frequency of the VFD output signal is that of the VFD line input frequency, or 60 Hz. The supply fan motor will operate at full speed when it receives a VFD output signal of 60 Hz.

Care should be taken when designing VAV systems that utilize DX refrigerant coils for cooling. Unless the refrigeration system is equipped with adequate capability to unload the refrigeration system capacity, freeze-up of the DX refrigerant cooling coil could occur under low airflow conditions.[8] Also, refrigeration systems having a capacity that is less than about 25 tons generally do not have the capability to adequately accommodate VAV operation. Unloading of the refrigeration system, adequate control of the discharge air temperature, and incorporation of a VFD into the unit cabinet are all issues that pose problems for these smaller-sized pieces of DX equipment. Therefore, equipment manufacturers do not currently offer VAV operation for air systems that utilize DX refrigerant cooling coils in sizes less than about 25 tons. However, recent energy-efficiency requirements in the industry are causing equipment manufacturers to develop technologies that will enable VAV capability for smaller-sized pieces of DX equipment. In the future, this lower limit of 25 tons for VAV operation in DX equipment may drop to as low as 10 tons or less.

The most common use of a VAV system is in serving multiple temperature zones. Therefore, we will discuss multiple-zone VAV systems first and then discuss the use of a VAV system for a single-zone application.

Multiple-Zone

VAV Terminal Units   Variable primary airflow is delivered to the zones through the modulation of the primary air damper in the VAV terminal units. As the zone temperature decreases, the primary air damper is modulated closed to supply less (55°F) primary air to the zone. Once the primary air damper reaches its predetermined minimum position (usually about 25% of maximum airflow[9]), upon a further drop in the zone temperature, VAV terminal units that have heating capabilities will position the primary air damper to the heating airflow and modulate the output of the heating coil to maintain the heating setpoint of the zone temperature sensor.[10] Fan-powered VAV terminal units are also equipped with a small fan that recirculates air (normally from the ceiling return air plenum) through the heating coil of the VAV terminal units. The various types of VAV terminal units are discussed in more detail in Chap. 7. Figure 5-5 is a schematic diagram of a VAV system serving multiple VAV terminal units.

Dual-Duct   Dual-duct VAV air systems are about as uncommon as dual-duct CAV air systems. The most likely time an HVAC system designer would encounter a dual-duct VAV air system would be in the case of a system that was originally designed as a dual-duct CAV system but was later renovated to function as a VAV system. The HVAC system designer may also have the task of designing the modifications to a dual-duct CAV system to convert it to a dual-duct VAV system.

Dual-duct VAV systems function much the same way as dual-duct CAV systems except the supply airflow to the zones is variable, not constant. The supply fan in the

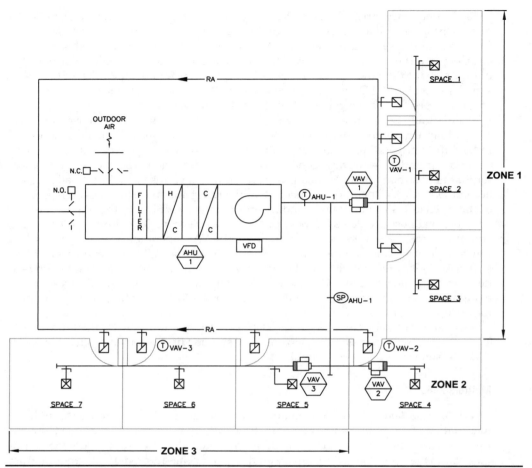

Figure 5-5 Schematic diagram of a variable air volume system.

dual-duct air handling unit has to be equipped with a means to modulate its airflow in response to static pressure in both the main hot and cold ducts. The energy efficiency of a dual-duct VAV system would be about the same as that of a conventional VAV system utilizing VAV terminal units. Care should be taken in converting a dual-duct CAV system to a VAV system to ensure that the zones do not require constant supply airflow to serve as makeup for constant exhaust airflow.

Single-Zone   Operation of a single-zone VAV system is similar to the operation of a VAV system serving multiple zones, except the supply airflow is modulated to maintain the cooling setpoint of the (single) zone temperature sensor rather than to maintain a constant primary air duct static pressure. The supply air temperature is maintained at 55°F as long as the zone temperature sensor is calling for cooling.

Once the zone temperature drops below the cooling setpoint (typically 75°F), the air handling unit will operate in the heating mode: cooling will be disabled, the supply fan will operate at the predetermined heating airflow, and the output of the heating coil within the unit will be modulated as required to maintain the heating setpoint of the zone temperature sensor (typically 70°F).[11]

Once the zone temperature rises above the cooling setpoint, the air handling unit will operate in the cooling mode: heating will be disabled, the output of the cooling coil will be modulated as required to maintain the supply air temperature at 55°F, and the supply air-flow will be modulated to maintain the cooling setpoint of the zone temperature sensor.

The advantage of a single-zone VAV over a single-zone CAV system is that during cooling operation, the supply air temperature will remain constant at approximately 55°F. This consistently cool supply air temperature will result in a lower space relative humidity than the same area served by a CAV system where the supply air temperature can vary anywhere between 55°F (full cooling load) and 75°F (no cooling load). The higher space relative humidity resulting from the use of a CAV system is exacerbated by outdoor air ventilation in moist climate zones and by a high density of occupants in the areas served by the unit.

Single-zone VAV systems that utilize DX refrigerant cooling coils have the same limitations on the minimum unit size as VAV systems serving multiple zones, that is, a minimum unit size of about 25 tons. In most cases, this results in an area that is too large to be practically served by one zone of temperature control. However, there are exceptions, such as an auditorium, gymnasium, or warehouse, where a single zone could require as much as 25 tons of cooling.

Ventilation Air Systems

Air systems that provide ventilation only and do not provide any heating or cooling to the spaces they serve include outdoor air and exhaust air systems, which are summarized as follows:

- Outdoor air systems operate under a positive air pressure to provide ventilation for the building occupants, keep certain types of equipment rooms at a maximum temperature during the summer, or provide outdoor air makeup for exhaust air systems or equipment.

- Exhaust air systems operate under a negative air pressure to remove air from the spaces or equipment they serve and discharge this air to the outdoors.

Outdoor Air Systems

The two types of outdoor air systems we will discuss are ventilation air systems and makeup air systems.

Ventilation Air Systems Ventilation air systems are used to either meet the outdoor air ventilation requirements of the building occupants, or to use outdoor air to keep certain types of equipment rooms under a positive air pressurization and keep them at a maximum temperature during the summer.

Dedicated Outdoor Air Systems Dedicated outdoor air systems (DOASs) are one part of what is known as a dual-path HVAC system. In a dual-path HVAC system, outdoor air for occupant ventilation and/or exhaust air makeup is handled by one air system while heating and cooling is handled by a separate air system (Fig. 5-6). In some situations, the outdoor air is delivered by the DOAS directly to the spaces served at a temperature and relative humidity that are approximately equal to the space temperature

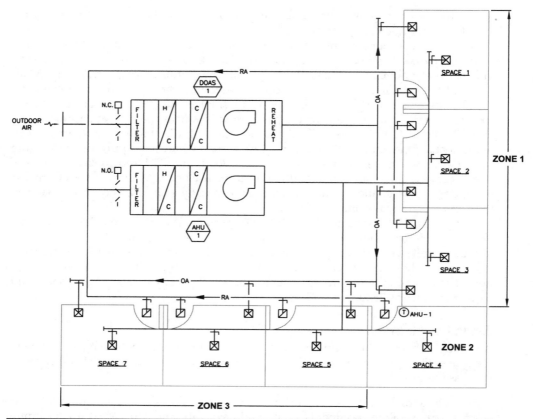

Figure 5-6   Schematic diagram of a dual-path HVAC system where the DOAS unit utilizes a reheat coil.

and relative humidity. In other situations, the outdoor air is delivered by the DOAS to the return air duct for the heating and cooling equipment in a filtered but uncondi- tioned or partially conditioned state. Partial conditioning could include heating only or heating and cooling with no humidity control.

A DOAS unit normally has a heating coil, cooling coil, and reheat coil so that the 100% outdoor air that it delivers can be heated and cooled, or cooled and reheated (which is required to dehumidify the air). A humidifier may also be part of a DOAS unit, but humidification is not as often required as heating, cooling, and dehumidifica- tion. The heating and reheat coils can be hot water, steam, gas, or electric, and the cool- ing coil can be chilled water or DX refrigerant.

However, utilizing hot water, steam, gas, or electricity for reheat requires more energy than utilizing a form of energy recovery for reheat. An example of an energy recovery reheat coil is a hot gas reheat coil that is part of a self-contained (or packaged) refrigeration system. When dehumidification is required for this type of system, heat, which is normally rejected to the outdoors through an air- or water-cooled condenser, is rejected to the airstream by a hot gas reheat coil located downstream of the cooling coil. Hot gas reheat can be incorporated into air-cooled or water-cooled refrigeration equipment, such as a packaged air-conditioning unit or a water-source heat pump unit. Packaged DOAS units with hot gas reheat still require an air- or water-cooled condenser for heat rejection when cooling (but not reheat) is required. However, these units have the added piping, controls, and hot gas reheat coil that enable them to utilize the hot gas reheat coil in the dehumidifying mode of operation. The capabilities and configurations of each type of DOAS unit with hot gas reheat vary depending upon the equipment manufacturer. Normally, manufacturers that specialize in constructing dehumidification equipment for indoor swimming pools (natatoriums) have the best selection of DOAS equipment.

Another example of a DOAS unit is a 100% outdoor air, modular central station air handling unit with a heating coil and a heat pipe refrigerant coil wrapped around a chilled water cooling coil. In this type of system, precooling of the outdoor air upstream of the cooling coil and reheating of the outdoor air downstream of the cooling coil is accomplished by the wrap-around heat pipe refrigerant coil.

The wrap-around heat pipe refrigerant coil consists of two coils filled with refrigerant that are connected by refrigerant pipes. One refrigerant coil is mounted upstream of the chilled water cooling coil and the other refrigerant coil is mounted downstream of the refrigerant coil, and the interconnecting refrigerant pipes wrap around the end of the chilled water cooling coil. When the air entering the upstream refrigerant coil is above the boiling point of the refrigerant (approximately 45°F), the refrigerant is evaporated and, as a result, precools the air by up to 14°F (depending upon the entering air temperature) before it enters the chilled water cooling coil. This precooling of the air is a completely sensible cooling process (i.e., no moisture is condensed). The precooled air enters the chilled water coil, and the capacity of the chilled water coil is modulated to further cool and dehumidify the air to achieve a 55°F dew point,[12] which corresponds to an approximately 55°F dry bulb temperature leaving the chilled water cooling coil. In the downstream refrigerant coil, the refrigerant that was evaporated in the upstream coil is condensed and flows by gravity back to the upstream refrigerant coil. The heat rejected by the condensation of the refrigerant reheats the air the same amount that it was precooled by the upstream refrigerant coil (up to 14°F of reheat, depending upon the amount of refrigerant that was evaporated in the upstream coil). The result is that the outdoor air delivered to the building by the DOAS unit has a dry bulb temperature of 69°F and a dew point of 55°F.

Using a wrap-around heat pipe refrigerant coil for precooling and reheat in a DOAS unit requires very little energy input. The only energy required to accomplish this heat transfer is the added fan energy that is required to overcome the additional (air) static pressure losses through the upstream and downstream refrigerant coils. There is no compressor in the refrigerant circuit because it is not required to circulate the refrigerant through the coils. The refrigerant flows between coils in the liquid and gaseous phases strictly as a result of gravity and the heat that is absorbed by the refrigerant in the upstream refrigerant coil. Using a wrap-around heat pipe refrigerant coil has the added benefit of decreasing the required cooling capacity of the chilled water cooling coil.

When the entering air temperature is below the boiling point of the refrigerant (approximately 45°F), there is no refrigerant flow and no heat transfer associated with the wrap-around heat pipe refrigerant coil. The heat output of the heating coil must be modulated to ensure that the air supplied by the DOAS unit does not drop below 69°F.

Figure 5-7 is schematic diagram of a DOAS unit that utilizes a wrap-around heat pipe refrigerant coil. Typical dry bulb and wet bulb air temperatures and air humidity ratios are given for the entering and leaving air conditions of the heat pipe and chilled

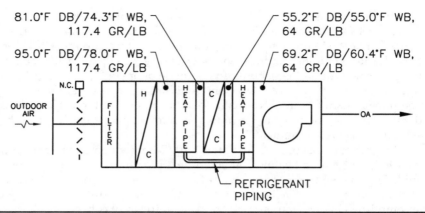

FIGURE 5-7   Schematic diagram of a DOAS unit with a wrap-around heat pipe refrigerant coil.

water cooling coils to give an indication of the sensible and latent cooling and heating that occurs at each point within the process.

Positive-Pressure Equipment Room Ventilation   Equipment rooms, including rooms that contain mechanical and/or electrical equipment, are commonly heated and ventilated only. Heat is provided during the winter typically through fan-forced unit heaters which recirculate room air to keep the rooms at a minimum of about 60°F (refer to Chap. 7 for a discussion of unit heaters). Ventilation is normally not provided in the winter. When the room temperature rises, normally above 85°F, unconditioned outdoor air is used to prevent the rooms from becoming excessively hot. This typically occurs during the summer but could occur at other times of the year depending upon the internal heat gains to the room from the equipment. The maximum acceptable temperature within an equipment room is normally about 100°F. However, equipment rooms that contain electronic components, such as elevator controls or computer systems, cannot be warmer than 85°F. In this case, the room would require mechanical cooling either from the building HVAC system or from a separate HVAC system. For equipment rooms that can tolerate a space temperature as high as 100°F, an outdoor air ventilation system that utilizes unconditioned outdoor air to keep the space temperature below this maximum acceptable temperature should be designed. HVAC calculations need to be performed to determine both the heat gains to the room at the design summer outdoor temperature and the outdoor airflow required to keep the room at 100°F, which is normally 5 to 10°F higher than the design summer outdoor temperature. However, the temperature within equipment room is not always this maximum temperature. The space temperature will vary between 85 and 100°F when ventilation is operating. When the space temperature within the equipment rooms is below 85°F and above 60°F, neither ventilation nor heating will be provided by the heating and ventilating system.

Most equipment rooms are ventilated by an exhaust air system in order to place them under a negative air pressurization, which is necessary to prevent odors within the equipment rooms from migrating to adjacent spaces (refer to the Negative-Pressure Equipment Room Ventilation section later). However, equipment rooms containing gas-fired or fuel-burning appliances that do not have a direct connection to the outdoors for combustion air (i.e., the appliances utilize room air for combustion) should

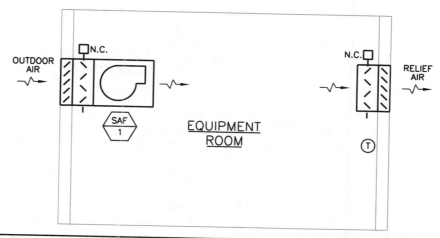

FIGURE 5-8 Schematic diagram of a positive-pressure equipment room ventilation system.

not be placed under a negative air pressurization by the ventilation system. In these situations, it is necessary for the ventilation system to positively pressurize the equipment room by blowing outdoor air into the room and allowing the excess air that is not used for combustion, if any, to be relieved from the room. Figure 5-8 is schematic diagram of a positive-pressure equipment room ventilation system.

Makeup Air Systems   Makeup air systems are utilized to provide the outdoor airflow required by certain types of equipment where it is not feasible or economical to condition this outdoor airflow through the HVAC system. Makeup air systems for kitchen exhaust hoods and combustion air systems for gas-fired or fuel-burning appliances are two examples of makeup air systems.

Kitchen Exhaust Hoods   *NFPA Standard 96—Standard for Ventilation Control and Fire Protection of Commercial Cooking Operations*, the *International Mechanical Code*, and the authority having jurisdiction require certain kitchen appliances that produce grease-laden vapor, smoke, or steam to be installed under one or more commercial kitchen exhaust hoods to remove these contaminants from the kitchen environment (Fig. 5-9). The exhaust airflow required to effectively remove these contaminants is typically in the range of 50 to 60 cfm per square foot of exhaust hood face area, although airflows as high as 125 cfm per square foot of exhaust hood face area are possible (depending upon the type and heat output of the appliances installed under the hood). This exhaust airflow must be replaced with outdoor airflow to prevent the negative air pressurization within the kitchen from exceeding 0.02 in. w.c. Because exhaust airflow from a commercial kitchen may be as much as double the supply airflow that is required to maintain the space temperature, special consideration must be given to conditioning the makeup outdoor airflow through a system that is separate from the HVAC system, particularly if cooling is provided for the kitchen in addition to heating.

   If the kitchen is heated and ventilated only through an H&V air system, the outdoor airflow required by the exhaust hoods is commonly provided through the H&V air system. The H&V system will normally position the outdoor air and return air dampers to deliver the required outdoor airflow when the exhaust hoods are operating and reduce

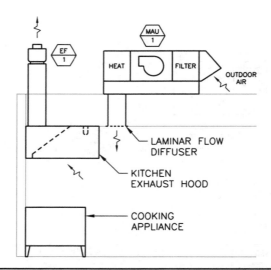

FIGURE 5-9 Schematic diagram of a kitchen exhaust hood makeup air system.

the outdoor airflow when the exhaust hoods are not operating. However, if cooling is provided for the kitchen, it is usually infeasible to deliver the required outdoor airflow through the HVAC system. Delivering a high percentage of outdoor air through the HVAC system would require the use of a reheat system to adequately dehumidify the air during cooling operation to prevent unacceptably high relative humidity within the kitchen. The use of reheat is costly from an energy standpoint because of the simultaneous cooling and heating that is required. In the case of a commercial kitchen, approximately 70% of the outdoor airflow required by the kitchen exhaust hoods can be delivered by a makeup air system that is dedicated to this purpose.

Typically, the makeup air system will heat the outdoor airflow required by the exhaust hoods and deliver this air through a laminar flow diffuser located within the kitchen near the exhaust hoods or, if makeup-air-type hoods are utilized, deliver this air to the makeup air connection on each of the exhaust hoods.[13] The goal is for the makeup air to be introduced at a low velocity into the hood capture zone so that it does not affect the capture and containment capabilities of the hood. The heating coil in the makeup air unit is commonly direct or indirect gas-fired or electric because the unit is often mounted on the roof of the building. If the heating coil is hot water or steam, and the makeup air unit is mounted on the roof, measures must be taken to protect the heating coil from freezing.

Combustion Air   As mentioned in Chap. 4, a source of combustion air must be provided for gas-fired or fuel-burning appliances. *NFPA Standard 54—National Fuel Gas Code* describes the combustion air requirements for gas-fired appliances and *NFPA Standard 31—Standard for the Installation of Oil-Burning Equipment* describes the combustion air requirements for fuel-burning appliances.

If it is not feasible to obtain the required combustion air from the room in which the appliances are installed, transfer the combustion air from adjacent spaces, provide adequate openings to the outdoors, or provide a direct connection of outdoor combustion air to each appliance, it will be necessary to provide combustion air for the appliances through a mechanical forced-air system. We will refer to this as a combustion air makeup system.

A combustion air makeup system consists mainly of a fan that blows outdoor air into the equipment room in which the appliances are installed. The combustion air makeup system must be sized to deliver the amount of outdoor combustion air required by all of the appliances installed within the room. It is recommended that the combustion air be filtered and heated to at least 50°F to prevent freezing conditions within the equipment room. Furthermore, operation of the combustion air unit should be interlocked with the operation of the appliances within the room so that combustion air is only provided when one or more of the appliances is firing.

The combustion air makeup system can also be used to ventilate the equipment room during the summertime in a manner similar to that which is discussed in the Positive-Pressure Equipment Room Ventilation section earlier. In this case, the combustion air makeup system would be controlled by a space thermostat in addition to being interlocked with the appliances. If there are multiple appliances within the equipment room and/or if the combustion air makeup system is used for summertime ventilation, provisions must be made in the room to relieve excess air to the outdoors that is not used for combustion.

Figure 5-10 includes a schematic diagram and sequence of operation for a combustion air makeup system that serves multiple gas-fired boilers and also provides positive-pressure ventilation of the equipment room in the summertime. The combustion air makeup system is designed to provide combustion air for six 1,800 thousand British thermal units per hour (MBH) input boilers.[14]

Exhaust Air Systems

Exhaust air systems remove air from within a building and discharge it to the outdoors. For commercial buildings, exhaust air systems are generally used to:

- Provide general exhaust for odor-producing areas to prevent odors from migrating to other areas of the building.

- Provide ventilation for spaces that do not require cooling but for which it is desired to keep the space temperatures below a maximum limit and also to keep the spaces under a negative air pressurization to prevent odor migration. Because this type of exhaust air system is normally utilized for equipment rooms, we will refer to this as negative-pressure equipment room ventilation.

- Provide the exhaust airflow required for certain types of equipment.

General Exhaust   Applicable codes and the authority having jurisdiction may require exhaust from certain areas within commercial buildings in order to prevent odor migration. For example, Table 403.3 in the 2009 *International Mechanical Code* requires public toilet rooms to be exhausted at a rate of 50 cubic feet per minute (cfm) per water closet[15] or urinal for normal use and 70 cfm per water closet or urinal for heavy use (theaters, schools, sports facilities, etc.). The code also requires sports locker rooms to be exhausted at a rate of 0.5 cfm/ft^2 of net occupiable floor area. In addition to exhausting these spaces with the code-required airflow, sufficient negative air pressurization must be established within these spaces with respect to adjacent spaces to ensure proper containment of odors, which is achieved by maintaining the space under a 0.02-in. w.c. negative air pressurization with respect to adjacent spaces. In situations where the space under concern is separated from adjacent spaces by one or more (closed) doors, the required 0.02-in. w.c. negative air pressurization can be achieved by exhausting

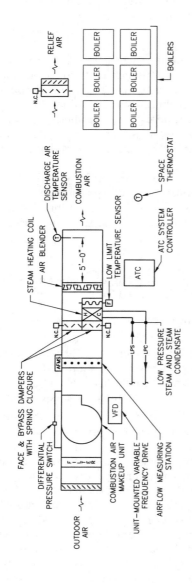

SEQUENCE OF OPERATION:

GENERAL

PROVIDE CURRENT SENSING RELAYS OR AUXILIARY CONTACTS IN THE CONTROL PANELS OF ALL SIX BOILERS WHICH WILL PROVIDE THE STATUS OF EACH BOILER TO THE ATC SYSTEM CONTROLLER.

COMBUSTION AIR MODE

THE COMBUSTION AIR MAKEUP UNIT SHALL BE INTERLOCKED WITH ALL SIX BOILERS SO THAT THE UNIT OPERATES WHENEVER ANY OF THE BOILERS IS FIRING.

THE MAKEUP AIR FAN SHALL BE CONTROLLED THROUGH THE VFD TO DELIVER THE FOLLOWING AIRFLOWS (AS SENSED BY THE AIRFLOW MONITORING STATION) DEPENDING UPON THE NUMBER OF BOILERS FIRING:

NO. OF BOILERS FIRING	AIRFLOW (CFM)
1	630
2	1,260
3	1,890
4	2,520
5	3,150
6	3,780

THE DISCHARGE AIR TEMPERATURE SENSOR SHALL MODULATE THE FACE AND BYPASS DAMPERS TO MAINTAIN A DISCHARGE AIR TEMPERATURE OF 50°F (ADJUSTABLE).

VENTILATION MODE

THE SPACE THERMOSTAT SHALL OPEN THE RELIEF AIR DAMPER AND ENERGIZE THE MAKEUP AIR FAN AT FULL SPEED WHENEVER ITS SETPOINT OF 85°F (ADJUSTABLE) IS EXCEEDED. UPON A DROP IN SPACE TEMPERATURE BELOW THE SETPOINT OF THE SPACE THERMOSTAT, THE RELIEF AIR DAMPER SHALL CLOSE AND THE MAKE-UP AIR FAN SHALL OPERATE IN THE COMBUSTION AIR MODE.

FILTER STATUS

A VISUAL ALARM SHALL SIGNAL AT THE ATC PANEL WHENEVER THE 0.50" W.C. (ADJUSTABLE) SETPOINT OF THE FILTER DIFFERENTIAL PRESSURE SWITCH IS EXCEEDED.

SAFETIES

UPON A DROP IN THE HEATING COIL LEAVING AIR TEMPERATURE BELOW 38°F AS SENSED BY THE LOW LIMIT TEMPERATURE SENSOR SERPENTINED ACROSS THE DOWNSTREAM FACE OF THE HEATING COIL, THE MAKEUP AIR FAN SHALL BE DE-ENERGIZED, THE FACE DAMPER SHALL CLOSE, THE BYPASS AIR DAMPER SHALL OPEN, AND AN AUDIBLE/VISUAL ALARM SHALL SIGNAL AT THE ATC PANEL. THE LOW LIMIT TEMPERATURE SENSOR SHALL REQUIRE A MANUAL RESET FOR THE FACE AND BYPASS DAMPERS TO FUNCTION IN THE NORMAL MODE.

AN AUDIBLE/VISUAL ALARM SHALL SIGNAL AT THE ATC PANEL WHENEVER THE MAKE-UP AIR FAN FAILS TO START WHEN COMMANDED. FAN STATUS SHALL BE DETERMINED BY THE DIFFERENTIAL PRESSURE SWITCH PIPED BETWEEN THE INLET AND DISCHARGE OF THE MAKEUP AIR FAN.

FIGURE 5-10 Schematic diagram of a combustion air makeup system.

approximately 100 cfm per 3-ft-wide (closed) door[16] more from the space than what is supplied to the space.

There may also be other spaces within the building for which exhaust is desired for odor containment but for which it is not required by the applicable codes or the authority having jurisdiction. For example, the 2009 *International Mechanical Code* does not require exhaust from janitor's closets. However, it is desirable to keep janitor's closets under a negative air pressurization to contain the odors produced by the cleaning agents that are often stored in these rooms. Therefore, each janitor's closet should be designed with an exhaust airflow of 100 cfm in order to properly contain these odors (supply airflow is not required for janitor's closets).

Often, the air systems providing general exhaust for commercial buildings serve multiple spaces and multiple occupancy classifications. For example, a four-story office building that has toilet rooms and janitor's closets in the same location on each floor would normally have one exhaust system to serve all of these toilet rooms and janitor's closets. However, physical separation of the spaces requiring general exhaust, differing hours of operation, and multiple tenants are all reasons why separate exhaust systems may be required to meet a building's general exhaust requirements.

Makeup air for public toilet rooms and janitor's closets is normally transferred from adjacent spaces that are served by the building HVAC systems. Typically, the outdoor air supplied by these HVAC systems to meet the occupant ventilation requirements is sufficient to provide makeup air for the toilet room and janitor's closet exhaust air systems.

There may be some spaces within a building for which the general exhaust airflow is a significant percentage of the supply airflow that is required to maintain the space temperature. One example is a large sports locker room. The 0.5-cfm/ft² exhaust airflow required for this occupancy may be as high as 50% of the supply airflow required for cooling. For this reason, sports locker rooms are normally heated and ventilated only. However, if cooling is required in addition to heating, the HVAC unit must have dehumidification capabilities in order to properly dehumidify the high percentage of outdoor airflow that will be required. Another alternative would be to design the HVAC system serving the locker rooms to utilize a maximum of 15% outdoor airflow and transfer the remaining airflow required for exhaust to the locker rooms from adjacent conditioned spaces.

General exhaust air systems normally operate when the building is occupied and are shut off when the building is unoccupied. It is common for exhaust fans serving general exhaust to be electrically interlocked with the operation of the HVAC systems serving the adjacent areas so that the exhaust fans can only operate when the HVAC systems are operating. If the building is designed with a computerized building automation system (BAS), the exhaust fans would be controlled by the BAS and can be assigned a different operating schedule than the associated HVAC systems, if desired.

Negative-Pressure Equipment Room Ventilation   Negative-pressure equipment room ventilation is provided for spaces that do not require cooling but for which it is desired to keep the space temperature at a maximum of about 100°F, which is 5 to 10°F higher than the summer design outdoor temperature (Figs. 5-11 and 5-12). Equipment room ventilation systems can positively pressurize the space they serve, which is the requirement for equipment rooms containing gas-fired or fuel-burning appliances that use the room air for combustion (refer to the Positive-Pressure Equipment Room Ventilation section earlier). However, it is more common for equipment room ventilation systems to place the spaces they serve under a negative air pressurization to prevent odors from migrating to adjacent

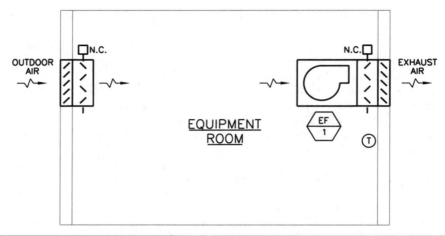

FIGURE 5-11   Schematic diagram of a negative-pressure equipment room ventilation system.

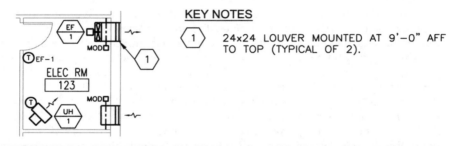

KEY NOTES

1   24x24 LOUVER MOUNTED AT 9'-0" AFF TO TOP (TYPICAL OF 2).

FIGURE 5-12   Floor plan representation of a negative-pressure equipment room ventilation system.

spaces. In these situations, an exhaust air system that draws makeup air from an opening in the building envelope, such as a wall louver or roof-mounted intake air hood, would be designed. The exhaust air system would be controlled by a space thermostat and be energized whenever the space temperature rises above the setpoint of the space thermostat, which is normally about 85°F. It is necessary for equipment room ventilation systems to incorporate a normally closed motor-operated damper on the outdoor air opening so that the damper closes whenever the ventilation system is off. This prevents outdoor air infiltration to the space when ventilation is not required.

One example of this type of exhaust air system is one that serves equipment rooms without gas-fired or fuel-burning appliances (or the appliances have direct connections to the outdoors for combustion air). Another example is the exhaust air system serving an electrical equipment room, particularly one that contains heat-generating equipment, such as transformers or uninterruptible power supplies (UPSs).

Equipment rooms can also be ventilated by transferring makeup air for the exhaust air systems from adjacent spaces that are conditioned by the building HVAC systems. In this situation, the maximum allowable space temperature within the rooms would still be 100°F,[17] but the makeup air would be the temperature of the air within the adjacent conditioned spaces, which is typically 75°F during occupied operation of the building HVAC systems. Because the temperature rise of the transfer air is 25°F, the exhaust airflow required to maintain the equipment rooms at 100°F is 2.5 to 5 times less than the exhaust airflow required to maintain the equipment rooms at 100°F when outdoor air at the design summer outdoor temperature is used to make up for the exhaust airflow.

Refer to Fig. 7-5 in Chap. 7 for a graphical representation of the exhaust fan, exhaust air louver, outdoor air louver, and motor-operated dampers for a negative-pressure equipment room ventilation system as they would be shown on a floor plan drawing.

Equipment Exhaust   Finally, exhaust systems are used to remove undesirable or potentially harmful contaminants from the building at the point where these contaminants are generated. An example of this type of exhaust air system is a kitchen hood exhaust system. The size, quantity, and location(s) of kitchen exhaust hood(s) are usually determined by the kitchen equipment consultant. The exhaust airflow through each exhaust hood depends upon the type and input rating of the appliances located under the hood. The kitchen hood exhaust system must be designed in accordance with *NFPA Standard 96—Standard for Ventilation Control and Fire Protection of Commercial Cooking Operations*.

Air Pressurization Calculations

Once all of the air systems have been defined for a building, it is necessary to perform air pressurization calculations for the overall building as well as for individual spaces within the building in order to:

- Ensure that sufficient outdoor makeup air is available to meet all exhaust airflow requirements within the building without placing the building under an overall negative air pressurization.
- Ensure that appropriate air pressure relationships exist between adjacent spaces within the building.

Generally, commercial buildings should have about 5% positive air pressurization; that is, the sum of the outdoor airflows supplied by the various HVAC and ventilation air systems to the building should be 5% greater than the sum of the exhaust airflows removed from the building. However, some judgment needs to be applied in the building air pressurization calculation, such as in the case where an equipment room is exhausted by an air system and makeup air is supplied to this room through an opening in the building envelope. In this case, the equipment room exhaust airflow would not be included in the air pressurization calculation for the overall building since the airflow through this room is actually separate from the rest of the building.

Also, appropriate air pressure relationships need to exist between adjacent spaces within the building. The air pressurization of a particular space within the building is equal to the sum of the airflows supplied to that space by the HVAC and ventilation air systems minus the sum of the airflows removed from that space. The air supplied to each space in the building includes both the supply air from the HVAC system(s) serving the space plus outdoor air delivered directly to the space by the outdoor air ventilation system(s). The air removed from the space includes both the return air to the HVAC system(s) serving the space plus exhaust air removed from the space by the exhaust air system(s).

Because some spaces of the building will have a more positive (relative) air pressurization than other spaces, there will be a transfer of air from spaces of greater air pressurization to spaces of lesser air pressurization. The transfer of air from one space to another needs to be evaluated to ensure that air is flowing in the proper direction. Typically, it is desirable for air to flow from spaces that are clean to spaces that are less clean or from spaces that are odor-free to spaces that are odor-producing. Section 403 of the 2009 *International Mechanical Code* defines the requirements for pressure relationships between various occupancy classifications within buildings.

It is very helpful in this regard to prepare an air pressurization schematic diagram of each floor of a building showing (with airflow arrows) the directions that air flows between the spaces of the building. The diagram should also show the resultant air pressurization of each space and the overall air pressurization of the floor, both in terms of cubic feet per minute.

Air System Equipment

In the first section of this chapter, we discussed the various types of air systems commonly used in commercial buildings, focusing on the overall systems and the ways in which the various systems meet the needs of the areas served. In this section, we will discuss the physical characteristics, connections, and design considerations common to air system equipment.

Physical Characteristics

HVAC Air Systems
As discussed earlier in this chapter, HVAC air systems circulate air primarily to meet the heating and/or cooling needs of the spaces they serve.

Centralized Equipment   An example of a centralized air system is a modular central station air handling unit[18] that contains a hot water heating coil and a chilled water cooling coil that receive their heating and cooling energy from a remote central plant. Figure 5-13 is a graphical representation of an air handling unit on a floor plan drawing.

Decentralized Equipment   An example of a decentralized air system is a packaged[19] rooftop unit (RTU) that contains an indirect gas-fired heater and a complete refrigeration system for cooling (Figs. 5-14 and 5-15).

Another example of a decentralized air system is an air handling unit with a split refrigeration system that provides DX cooling or heat pump operation. This type of system is called a split system. In a split system, the compressor and condenser are contained within an outdoor air-cooled unit called a condensing unit. The outdoor air-cooled condensing unit is connected to the evaporator (DX cooling coil) in the air handling unit by refrigerant liquid and refrigerant suction piping. The refrigerant liquid piping carries the high-pressure refrigerant liquid from the outlet of the condenser to the thermostatic expansion valve on the inlet of the DX cooling coil. The refrigerant suction piping carries the low-pressure refrigerant gas from the outlet of the DX cooling coil to the suction of the compressor.

In order to increase the unloading capabilities of the refrigeration system, a hot gas bypass line can be connected to the high-pressure (hot gas) side of the compressor, which carries a portion of the hot gas to the DX cooling coil, bypassing the condenser.

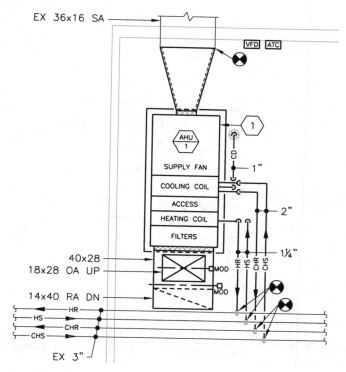

EX 36x16 SA

VFD ATC

1

AHU 1

SUPPLY FAN

COOLING COIL

ACCESS

HEATING COIL

FILTERS

CD

1"

2"

1¼"

HR HS CHR CHS

40x28

18x28 OA UP

14x40 RA DN

MOD

MOD

HR

HS   CHR

CHS

EX 3"

KEY NOTES

1   REPLACEMENT AIR HANDLING UNIT MOUNTED ON 4"
HOUSEKEEPING PAD. CONNECT TO EXISTING
OUTDOOR, RETURN, AND SUPPLY AIR DUCTWORK,
AND EXISTING PIPING AS INDICATED.

FIGURE 5-13 Floor plan representation of a modular central station air handling unit with hot water heating coil and chilled water cooling coil.

The hot gas adds load to the DX cooling coil during low-load conditions and keeps the compressor running.

One limitation of a split system is the distance between the air handling unit and the air-cooled condensing unit. Special sizing of the refrigerant lines must be performed by the equipment manufacturer if the distance between the two units exceeds the allowable separation distance, which is usually 75 to 100 ft. Furthermore, the allowable separation distance between the air handling unit and the condensing unit may be decreased if the condensing unit is installed at a higher elevation than the air handling unit.

Figure 5-16 illustrates a rooftop, cooling-only air handling unit with a split-system DX cooling coil. The refrigeration system has two separate refrigeration circuits and a hot gas bypass line for each circuit. This explains why there are six refrigerant lines connecting the two units.

Another example of a decentralized air system is a computer room air-conditioning (CRAC) unit with an electric heating coil, self-contained steam humidifier, and outdoor air-cooled condenser (Figs. 5-17 through 5-19).

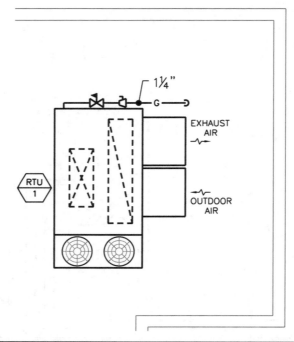

1¼"

EXHAUST AIR

OUTDOOR AIR

RTU 1

G

FIGURE 5-14   Roof plan representation of a packaged RTU with indirect gas-fired heater and DX cooling coil.

FIGURE 5-15   Photograph of a packaged RTU with indirect gas-fired heater and DX cooling coil.

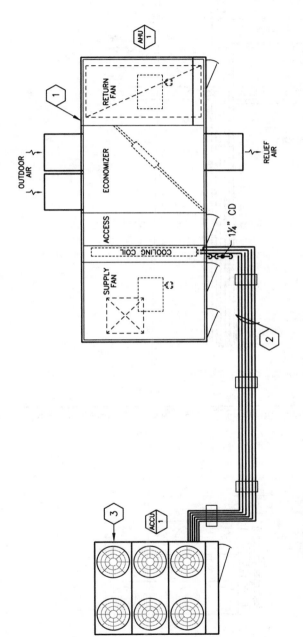

KEY NOTES

1. MOUNT ROOFTOP AIR HANDLING UNIT ON ROOF CURB. TRANSITION AND MAKE CONNECTIONS TO SUPPLY AND RETURN AIR DUCTWORK BELOW ROOF WITH FLEXIBLE DUCT CONNECTORS.

2. MOUNT REFRIGERANT SUCTION, REFRIGERANT LIQUID, AND HOT GAS PIPING FOR BOTH REFRIGERATION CIRCUITS ON PIPE SUPPORTS.

3. MOUNT AIR–COOLED CONDENSING UNIT ON STRUCTURAL STEEL SUPPORT (REFER TO STRUCTURAL DRAWINGS). PROVIDE ¾" THICK NEOPRENE VIBRATION ISOLATION PADS BETWEEN UNIT AND STRUCTURAL STEEL SUPPORT.

FIGURE 5-16 Roof plan representation of a rooftop, cooling-only air handling unit with a split-system DX cooling coil.

175

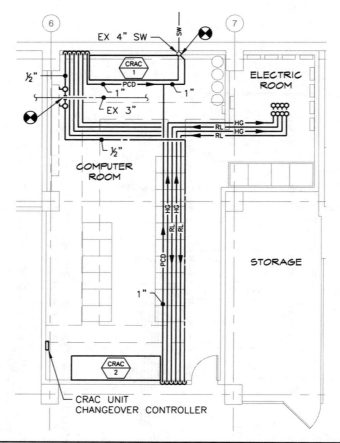

FIGURE 5-17 Floor plan representation of two CRAC units with electric heating coils and self-contained steam humidifiers.

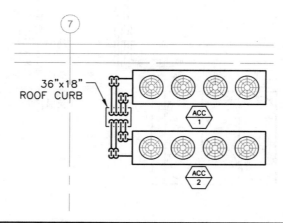

FIGURE 5-18 Roof plan representation of two outdoor air-cooled condensers.

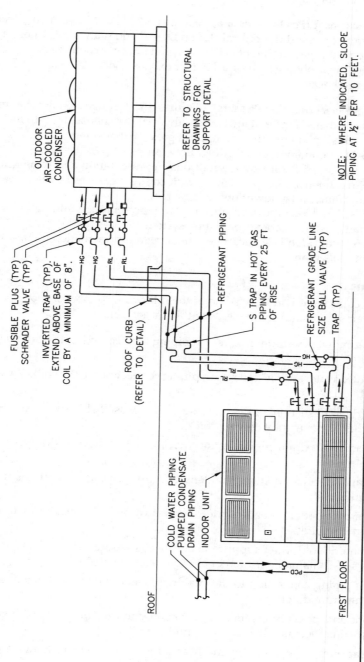

OUTDOOR AIR-COOLED CONDENSER

REFER TO STRUCTURAL DRAWINGS FOR SUPPORT DETAIL

FUSIBLE PLUG (TYP) SCHRADER VALVE (TYP)

INVERTED TRAP (TYP). EXTEND ABOVE BASE OF COIL BY A MINIMUM OF 8".

HG
HG
RL
RL

REFRIGERANT PIPING

ROOF CURB (REFER TO DETAIL)

S TRAP IN HOT GAS PIPING EVERY 25 FT OF RISE

REFRIGERANT GRADE LINE SIZE BALL VALVE (TYP)

TRAP (TYP)

HG
HG
RL
RL

COLD WATER PIPING PUMPED CONDENSATE DRAIN PIPING

INDOOR UNIT

PCD
PCD

ROOF

FIRST FLOOR

NOTE: WHERE INDICATED, SLOPE PIPING AT ½" PER 10 FEET.

FIGURE 5-19 CRAC unit connection detail.

177

Ventilation Air Systems

Ventilation air systems provide ventilation only and do not provide any heating or cooling to the spaces they serve.

Outdoor Air Equipment (Positive Pressure)   Figures 5-20 and 5-21 show the section view and photograph of a modular central station DOAS unit with a hot water heating coil, wrap-around heat pipe refrigerant coil, and a chilled water cooling coil.

Figure 5-22 is a photograph of a kitchen exhaust hood makeup air unit with an indirect gas-fired heater.

Exhaust Air Equipment (Negative Pressure)   Figures 5-23 through 5-31 are examples of different types of exhaust fans that are commonly used in exhaust air systems.

Figure 5-23 is a photograph of an upblast grease exhaust fan.

Figure 5-24 is a photograph of a downflow fan.

Figures 5-25 and 5-26 illustrate a roof plan representation and photograph of two kitchen exhaust hood makeup air units with indirect gas-fired heaters, two upblast grease exhaust fans, and a downflow exhaust fan.

Figures 5-27 and 5-28 illustrate a floor plan representation and detail of an in-line centrifugal fan used in a garage ventilation system.

Figures 5-29 through 5-31 illustrate a floor plan representation, detail, and photograph of a propeller fan.

Connections

Air system equipment will have some of the following connections, depending upon the type of air system and components within the system:

- Heating water supply and return piping connections to the hot water heating coil
- Chilled water supply and return piping connections to the chilled water cooling coil
- Refrigerant suction, refrigerant liquid, and possibly hot gas piping connections to the DX refrigerant cooling coil
- Condensate drain piping connection to the condensate drain pan under the cooling coil
- Steam supply and steam condensate return piping connections to the steam heating coil
- Steam supply and steam condensate drain piping connections to the centralized steam humidifier
- Domestic cold water supply and steam condensate drain piping connections to the self-contained gas-fired or electric steam humidifier
- Gas piping connection to the gas-fired heater and/or self-contained gas-fired steam humidifier
- Vent and possibly combustion air connection to the gas-fired heater and/or self-contained gas-fired steam humidifier
- Heat pump water supply and heat pump water return piping connections to water-source heat pump units

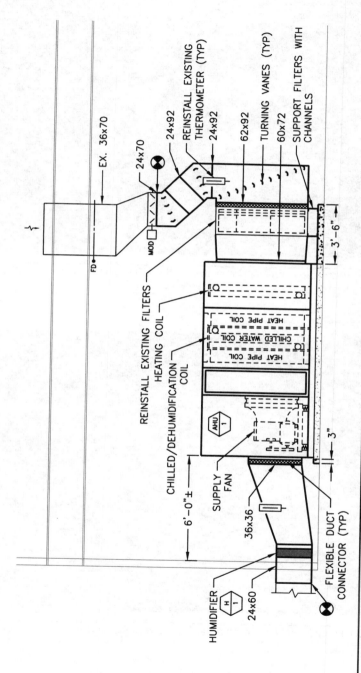

Figure 5-20 Section view of a modular central station DOAS with a hot water heating coil, wrap-around heat pipe refrigerant coil, and a chilled water coil.

FIGURE 5-21 Photograph of a modular central station DOAS with a hot water heating coil, wraparound heat pipe refrigerant coil, and a chilled water coil.

FIGURE 5-22 Photograph of a kitchen exhaust hood makeup air unit with an indirect gas-fired heater.

Figure 5-23 Photograph of an upblast grease exhaust fan.

Figure 5-24 Photograph of a downflow exhaust fan.

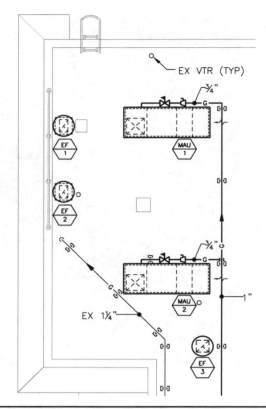

FIGURE 5-25 Roof plan representation of two kitchen exhaust hood makeup air units with indirect gas-fired heaters, two upblast grease exhaust fans, and a downflow exhaust fan.

FIGURE 5-26 Photograph of two kitchen exhaust hood makeup air units with indirect gas-fired heaters, two upblast grease exhaust fans, and a downflow exhaust fan.

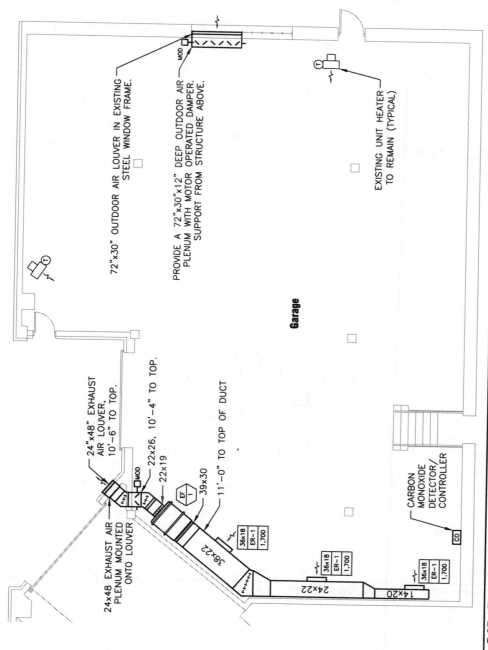

24"x48" EXHAUST AIR LOUVER, 10'-6" TO TOP.

22x26, 10'-4" TO TOP.

22x19

11'-0" TO TOP OF DUCT

24x48 EXHAUST AIR PLENUM MOUNTED ONTO LOUVER

39x30

EF 1

38x22

24x22

14x20

36x18
ER-1
1,700

36x18
ER-1
1,700

36x18
ER-1
1,700

CARBON MONOXIDE DETECTOR/ CONTROLLER

CO

72"x30" OUTDOOR AIR LOUVER IN EXISTING STEEL WINDOW FRAME.

PROVIDE A 72"x30"x12" DEEP OUTDOOR AIR PLENUM WITH MOTOR OPERATED DAMPER. SUPPORT FROM STRUCTURE ABOVE.

MOD

EXISTING UNIT HEATER TO REMAIN (TYPICAL)

T

Garage

FIGURE 5-27 Floor plan representation of an in-line centrifugal fan.

183

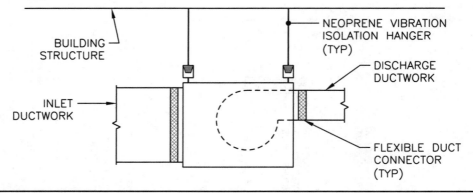

FIGURE 5-28 Detail of an in-line centrifugal fan.

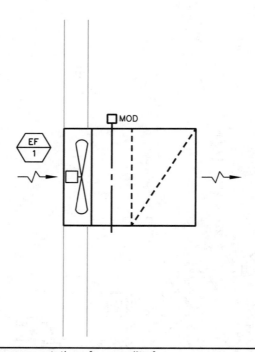

FIGURE 5-29 Floor plan representation of a propeller fan.

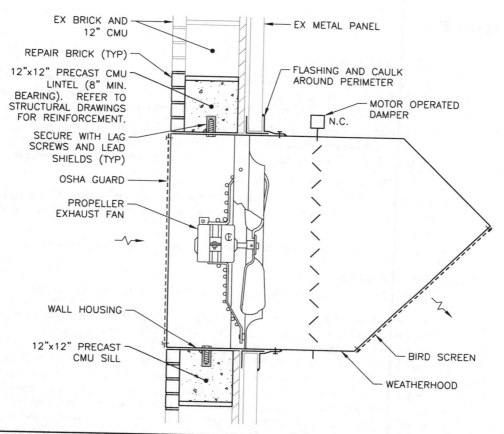

EX BRICK AND 12" CMU

REPAIR BRICK (TYP)

12"x12" PRECAST CMU LINTEL (8" MIN. BEARING). REFER TO STRUCTURAL DRAWINGS FOR REINFORCEMENT.

SECURE WITH LAG SCREWS AND LEAD SHIELDS (TYP)

OSHA GUARD

PROPELLER EXHAUST FAN

WALL HOUSING

12"x12" PRECAST CMU SILL

EX METAL PANEL

FLASHING AND CAULK AROUND PERIMETER

MOTOR OPERATED DAMPER

N.C.

BIRD SCREEN

WEATHERHOOD

FIGURE 5-30 Detail of a propeller fan.

FIGURE 5-31 Photograph of a propeller fan.

- Duct connections
  - HVAC equipment will have return air and supply air duct connections. Units may have an outdoor air connection to the mixing box section and may also have a relief air connection if a return fan is used.
  - Ventilation equipment will have outdoor air duct inlet and supply air duct discharge connections for outdoor air systems, and exhaust air duct inlet and (possibly) exhaust air duct discharge connections for exhaust air systems.
- Automatic temperature controls
- Electrical connections to each motor and electric coil (if equipped) or single-point electrical connection for all electrical components that are part of the air system
- Electrical connection to the self-contained gas-fired or electric steam humidifier

Design Considerations

The following are some important items to consider when designing air systems:

- The width and height of an air handling unit is proportional to the airflow through the unit. The velocity of air through the cooling coil (or face velocity) is typically what determines the width and height dimensions of an air handling unit. Cooling coils are normally selected for a maximum of 500 feet per minute (fpm) face velocity.[20] If 10% future capacity is required, the cooling coil should be selected with the additional capacity but with a face velocity of 450 fpm for the present-day airflow so that when the future capacity is required and the fan speed is adjusted for the increased airflow, the face velocity will be 495 fpm, which is below the maximum 500 fpm face velocity. Equipment manufacturers offer some flexibility in the width and height dimensions for air handling units.
- The outdoor air/return air mixing box section is normally furnished with dampers integral to the box. Field installation of the damper actuators is required.
- An angled filter section is recommended in lieu of a flat filter section if sufficient airway length is available for the unit. The angled arrangement of the filters provides a greater filter surface area, a lower face velocity and static pressure loss, and a greater dust-holding capacity.
- A pressure gauge mounted on the exterior of the unit piped across the filter section is recommended because the differential pressure across the filter section provides maintenance personnel with information regarding the loading of the filters. In addition to (or as an alternative to) a pressure gauge, a differential pressure switch piped across the filter section that provides a "dirty filter" alarm at the BAS whenever the (adjustable) setpoint of the differential pressure switch is exceeded can be specified.
- Access sections are recommended both upstream and downstream of the heating and cooling coils to facilitate routine inspection and cleaning of the coils. Access sections are typically 12 or 24 in. in airway length.
- The required capacity for the heating and cooling coils is what determines the number of rows[21] and fin spacing for the coils.

- Heating coils are typically one or two rows deep.

- Cooling coils are typically four to eight rows deep. A higher number of rows for a cooling coil increases the dehumidification capabilities of the cooling coil because the air is in contact with the cold surface of the coil for a longer period of time.

- The fin spacing on air coils is given in terms of fins per inch (fpi). Air coils typically have a fin spacing of 8 to 12 fpi. Closer fin spacing increases the heat transfer efficiency and air pressure drop of the air coil; however, coils with a fin spacing greater than 12 fpi are difficult to clean.

- The condensate drain pan below the cooling coil should be double-sloped; that is, it should be sloped in two planes to drain condensate within the pan to the drain pipe. If available as an option, double-wall construction with foam insulation between the walls should be specified for the drain pan because it prevents condensation of moisture on the outer surfaces of the drain pan. If available as an option, stainless steel construction for the drain pan is also recommended (in lieu of galvanized steel) because it provides better corrosion resistance.

- Double width, double-inlet (DWDI)[22] centrifugal fans are typically used for air handling units. The fan blades can be forward-curved (FC), backward-inclined (BI), or airfoil. FC fans are typically used for low total static pressure applications. BI and airfoil fans are typically used for high total static pressure applications.

- Plenum fans can also be used in air handling units. Plenum fans provide some flexibility in the location of the supply fan outlet connection(s) [or return fan inlet connection(s)] to the air handling unit and are also shorter in airway length than centrifugal fans.

- Air handling units are normally internally vibration isolated; that is, the fan(s) is mounted on spring vibration isolators inside of the unit. External spring vibration isolators are not necessary for floor-mounted air handling units. Neoprene vibration isolation hangers should be used if the air handling unit is suspended from the building structure.

- Lights are typically installed in selected sections (such as filter, access, and fan sections) of larger air handling units. The units can be specified with lights that are factory-mounted and wired to a switch on the outside of the unit.

- The stiffness of the panels used to construct air handling units is usually given in terms of the deflection of the panel at mid-span when the unit is subjected to a test static pressure that is 125% of the design static pressure, or a maximum of 5-in. w.c. positive or 6-in. w.c. negative static pressure. The deflection at mid-span is described as a ratio of the length of the span divided by a constant. A good panel stiffness of L/240 defines a panel deflection at mid-span that is 1/240 of the length of the panel.

- The air leakage of an air handling unit is normally given in terms of cubic feet per minute per square foot of cabinet area at a test pressure of 5-in. w.c. positive or 6-in. w.c. negative static pressure. Tight air handling unit construction is 0.5-cfm/ft² air leakage.

- Factory mounting and wiring of electrical components, such as motor starters or VFDs, are recommended whenever available as an option because it reduces field labor required for installation and improves coordination between trades.

- If possible, a single-point electrical connection[23] should be specified for air handling units that have multiple electrical components. This simplifies the installation procedure and also provides factory quality control of the installation and wiring of the electrical components within the unit.

- Premium-efficiency motors complying with the 1992 Energy Policy Act (EPAct) should always be specified.

- In addition, motors controlled by VFDs must be specified as inverter-duty motors.[24]

- A concrete housekeeping pad is normally designed for floor-mounted air handling units. Housekeeping pads are normally 4 in. high and 4 to 6 in. larger than the equipment base frame on all sides. A housekeeping pad higher than 4 in. may be required for air handling units with a high negative static pressure in the cooling coil drain pan in order to allow sufficient height to maintain the water seal in the condensate drain trap.

- The entering and leaving water temperatures, type of brine (if used), flow rates, maximum water pressure drop, maximum air pressure drop, minimum number of rows, and minimum fin spacing (maximum number of fins per inch) for heating and chilled water coils should be specified.

- Whether 2-way or 3-way control valves will be used for the heating and chilled water coils should be specified.

- A primary-secondary pumping arrangement may be required for the heating and/or chilled water coils.

- A circulating pump for freeze protection on the heating water coil should be specified if the mixed air temperature is less than 40°F.

- The heating coil should be located upstream of the chilled water coil to protect the chilled water coil from freezing.

- A low-limit temperature sensor mounted on the downstream side of hot water or steam heating coils is recommended. The low-limit temperature sensor de-energizes the unit and closes the outdoor air damper if the coldest point along the sensor drops below the (adjustable) setpoint of the sensor, which is usually set at 38°F.

- If steam coils are used, the steam pressure and flow rate should be specified.

- Hot gas bypass for DX refrigerant cooling coils used in VAV air handling units should be specified.

- The minimum outdoor airflow that is required to be delivered by the unit should be specified.

- If the static pressure loss through the return air duct system at the design airflow exceeds 0.25 in. w.c., a return fan should be specified. The return fan should be selected to deliver the return airflow (supply airflow minus outdoor airflow) at a static pressure that is sufficient to overcome the (external) static pressure loss of the longest run of the return air duct system, including the pressure loss through the return air inlet at the end of this run.

- The airflow and external static pressure for supply and return fans should be specified. The supply air fan should be selected to deliver the maximum design supply airflow at a static pressure that is equal to the (external) static pressure loss of the longest run of the supply air duct system, including the pressure loss through the air outlet at the end of this run, plus the (internal) static pressure loss of all the components within the air handling unit (mixing box, filters, coils, etc.).

- The design criteria for the air handling unit should be provided to the manufacturer's representative for the selection of the equipment that forms the basis of design. The actual equipment selection data will include the internal pressure drop of all components within the air handling unit. This information is used to specify the maximum air pressure drop through the coils and air filters and also to specify a minimum number of coil rows and minimum fin spacing (maximum number of fins per inch).

- The minimum face area of all air coils should be specified to ensure that the face velocity does not exceed the desired limit (450 to 500 fpm) for the cooling coil.

- A duct-mounted smoke detector is required in return air systems that have a design airflow greater than 2,000 cfm (refer to the 2009 *International Mechanical Code*, Section 606).

- Duct-mounted smoke detectors are required at each floor in return air systems serving two or more stories that have a design airflow greater than 15,000 cfm (refer to the 2009 *International Mechanical Code*, Section 606).

- An airside economizer[25] should be employed if the spaces served by the air system require cooling when the outdoor temperature is below 55°F. Provisions to relieve the outdoor airflow that is in excess of the minimum outdoor airflow required for ventilation and/or exhaust air makeup need to be accommodated in the air system design.

- The dimensions of the condensate drain trap should be specified based on the static pressure that exists in the cooling coil section.

- A 1-in. air gap at the condensate pipe discharge to drain should be specified.

- Single- or double-wall construction for the unit cabinet should be specified. Double-wall construction is preferred but is also more costly.

- A roof curb is normally used for rooftop equipment.

- The maximum sound power levels for the unit, both radiated and discharge, should be specified.

- The connecting ductwork to the unit must be properly designed in order to minimize the fan system effect.

- All ductwork connections to the unit must be made with flexible duct connectors to reduce vibration transmission through the duct systems.

- It is necessary to allow adequate coil pull clearance equal to the width of the unit on the coil connection side of the unit.

- It is necessary to allow adequate filter replacement clearance on the proper side of the unit.

- It is necessary to allow sufficient clearance within the equipment room to install and remove the unit. Clearance may only need to be provided for the largest section of modular central station air handling units if the sections can be field-assembled.

- Doorways into equipment rooms need to be large enough to install and remove the unit. Individual sections of modular central station air handling units can be shipped separately for field assembly, if required.

- Hose kits are recommended for water-source heat pump units to simplify the final heat pump water supply and return connections to the units. Hose kits typically include 2- or 3-ft flexible, fire-rated hoses; shutoff valves; strainer; and a flow control valve (for water flow balancing).

- If installed above critical equipment, a drain pan should be included under units suspended from the building structure. A leak detector in the pan or a drain line from the pan should also be designed.

- The required openings in the building envelope need to be coordinated with the project architect and structural engineer.

- The size, location, and weight of all air system equipment need to be coordinated with the project architect and structural engineer.

- It is necessary to coordinate the electrical requirements of the air system equipment with the project electrical engineer.

- Coordination with the project electrical engineer is necessary in order to ensure that the ground fault circuit interrupter receptacle required by Section 210.63 of the *National Electric Code* for use by the maintenance personnel is designed within 25 ft of all HVAC units.

- It is important to ensure that all appropriate automatic temperature controls are designed for the air systems, including heating, cooling, occupied, unoccupied, and morning warm-up modes of operation, as well as safeties and interlocks with other air systems and connection to the BAS, if required.

- The location of outdoor air intake openings must comply with the 2009 *International Mechanical Code*, Section 401.4. One requirement of this section is for intake openings to be at least 10 ft away from noxious contaminant sources, such as sanitary plumbing system vents and loading docks. It is also good practice to locate outdoor air intake openings at least 10 ft away from building exhaust outlets.

- Care should be taken when locating exhaust air outlets to ensure that the discharged air does not create a nuisance, such as discharging onto a walkway or below operable windows.

Endnotes

1. HVAC air systems may also provide humidification and dehumidification for the spaces they serve.
2. HVAC air systems that serve laboratories or other spaces with high exhaust airflows may have a significantly higher percentage of outdoor air.
3. In an indirect gas-fired heater, fuel is burned within a combustion chamber that is separate from the conditioned airstream, and the products of combustion are vented

to the outdoors. In a direct gas-fired heater, the fuel is burned directly within the conditioned airstream; thus the products of combustion are entrained within the conditioned air and are delivered to the spaces served.

4. Depending upon the project, these spaces may require cooling as well as heating and ventilation.

5. The primary air duct is the portion of the supply air duct system between the VAV air handling unit and the VAV terminal units.

6. VAV terminal units without a fan are called single-duct VAV terminal units. Those that contain a fan are called fan-powered VAV terminal units.

7. Other methods of modulating fan airflow are available, such as the use of inlet guide vanes or discharge dampers. However, these methods are less energy efficient than the use of VFDs. Furthermore, recent improvements in VFD technology and reduced cost have made VFDs the first choice for modulating fan airflow for VAV systems.

8. DX refrigerant cooling coils are not as "forgiving" as chilled water coils. Without adequate unloading capabilities in the refrigeration system, the saturated suction pressure of the refrigerant will drop at low-load conditions, which could cause a decrease in the cooling coil temperature below the freezing point of water. If this occurs, any moisture on the cooling coil will freeze, leading to further freezing of the moisture contained within the air flowing through the coil. This could eventually lead to a complete freeze-up of the coil and blockage of airflow through the coil.

9. The purpose of the minimum position for the primary air damper in a VAV terminal unit is to provide minimum ventilation airflow to the zone. A minimum primary air damper position is necessary only for zones that have a heating load. For zones with a heating load, there could be times when the (external) heat losses from the zone equal the (internal) heat gains to the zone, producing a net zone cooling load of zero. With no zone cooling load, the primary air damper in the VAV terminal unit would close completely (even though the zone could be occupied and require ventilation) were it not for the minimum primary air damper position. A heating coil in the VAV terminal unit is required in this case to prevent the possibility of overcooling the zone when the primary air damper is at its minimum position.

10. The primary air damper in VAV terminal units serving zones that do not have any heating load (purely interior spaces) can have a minimum position that is fully closed because if there is no zone cooling load, the zone would have to be unoccupied. That is, there would be no internal heat gains from lights, equipment, or people. If any of these heat gains were present, there would be a cooling load, and the primary air damper in the VAV terminal unit would not close completely. A heating coil in the VAV terminal unit is not required in this case.

11. The temperature range between the 70°F heating setpoint and the 75°F cooling setpoint is called the deadband. Within this range of temperatures, neither heating nor cooling is provided to the zone; rather, the zone temperature is allowed to "float" between these two limits. If closer control of the zone temperature is required, the deadband can be reduced. However, a 5°F deadband is usually appropriate for commercial HVAC systems because it is acceptable for human comfort and it reduces the cycling from one mode of operation to the other.

12. Indoor air at 75°F/50% has a dew point of approximately 55°F. Thus, if the dew point of the outdoor air delivered by the DOAS is lowered to 55°F, it will not increase the total moisture content (humidity ratio) of the indoor air.

13. If a makeup-air-type exhaust hood is not used, the makeup air distribution system (particularly the location, size, type, and airflow delivered by the makeup air outlets) must be coordinated with the exhaust hood design to ensure a proper capture of the contaminated air by the exhaust hood.

14. For a mechanical forced-air system, the 2009 *National Fuel Gas Code* (*NFPA Standard 54*), Section 9.3.6, requires 0.35 cfm of combustion air per 1,000 Btuh input rating for all of the appliances located within the room.

15. A water closet is more commonly known as a toilet.

16. A 0.02-in. w.c. negative air pressurization is achieved with 105 cfm of airflow through a ¾-in. undercut on the bottom of each 3-ft-wide (closed) door. It is not feasible to establish space air pressurization through doors that remain open.

17. An equipment room ventilation system that utilizes transfer air from adjacent conditioned spaces is not recommended for equipment rooms containing electronic equipment for which the maximum space temperature is 85°F because the temperature in the conditioned spaces can rise above this temperature when the building HVAC systems are operating in the unoccupied mode.

18. A modular central station air handling unit is a built-up unit consisting of various individually specified components, such as the (outdoor/return air) mixing section, filter section, heating section, cooling section, access sections, and fan section.

19. Packaged equipment is only available with the standard components and capacities offered by the manufacturer. Packaged units offer limited flexibility in the selection of each individual component.

20. If the face velocity of a cooling coil is too high, moisture carryover from the coil will occur.

21. The number of rows in an air coil is given in terms of the airway length of the coil. For example, air passes through four rows of coils in a 4-row air coil.

22. A DWDI fan consists of two single-width single-inlet fans mounted on a common shaft, driven by a single motor.

23. A single-point electrical connection means the equipment requires only a single electrical connection to the building electrical power distribution system. All required wiring, conduit, starters, VFDs, overload, and overcurrent protection for the multiple electrical components within the equipment is factory-mounted and wired.

24. Inverter-duty motors are designed in accordance with the *National Electric Manufacturers Association (NEMA) Standard MG 1*, Part 31, Section IV.

25. An airside economizer utilizes an increased outdoor airflow for cooling whenever the outdoor conditions are appropriate and cooling is required by the air system. An airside economizer is sometimes referred to as "free cooling" because the cooling energy obtained from the outdoor air does not require energy associated with mechanical cooling.

CHAPTER 6

Piping and Ductwork Distribution Systems

In this chapter we will discuss the basic principles of designing HVAC piping and ductwork distribution systems for commercial buildings. In order to do this, the HVAC load calculations for the entire building must be complete so that the water, steam, and/or airflow rates required by the air systems and terminal equipment can be determined and the HVAC piping and ductwork distribution systems can be laid out and sized accordingly.

Equations and Conversion Factors

Before proceeding to the method of designing HVAC piping and ductwork distribution systems, it is necessary to present some basic equations that are used to calculate the rates of fluid flow for water, steam, and air, which are expressed in terms of gallons per minute (gpm) for water, pounds per hour (lb/h) for steam, and cubic feet per minute (cfm) for air, from the heating and cooling loads for the building, which are expressed in terms of heat flow [British thermal units per hour (Btu/h or Btuh[1])]. These fluid flow rates are necessary to properly size the HVAC piping and ductwork distribution systems within the building. We will also discuss some helpful conversion factors.

It is highly recommended that the HVAC system designer commit at least the following equations and conversion factors to memory:

Equations:

Water:

$$q = 500(Q)(\Delta t)$$

where

q = heat flow (Btuh)
Q = water flow (gpm)
Δt = temperature difference (°F)

Steam:

$$q = (Q)(h_{fg})$$

193

where

q = heat flow (Btuh)
Q = steam flow (lb/h)
h_{fg} = latent heat of vaporization (Btu/lb)

Air:

$$q_s = (1.08)(Q_s)(\Delta t)$$

where

q_s = sensible heat flow (Btuh)
Q_s = airflow (cfm) of standard air[2]
Δt = temperature difference (°F)

Conversion factors:

Water pressure:

$$1 \text{ psi} = 2.31 \text{ ft w.c.} = 28 \text{ in. w.c.}$$

Steam heat flow:

1. For low-pressure steam systems (0 to 15 psig), the heat flow of 1 lb/h of steam is approximately equal to 1,000 Btuh (or 1 MBH[3]). This figure can be used to quickly estimate the steam flow required to meet the heating needs of a particular space or piece of equipment. The heat flow for low pressure steam is more precisely given as follows:
   a. 966 Btuh for 2 psig steam
   b. 946 Btuh for 15 psig steam

2. 1 ft² EDR[4] = 240 Btuh

Electrical power:

$$1 \text{ watt (W)} = 3{,}413 \text{ Btuh}$$

Other equations and conversion factors that are not defined here are also used; these can be found in various reference books when needed.

Piping Systems

In this section, we will discuss general and specific guidelines for designing hydronic and low pressure steam systems.

General Design Guidelines

Piping Drawing Conventions

Piping systems are normally represented by single lines for pipe sizes up to about 12 in. Piping systems larger than 12 in. are normally represented by double lines.

Changes in elevation of piping systems are denoted on floor plans by "pipe up" and "pipe down" symbols. These symbols can be understood by defining the viewing plane for a plan drawing as a horizontal plane located just below the underside of the deck that is above the ceiling of the floor being presented on the drawing. If the pipe rises up through this plane, it is represented by a "pipe up" symbol. If the pipe drops down and does not rise up through this plane, it is represented by a "pipe down" symbol. Symbols for other piping components, such as valves, are also shown on floor plans. The conventions for pipe symbols should be given in the legend for each project. Although there are basic similarities, pipe symbol conventions vary somewhat from one company to another. Refer to the sample legend and abbreviations in Chap. 10 for commonly used pipe, valve, and specialty symbols.

Piping Layout

Generally, piping systems should be laid out parallel and perpendicular to the building structure. This should be done utilizing the fewest number of bends and offsets in the vertical and horizontal planes necessary to connect the heating and cooling sources to the loads.

Piping systems are suspended from the building structure with clevis hangers for single pipes or trapeze hangers for multiple pipes. Figures 6-1 through 6-4 illustrate these hangers.

The (horizontal) floor plan location and (vertical) elevation of HVAC piping systems above the floor must be coordinated with the systems of other disciplines, such as the building structure, ductwork, plumbing piping, sprinkler piping, and lighting, in the same area. If close coordination with the systems of other disciplines is required, a section view of the area should be drawn showing the work of all disciplines and the proposed locations and mounting heights of the systems within the available space.

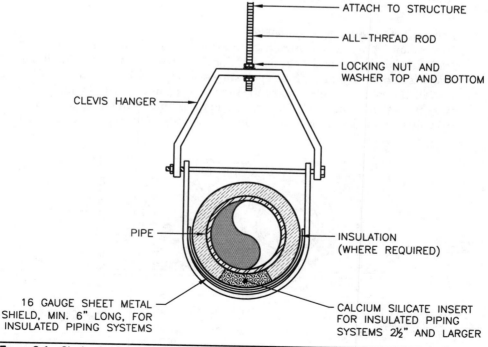

FIGURE 6-1 Clevis pipe hanger detail.

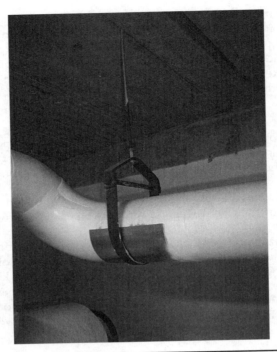

FIGURE 6-2   Photograph of a clevis pipe hanger.

Penetrations of Building Components

It is common practice in commercial construction to require sleeves for all pipe penetrations of the floors and both the interior and exterior walls of the building. A pipe sleeve is a short section of pipe or galvanized sheet metal that is embedded in the floor or wall. It is large enough to allow the pipe to pass through without interruption to the pipe insulation (if required). Sleeves isolate the piping from the building components, provide a finished appearance, and allow movement of the piping systems, which may

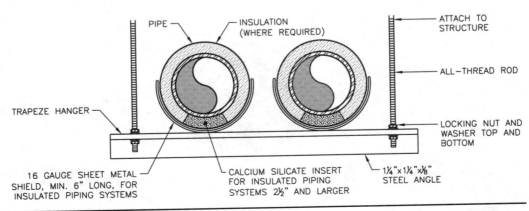

FIGURE 6-3   Trapeze pipe hanger detail.

FIGURE 6-4   Photograph of a trapeze pipe hanger.

be required for pipe expansion. Sleeves for pipes are typically Schedule 40 steel pipe; however, sleeves for large pipes may be constructed of galvanized sheet metal. Penetrations through fire-rated components must be firestopped. That is, the voids between the pipe (or pipe insulation) and the sleeve must be sealed with Underwriters Laboratories- (UL) listed materials or a UL-listed fire-stopping assembly. Sleeves through floors should extend 1 in. above the floor in dry locations and 2 in. above the floor in wet locations. A pipe clamp is required at the top of each pipe riser (and may be required at other locations along the riser, depending upon the length) to support the riser. Sleeves should be cast into concrete floor slabs and poured concrete foundation walls during the initial construction to eliminate core drilling after the concrete has been poured. Pipe penetrations through exterior walls and roofs must be waterproofed.

Although the specifics of the sleeves and waterproofing methods may vary from one project to another, the details shown in Figs. 6-5 through 6-7 provide some general guidelines for pipe penetrations through interior and exterior building components.

Pipe Expansion Compensation

Piping systems expand longitudinally (increase in length) when they are heated. For example, carbon steel pipe expands 1.53 in. per 100 ft of pipe length and copper tube expands 2.30 in. per 100 ft when they are heated from 0 to 200°F. If a straight section of pipe is restrained at both ends and the working fluid, such as heating water or steam, is heated to the operating temperature, the thermal expansion of the pipe will cause increased stress within the pipe wall and increased forces at the end supports of the

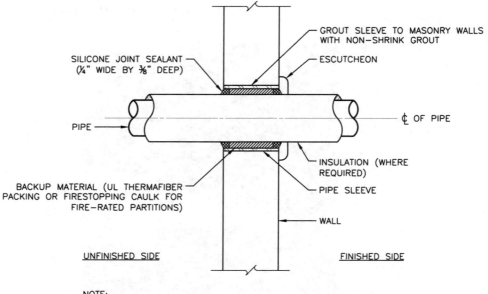

GROUT SLEEVE TO MASONRY WALLS WITH NON-SHRINK GROUT

ESCUTCHEON

SILICONE JOINT SEALANT (¼" WIDE BY ⅜" DEEP)

₵ OF PIPE

PIPE

INSULATION (WHERE REQUIRED)

PIPE SLEEVE

BACKUP MATERIAL (UL THERMAFIBER PACKING OR FIRESTOPPING CAULK FOR FIRE-RATED PARTITIONS)

WALL

UNFINISHED SIDE

FINISHED SIDE

NOTE:

1. AT THE CONTRACTOR'S OPTION, A UL LISTED FIRESTOPPING PIPE SLEEVE ASSEMBLY MAY BE UTILIZED.

FIGURE 6-5 Pipe penetration through interior walls.

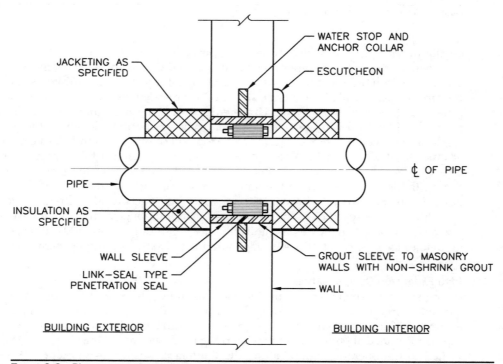

WATER STOP AND ANCHOR COLLAR

ESCUTCHEON

JACKETING AS SPECIFIED

₵ OF PIPE

PIPE

INSULATION AS SPECIFIED

WALL SLEEVE

LINK-SEAL TYPE PENETRATION SEAL

GROUT SLEEVE TO MASONRY WALLS WITH NON-SHRINK GROUT

WALL

BUILDING EXTERIOR

BUILDING INTERIOR

FIGURE 6-6 Pipe penetration through exterior walls.

198

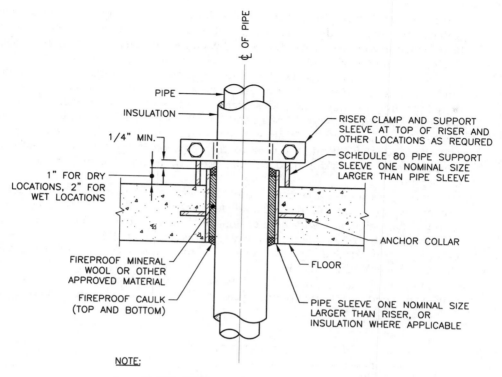

FIGURE 6-7   Pipe penetration through floors.

pipe section. If the pipe expansion is significant enough, the stress within the pipe wall can exceed acceptable limits and cause leakage in the joints, failure of the pipe, failure of the supports, and/or detrimental forces at the connections to equipment.

Therefore, it is necessary to design heating water, steam, steam condensate return, and even long sections of chilled and condenser water[5] piping to compensate for the thermal expansion of the piping within the various sections of the system. The first rule is not to anchor both ends of a straight section of pipe. Pipe anchors are different than standard pipe hangers, such as clevis hangers and trapeze hangers. Pipe anchors are designed to anchor a pipe at a certain point to the building structure, whereas pipe hangers allow for some movement of the pipe relative to the building structure. Pipe anchors are used to protect equipment connections and are also used in conjunction with pipe expansion compensation devices such as Z bends, U bends (expansion loops), or expansion joints (refer to the discussion of expansion compensation devices below).

Straight sections of pipe that serve heating water and low-pressure steam and condensate return systems that are less than 40 ft in length do not require any expansion compensation. Even so, it is good practice to design a 90° bend in the piping system every 25 ft or so to provide some expansion compensation for the piping through the flexure of the pipe bend. For straight sections of heating water and low-pressure steam and condensate return piping that exceed 40 ft in length and for chilled and condenser water piping that exceeds 100 ft in length, either a Z bend or an expansion loop should be designed to compensate for the thermal expansion of the piping section. As shown in Figs. 6-8 and 6-9, Z bends and expansion loops require pipe anchors to isolate the pipe section for expansion. Furthermore, expansion loops require pipe alignment guides in order to focus the flexure of the pipe section on the expansion loop (Z bends do not require pipe alignment guides). Refer to Figs. 6-8 and 6-9 and Tables 6-1 through 6-4 for the offset dimension "L" of Z bends and the height "H" and width "W" dimensions of expansion loops for heating, steam, and chilled or condenser water systems. The dimensions of Z bends and expansion loops depend upon the pipe material, the temperature of the working fluid, the distance between the pipe anchors, and the size of the pipe.

For example, a straight section of 4-in. steel heating water pipe that is 100 ft between anchors requires a Z bend that has an offset dimension of L = 16.5 ft or an expansion loop that has height and width dimensions of H = 7 ft and W = 3.5 ft. The pipe alignment guides are no closer than 14 ft away from the midpoint of the expansion loop. The locations of Z bends and expansion loops, and their associated dimensions, should be shown on the HVAC piping plans.

Expansion joints are normally not used in heating water, low-pressure steam, chilled, or condenser water piping systems for commercial buildings because pipe

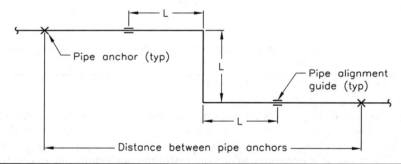

FIGURE 6-8   Diagram of a Z bend.

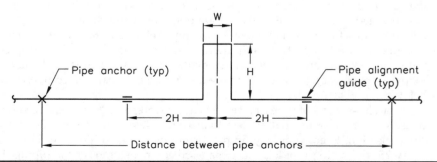

FIGURE 6-9   Diagram of an expansion loop.

Pipe Size, in.	Distance Between Anchors, ft								
	40 to 50			51 to 100			101 to 150		
	L, ft	W, ft	H, ft	L, ft	W, ft	H, ft	L, ft	W, ft	H, ft
¾	4.0	—	—	8.0	2.0	4.0	12.0	2.5	5.0
1	4.5	—	—	9.0	2.0	4.0	13.5	2.5	5.0
1¼	5.0	—	—	10.0	2.0	4.0	15.0	3.0	6.0
1½	5.5	—	—	10.5	2.0	4.0	16.0	3.5	7.0
2	6.0	—	—	12.0	2.5	5.0	18.0	3.5	7.0
2½	6.5	—	—	13.0	2.5	5.0	19.5	4.0	8.0
3	7.5	—	—	14.5	3.0	6.0	21.5	4.5	9.0
4	8.0	2.0	4.0	16.5	3.5	7.0	24.5	5.0	10.0
6	10.0	2.0	4.0	20.0	4.0	8.0	30.0	6.0	12.0
8	11.5	2.5	5.0	22.5	4.5	9.0	34.0	7.0	14.0
10	12.5	2.5	5.0	25.5	5.0	10.0	38.0	7.5	15.0
12	14.0	3.0	6.0	27.5	5.5	11.0	41.0	8.5	17.0

TABLE 6-1 Z-Bend and Expansion Loop Dimensions for Steel Heating Water Piping (200°F)

expansion can usually be accommodated through the use of Z bends and expansion loops. Furthermore, expansion joints are more costly than Z bends and expansion loops and also require routine maintenance. Therefore, expansion joints should only be designed when there is insufficient space to install a Z bend or an expansion loop. Refer to Chap. 45 of the 2008 *ASHRAE Handbook—HVAC Systems and Equipment* for more information regarding pipe expansion compensation.

Pipe Size, in.	Distance Between Anchors, ft								
	40 to 50			51 to 100			101 to 150		
	L, ft	W, ft	H, ft	L, ft	W, ft	H, ft	L, ft	W, ft	H, ft
¾	4.5	—	—	9.0	2.0	4.0	13.5	2.5	5.0
1	5.0	—	—	10.0	2.0	4.0	15.0	3.0	6.0
1¼	5.5	—	—	11.0	2.5	5.0	16.5	3.5	7.0
1½	6.0	—	—	12.0	2.5	5.0	18.0	3.5	7.0
2	7.0	—	—	14.0	3.0	6.0	20.5	4.0	8.0
2½	7.5	—	—	15.5	3.0	6.0	23.0	4.5	9.0
3	8.5	2.0	4.0	17.0	3.5	7.0	25.0	5.0	10.0
4	9.5	2.0	4.0	19.5	4.0	8.0	29.0	6.0	12.0

TABLE 6-2 Z-Bend and Expansion Loop Dimensions for Copper Heating Water Piping (200°F)

Pipe Size, in.	Distance Between Anchors, ft								
	40 to 50			51 to 100			101 to 150		
	L, ft	W, ft	H, ft	L, ft	W, ft	H, ft	L, ft	W, ft	H, ft
¾	4.0	—	—	8.5	2.0	4.0	12.5	2.5	5.0
1	4.5	—	—	9.5	2.0	4.0	14.0	3.0	6.0
1¼	5.5	—	—	10.5	2.0	4.0	15.5	3.0	6.0
1½	5.5	—	—	11.0	2.5	5.0	17.0	3.5	7.0
2	6.5	—	—	12.5	2.5	5.0	19.0	4.0	8.0
2½	7.0	—	—	14.0	3.0	6.0	20.5	4.0	8.0
3	7.5	—	—	15.0	3.0	6.0	22.5	4.5	9.0
4	8.5	2.0	4.0	17.0	3.5	7.0	26.0	5.0	10.0
6	10.5	2.0	4.0	21.0	4.5	9.0	31.5	6.5	13.0
8	12.0	2.5	5.0	24.0	5.0	10.0	35.5	7.0	14.0
10	13.5	2.5	5.0	26.5	5.5	11.0	40.0	8.0	16.0
12	14.5	3.0	6.0	29.0	6.0	12.0	43.5	8.5	17.0

TABLE 6-3 Z-Bend and Expansion Loop Dimensions for Steel 2-psig Steam and 0-psig Condensate Return Piping (219°F)

Pipe Size, in.	Distance Between Anchors, ft								
	100			101 to 200			201 to 300		
	L, ft	W, ft	H, ft	L, ft	W, ft	H, ft	L, ft	W, ft	H, ft
¾	5.5	—	—	11.1	2.5	5.0	16.7	3.5	7.0
1	6.5	—	—	12.4	2.5	5.0	18.7	4.0	8.0
1¼	7.0	—	—	14.0	3.0	6.0	21.0	4.5	9.0
1½	7.5	—	—	15.0	3.0	6.0	22.4	4.5	9.0
2	8.5	—	—	16.7	3.5	7.0	25.1	5.0	10.0
2½	9.5	2.0	4.0	18.4	4.0	8.0	27.6	5.5	11.0
3	10.5	2.0	4.0	20.3	4.0	8.0	30.5	6.0	12.0
4	11.5	2.5	5.0	23.0	4.5	9.0	34.5	7.0	14.0
6	14.0	3.0	6.0	27.9	5.5	11.0	41.9	8.5	17.0
8	16.0	3.5	7.0	31.9	6.5	13.0	47.8	9.5	19.0
10	18.0	3.5	7.0	35.6	7.0	14.0	53.4	11.0	22.0
12	19.5	4.0	8.0	38.8	8.0	16.0	58.1	11.5	23.0

TABLE 6-4 Z-Bend and Expansion Loop Dimensions for Steel Chilled and Condenser Water Piping (100°F)

Equipment Connections

It is not necessary to show the valves or specialties associated with the piping connections for hydronic or steam equipment on the floor plan drawings. These items will be shown in the connection details for each piece of equipment.

The following are guidelines for the valves and specialties that should be designed for each piece of hydronic or steam equipment:

1. Unions should be designed for:
   a. Equipment piping connections for pipe sizes 2 in. and smaller.
   b. Piping connections for valves or specialties that may require service, such as:
      (1) Control valves.
      (2) Steam traps.
      (3) Pressure reducing valves.

2. Flanges should be designed for equipment piping connections for pipe sizes 2½ in. and larger.

3. Flexible pipe connectors should be designed for:
   a. Piping connections to pumps.
   b. Piping connections to cooling towers.
   c. Piping connections to water-source heat pumps.
   d. Other pieces of motor-driven equipment where the noise and vibration generated by the equipment may be transmitted to the piping system.

4. Types of shutoff valves that should be designed include:
   a. Gate valves for steam and condensate return piping.
   b. Ball valves for hydronic pipe sizes 2 in. and smaller.
   c. Butterfly valves for hydronic pipe sizes 2½ in. and larger:.
      (1) Hand-lever actuators for valves 6 in. and smaller.
      (2) Gear actuators for valves 8 in. and larger.

5. Types of check valves that should be designed include:
   a. Silent check valve in pump discharge piping for pipe sizes 2½ in. and larger.
   b. Swing check valve:
      (1) Pump discharge piping for pipe sizes 2 in. and smaller.
      (2) Steam trap outlet piping.

6. A balancing valve should be designed in conjunction with a venturi flow meter in the pump discharge piping for pipe sizes 2½ in. and larger to balance the pump flow. The balancing valve can be either a globe valve or a butterfly valve with a memory stop.
   a. A globe valve provides more accurate balancing of flow, but at a higher cost than a butterfly valve with memory stop. The globe balancing valve should not be used for shutoff duty because it is a multiturn valve and its balancing position cannot be easily restored. Thus, a discharge shutoff valve is required in addition to the globe balancing valve.
   b. A butterfly valve with a memory stop is the lower-cost alternative to the globe valve. It can perform both shutoff and balancing duty, eliminating the requirement for a discharge shutoff valve in addition to the balancing valve.

7. A venturi flow meter should be designed in the pump discharge piping for pipe sizes 2½ in. and larger. Appropriate lengths of straight pipe upstream and downstream of the flow meter must be provided (refer to the flow meter manufacturer's product data).

8. A calibrated balancing valve should be designed for:
   a. Pump discharge piping for pipe sizes 2 in. and smaller.
   b. All pieces of hydronic equipment, such as coils, where multiple pieces of equipment are served by a single pump.
   c. Bypass pipe for equipment controlled by a 3-way valve.

9. A multipurpose valve may be designed for the pump discharge piping for pipe sizes 2½ in. and larger in place of the discharge shutoff valve, balancing valve, flow meter, and check valve.

10. A suction diffuser should be designed for the pump suction connection to end-suction pumps where there is insufficient space for the recommended five pipe diameters of straight pipe upstream of the pump suction connection.

11. Globe valve bypass:
    a. A globe valve bypass is a pipe in which a globe valve is installed that is connected to the inlet and outlet of a valve or specialty that serves a piece of equipment or a system that must remain in continuous operation, even when the valve or specialty requires maintenance (or replacement). Shutoff valves on the inlet and outlet pipes of the valve or specialty being bypassed are required to isolate the valve or specialty when maintenance is required. When the shutoff valves are closed, the globe valve in the bypass pipe is opened as required to provide the appropriate throttling of the flow while the maintenance is being performed. Once the maintenance is complete, the shutoff valves are reopened and the bypass valve is closed. Refer to the detail of the air separator/makeup water assembly (Fig. 4-21) in Chap. 4 for an example of a globe valve bypass around the makeup water pressure reducing valve.
    b. A globe valve bypass is commonly designed for:
       (1) Control valves for major pieces of equipment.
       (2) Steam traps for major pieces of equipment.
       (3) Pressure reducing valves.

12. Strainers:
    a. A strainer should be designed upstream of:
       (1) Control valves.
       (2) Steam traps.
       (3) Pressure reducing valves.
       (4) Pumps.
    b. A blow-down valve is usually designed for strainers that are 1¼ in. and larger.

13. A control valve should be designed where automatic control of the fluid flow is required. Control valves can be either two-position or modulating, depending upon the control requirements.
    a. A 2-way control valve is typically used for variable flow systems.
    b. A 3-way control valve is used for constant flow systems.

14. A steam trap should be designed for all steam equipment.

15. A pressure relief valve should be designed for:
    a. Boilers.
    b. Shell side of steam to hot water shell and tube heat exchanger.
    c. Other locations where pressurization of the equipment may exceed the maximum pressure rating of the equipment.
16. Pressure gauges should be designed for major pieces of equipment where frequent pressure readings are required.
17. Thermometers should be designed for major pieces of equipment where frequent temperature readings are required. Thermometers are not required in steam or condensate return systems.
18. Test plugs should be designed for minor pieces of equipment where infrequent temperature and/or pressure readings are required.
19. The types and locations of equipment air vents that should be designed are as follows:
    a. Manual air vent at high point of hydronic equipment connection.
    b. Automatic air vent at hydronic system air separator.
    c. Automatic air vent at hydronic system expansion tank.
20. A drain should be designed in the low point of hydronic or steam equipment piping connections.
21. A vacuum breaker should be designed for all steam equipment where the steam flow is modulated by a control valve.
22. A condensate drain should be designed for:
    a. Cooling coil drain pan.
    b. Humidifier drain pan.

Hydronic System Design Fundamentals

In a hydronic system, the system pump must be capable of circulating the design flow rate of water (or brine) at the pressure (or head) necessary to overcome the pressure losses in the system from the pump discharge connection to the farthest point in the system and back to the pump suction connection. This includes the pressure losses associated with the supply piping, farthest piece of equipment, return piping, and all of the fittings, valves, and specialties that lie along that path. Because hydronic systems are normally closed systems, there is no elevation head that must be overcome by the system pump.

Pipe Sizing Criteria

In order to perform an accurate estimate of the hydronic system pressure losses and size the hydronic system piping, it is necessary to establish the pressure drop per unit length criterion that will be used [expressed in terms of feet of water column (ft w.c.) per 100 ft of piping] and the maximum fluid velocity [expressed in terms of feet per second (fps)] within the piping that will be allowed. A common sizing criterion for hydronic piping systems is 4 ft w.c. pressure drop per 100 ft of piping, with a maximum velocity of 4 fps for pipe sizes 2 in. and smaller and 6 fps for pipe sizes 2½ in. and larger. Table 6-5 lists the allowable water flow rates for hydronic piping between ¾ and 12 in. based on this criterion. Pipe sizing charts and software programs are available to assist in sizing piping if a different criterion or a brine is utilized.

Pipe Size, in.	Water Flow,[1] gpm	P.D., ft w.c./ 100 ft of Pipe	Velocity, fps
¾	3	2.5	1.8
1	7	3.6	2.6
1¼	15	3.7	3.2
1½	23	3.8	3.6
2	42	3.4	4.0
2½	74	4.0	5.0
3	132	4.0	5.7
4	240	3.2	6.0
6	540	1.9	6.0
8	940	1.4	6.0
10	1,480	1.0	6.0
12	2,100	0.8	6.0

[1]Based on a maximum P.D. of 4 ft w.c./100 ft of pipe and a maximum velocity of 4 fps for piping 2 in. and smaller, and 6 fps for piping 2½ in. and larger.
Note: P.D. = pressure drop

TABLE 6-5   Allowable Water Flow Rates for Hydronic Piping Between ¾ and 12 in.

Refer to the sample 4-pipe system flow diagram and heating water riser diagram (Figs. 10-19 and 10-20) in Chap. 10 for applications of this pipe sizing criterion. Piping for open systems, such as a condenser water system that utilizes an open cooling tower for heat rejection, can also be sized based on this pipe sizing criterion.

Piping Design

The following are some basic guidelines for designing hydronic piping systems:

- Manual air vents[6] are required at all high points in the piping distribution system. These vents are used to release the air from the system during start-up to prevent the system from becoming air bound. This occurs when a pocket of air within the piping system prevents the flow of water through that section of pipe.

- Drains are required at all low points in the system in order to drain the system, or parts of the system, for maintenance, repair, or replacement.

- Air vents and drains do not need to be shown on the floor plan drawings. The requirement for these specialties should be included in the project specifications.

- Supply and return mains should be pitched at ¼ in./10 ft toward the drains in the low points of the system.

- Pipe expansion compensation should be designed as described in the Pipe Expansion Compensation section earlier.
- Shutoff valves, a balancing valve,[7] and a calibrated balancing valve (for pipe sizes 2 in. and smaller) or an orifice flow meter (for pipe sizes 2½ in. and larger) should be designed for the hydronic piping mains serving each floor of a building.
- Shutoff valves should be designed at all major branches in the piping system.
- The minimum size for hydronic supply and return pipes is ¾ in.
- Takeoffs[8] to upfeed risers and equipment should be made at a 45° angle above the horizontal centerline of the main pipe (Fig. 6-10).
- Takeoffs to downfeed risers should be made at a 45° angle below the horizontal centerline of the main pipe (Fig. 6-11).

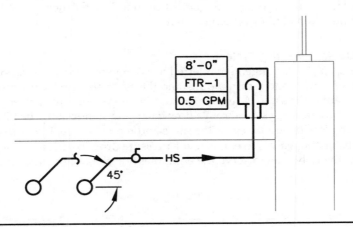

FIGURE 6-10   Top takeoff from main piping.

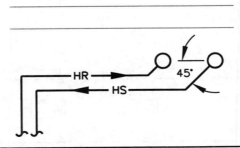

FIGURE 6-11   Bottom takeoff from main piping.

Direct Return Piping System

Typically, the supply and return pipe mains serving a particular floor of a building are routed parallel to each other from the piping risers to the farthest piece of equipment on the floor. The supply and return pipe takeoffs are routed from these mains to each piece of equipment served. Thus, the closest piece of equipment has the shortest run of supply piping and the shortest run of return piping between the equipment and the piping risers. Similarly, the farthest piece of equipment served by the mains has the longest run of supply piping and the longest run of return piping between the equipment and the piping risers. This piping arrangement is called a direct return piping system.

The difficulty with this type of system is that the pressure drop through the supply and return piping to the closest piece of equipment is less than the pressure drop through the supply and return piping to the farthest piece of equipment. As a result, the balancing valves on the closer pieces of equipment need to be throttled more than the balancing valves on the farther pieces of equipment in order to equalize the pressure drop to each piece of equipment. This is possible and it is done frequently. However, it is not the optimal piping arrangement from a flow balancing standpoint. Figure 6-12 illustrates a direct return piping system.

Reverse Return Piping System

There is a piping arrangement referred to as a reverse return piping system in which the length of supply and return piping (and the associated pressure drop) for each piece of equipment is the same. A reverse return piping system is the preferred piping arrangement for hydronic systems from a flow balancing standpoint. The piping pressure drop for each piece of equipment is the same and requires very little throttling of the equipment balancing valves to equalize the pressure drop to each piece of equipment. However, if the only way to serve the equipment on the floor is in a straight line from the

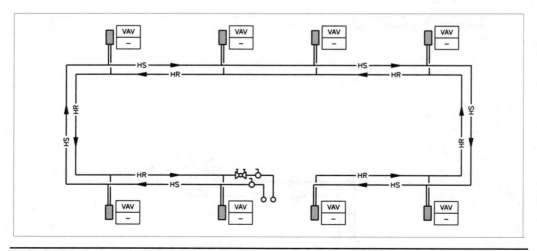

FIGURE 6-12   Direct return piping system.

piping risers to the farthest piece of equipment, a reverse return piping system is not economical. A second return pipe main that is the full length of the run to the farthest piece of equipment must be added to equalize the supply and return pipe lengths for each piece of equipment on the floor. If, however, it is possible to serve the equipment in a looped arrangement, there will be virtually no additional piping required to accommodate a reverse return system. These two configurations of a reverse return piping system are illustrated in Figs. 6-13 and Figs. 6-14.

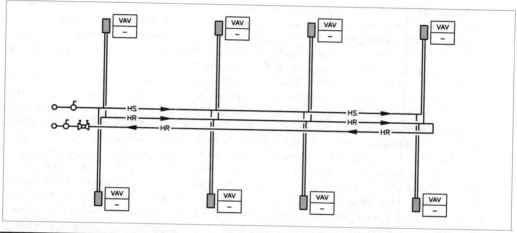

FIGURE 6-13   Straight-line reverse return system.

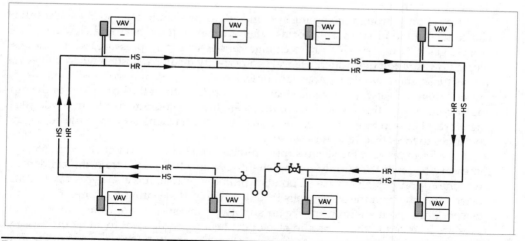

FIGURE 6-14   Looped reverse return system.

Low-Pressure Steam System Design Fundamentals

Steam has many qualities that make it the preferred fluid for transferring heating energy from the source to the load in industrial processes and commercial heating and plumbing systems. However, steam also has qualities that must be respected in the system design in order to avoid undesirable, and potentially harmful, consequences.

Steam can be generated at whatever pressure the system is designed for, even pressures exceeding 1,000 psig. However, steam pressures in excess of 100 psig are typically used for industrial processes or for district steam systems where the piping network is large. Within commercial buildings, where the length of steam piping is less than 1,000 ft, low-pressure steam systems, which are defined by ASME as those operating at 15 psig or less, are utilized. Because steam systems utilizing an operating pressure of 2 psig are used extensively in commercial buildings, we will limit our discussion to the design principles associated with this operating pressure. However, many of the same principles that apply to 2-psig steam systems can also be applied to steam systems operating at higher steam pressures.

Before discussing the design of steam distribution systems, it is necessary to discuss the basics of using steam as a heating fluid. Steam, as mentioned in Chap. 4, is generated within a steam boiler. This steam is distributed through the steam piping system to the heating equipment within the building. The condensate that is formed as the steam changes phase from a gas to a liquid is returned to the boiler, thus completing the cycle.

There are two types of steam heating systems: a 1-pipe system and a 2-pipe system. In a 1-pipe system, each piece of heating equipment has only one pipe connection. Steam flows to the heating equipment and the condensate returns (against the steam flow) from the heating equipment back to the boiler, all within the same pipe. One-pipe steam systems are not commonly used in the heating systems for commercial buildings; therefore, we will not discuss them in this book. In a 2-pipe steam heating system (not to be confused with a 2-pipe hydronic system), each piece of heating equipment has two connections, one for the steam supply and one for the condensate return. Steam flows from the boiler through the supply piping system to the heating equipment and is returned to the boiler through the condensate return piping system and the condensate recovery equipment.

Steam flows through the piping system and heating equipment solely as a result of the condensation of steam that forms when the steam loses heat and changes phase from a gas to a liquid. This phase change causes a drop in pressure, which induces steam flow through the system. Hence, in a perfectly insulated system where there was no heat loss, there would be no steam flow. However, even in a well-insulated piping system, there will always be some steam flow resulting from the heat losses of the piping system, even if there is no load on the steam heating equipment. Steam condensate, therefore, begins to form at the connection of the steam piping system to the boiler and continues to form throughout the system.

In order to prevent the steam supply piping from being filled with condensate, it is necessary to pitch the supply piping slightly (½ in./10 ft) and drain the condensate from the supply piping at various points within the system. It is also necessary to drain the condensate from the steam coils installed within the heating equipment. The condensate drain points within the steam supply piping system are called drips. Steam traps are installed at the drips and also at the heating equipment to allow the passage of condensate into the condensate return piping. The steam traps also prevent the flow of steam into the condensate return piping.

The condensate that is collected from the drips and the steam heating equipment is carried through the condensate return piping system to one or more receivers in the condensate recovery equipment. The condensate return piping can be either open (gravity motivated) or closed (steam pressure motivated). Open condensate return systems are vented at various locations in the piping system, ensuring that the pressure at all points within the return piping system is atmospheric pressure. The condensate flows through an open return piping system strictly due to gravity. Open condensate return piping systems must be pitched in the direction of the condensate flow at a minimum of 1/16 in./ft. However, we will not discuss open return piping systems in this book because they are not commonly used in commercial buildings. Refer to Chap. 22 of the 2009 *ASHRAE Handbook—Fundamentals* for information on designing open condensate return systems.

We will discuss closed condensate return systems in this book because this type of return system is often used in commercial buildings. In particular, we will discuss closed return systems that are vented to the atmosphere only at the receiver in the condensate recovery equipment (the point where the condensate leaves the return piping system). Furthermore, we will only discuss the principles of designing dry, closed return piping systems (those in which the return piping is installed above the waterline of the downstream receiver) because this is the type of closed return system most commonly used commercial heating systems.

In a closed return piping system, there is a continual difference in pressure within the condensate return piping between the point where the condensate enters the line (at the steam traps) and where it leaves the line (at the vented receiver). As a result, the condensate flow in a dry, closed return system is steam pressure motivated. The pressure within the return system is equal to the sum of:

- The pressure at the end of the return line (at the receiver in the condensate recovery equipment), which for our purposes is atmospheric pressure, or 0 psig (because the receiver is vented to the atmosphere).

- The hydrostatic head required to lift the condensate to a higher elevation. However, because lifting of condensate is not recommended for 2-psig steam systems, the hydrostatic head is also equal to zero.[9]

- The pressure drop due to friction losses in the return piping system.

Therefore, the pressure at any point in the dry, closed condensate return system for a 2-psig steam system is equal to the pressure drop due to friction of the steam condensate flow from that point in the return line to the end of the return line at the receiver in the condensate recovery equipment.

Although there is a slight positive pressure throughout the return piping system, for the purposes of determining the condensate carrying capacity of the return piping, the return pressure is assumed to be atmospheric pressure (0 psig). The difference between the supply pressure of the saturated steam condensate at the inlet of the steam traps and the nearly atmospheric pressure of the return piping causes a small percentage (by mass) of the condensate to flash to steam.[10] Although the flash steam percentage is low on a mass basis, it is very high on a volume basis (91.9% by volume for 2 psig supply pressure). Thus, the overwhelming volume of the two-phase fluid within the condensate return piping consists of steam at 0 psig. For this reason, the condensate return piping can be sized as if it were flowing low-pressure steam at 0 psig. We will discuss the principles of sizing the steam supply and condensate return piping later in this chapter.

The condensate return piping is also pitched at ½ in./10 ft in the direction of the return flow. Sometimes there is sufficient elevation to route the condensate return piping all the way back to the boiler feedwater receiver located in the central plant. However, in most cases, it is necessary to collect the condensate within multiple receivers located throughout the building. The condensate is pumped from these receivers back to the boiler feedwater receiver through pumped condensate return piping.

Once the condensate is returned to the boiler feedwater receiver, makeup water is added to account for any fluid losses in the system in order to maintain the appropriate water level in the receiver. Feedwater is pumped from the feedwater receiver to the steam boiler by the feedwater pumps at a rate necessary to maintain the operating water level within the boiler.

Although we are dealing strictly with 2-pipe steam systems where the condensate return is separated from the steam supply, there are certain parts of a 2-pipe steam system where the condensate flows within the same pipe as the steam supply and in the reverse direction as the steam flow. The steam supply piping in these portions of the system must be sized differently than the rest of the steam piping. Two examples of this situation are the vertical sections of steam piping, called risers, where the steam is flowing in the upward direction and the horizontal runouts to steam risers or steam heating equipment, which are pitched back toward the main and are not dripped.

In short, 2-pipe steam heating systems consist of the following components:

1. Steam boiler
2. Steam supply piping
3. Traps
   a. Steam main drips
   b. Steam heating equipment
4. Condensate return piping
5. Condensate transfer equipment
   a. Condensate receiver
   b. Condensate transfer pump
6. Boiler feedwater equipment
   a. Feedwater receiver
   b. Makeup water
   c. Feedwater pump(s)

In Chap. 4 we discussed the fundamentals of steam boilers and the boiler feedwater equipment. In the following sections, we will discuss the basics of designing a 2-pipe steam and condensate return system that utilizes 2 psig steam.[11] Our discussion will be limited to systems that have a maximum steam supply (or condensate return) piping length of:

- 400 ft to the farthest point: Allowing 100% of this length for pipe fittings and valves yields a piping system with an equivalent length[12] of 800 ft. This criterion of 800 equivalent ft describes systems that are common for commercial buildings up to about 50,000 ft².
- 50 ft to the farthest point: This would include the steam and condensate return piping that is limited to the equipment room in which the boiler is installed.

Allowing 100% of this length for pipe fittings and valves yields a piping system with an equivalent length of 100 ft.

For larger buildings, a higher steam supply pressure and different steam and condensate return pipe sizing criteria would be utilized.

Pipe and Trap Sizing Criteria

In a 2-psig steam system, there is only 2 psi of steam pressure available to overcome all of the pressure losses in the system, that is, the losses associated with the steam supply piping, condensate return piping, and steam traps. Therefore, the following criteria should be followed for sizing the steam supply and condensate return piping and the steam traps in a 2-psig steam heating system:

1. Steam supply piping
   a. The steam supply piping should be sized for a total pressure drop (P.D.) of 0.5 psi from the steam boiler to the farthest point in the steam supply piping system. Furthermore, a maximum steam velocity of 6,000 fpm within the steam piping is recommended for quiet operation of the steam system.
2. Condensate return piping
   a. The condensate return piping should be sized for a total P.D. of 0.5 psi from the farthest point in the system to the condensate receiver.
3. Steam traps
   a. The steam traps should be sized for a total P.D. of 0.5 psi. Float and thermostatic (F&T) traps should be used for the steam main drips and most pieces of heating equipment in 2-psig steam systems.

Following the above sizing criteria leaves a safety factor of 0.5 psi steam pressure.

Thus, the steam supply and condensate return piping for a system with 800 equivalent ft of steam or condensate return piping should be sized for a friction loss of 1/16 psi/100 ft of pipe to yield a pressure drop of 0.5 psi in the steam supply piping and condensate return piping to the farthest points in these systems. The steam supply piping for a system with 100 equivalent ft of steam piping (such as piping limited to the equipment room) can be sized for a friction loss of 0.5 psi/100 ft of pipe and still have a pressure drop of 0.5 psi in the steam supply piping. The condensate return piping, even for systems with only 100 equivalent ft of condensate return piping, should be sized for a friction loss of 1/16 psi/100 ft of pipe to keep the pressure drop in the condensate return piping to a minimum.

The final sizing criterion applies to the sizing of the F&T steam traps in the system. Each F&T steam trap should be selected to handle the full condensing load at a differential pressure (steam supply inlet pressure minus the condensate return back pressure) of 0.5 psi.

Tables found in the 2009 *ASHRAE Handbook—Fundamentals* can be used for sizing the steam supply and condensate return piping for 2-psig steam systems as follows:

- 2-psig steam supply piping (400 ft longest run): Chap. 22, Table 18, 1/16 psi pressure drop/100 ft of length, saturation pressure of 3.5 psig.
- Upfeed steam risers and undripped horizontal steam runouts: Chap. 22, Table 19, columns B and C.

- Dry, closed condensate return piping: Chap. 22, Table 20, column O.
- 2-psig steam supply piping (50 ft longest run): Chap. 22, Table 18, ½ psi pressure drop/100 ft of length; saturation pressure of 3.5 psig can be used for pipe sizes 2 in. and smaller. For pipe sizes 2½ in. and larger, use Table 6-6, which limits the steam velocity in these pipe sizes to 6,000 fpm.

Table 6-7 can be used to size the F&T steam traps for the steam main drips in 2-psig steam systems. Table 6-8 can be used to size the F&T steam traps for the steam heating equipment in 2-psig steam systems.

Pipe Size, in.	2 psig Steam Flow,[3] lb/h
2½	480
3	750
4	1,500
6	2,900
8	5,000
10	8,000
12	11,500

[1]Dripped horizontal mains.
[2]Flow of condensate does not inhibit flow of steam.
[3]Based on a velocity of 6,000 fpm.

TABLE 6-6   2-psig Steam Supply Piping[1,2] (50 ft Longest Run)

Steam Main Size, in.	Condensate Load,[2] lb/h	Drain Pipe to Trap,[3] in.	F&T Trap Size,[4,5] in.	Trap Discharge Pipe Size,[6] in.
2	84	¾	¾	1¼
2½	129	1	1	1¼
3	170	1	1	1½
4	240	1	1	1½
6	418	1¼	1¼	2
8	596	1¼	1¼	2½
10	888	1½	1½	2½
12	1,157	1½	1½	3

[1]Based on 100 ft of pipe.
[2]Includes warm-up load (5 min warm-up) and running load of pipe.
[3]Same size as trap size. Pitch pipe at 1 in./10 ft.
[4]Based on Steam Heating Equipment Manufactures Association rating (built-in safety factor).
[5]Based on a differential pressure of 0.5 psi.
[6]Same size as dry, closed condensate return main size.
Note: F&T = float and thermostatic

TABLE 6-7   Low-Pressure (up to 15 psig) F&T Steam Trap Sizing (Steam Main Drips[1])

The condensate loads for the steam main drips given in Table 6-7 are based on 100 ft of pipe. If the actual length of steam main to be dripped is different than 100 ft, the actual condensate load can be calculated by multiplying the condensate load given in Table 6-7 by a factor equal to the actual pipe length divided by 100 ft. For example, if the actual length of a 4-in. steam main to be dripped is 50 ft, the condensate load would be:

$$\text{Condensate load} = (50\ \text{ft}/100\ \text{ft}) \times (240\ \text{lb/h}) = 120\ \text{lb/h}$$

The float and thermostatic trap size can be selected for the actual condensate load based on the capacities given for the steam heating equipment listed in Table 6-8. For the example above, a 1-in. float and thermostatic trap would be required for the steam main drip. In Table 6-8, the F&T trap size incorporates the industry-recommended safety factor of 3 on the condensate load.

Piping Design

2-psig Steam Supply Piping   The following are some basic guidelines for designing 2-psig steam supply piping systems:

1. Connection(s) of the main steam supply pipe(s) to the building from the steam header at the steam boiler(s) should be made on the top of the header.
2. Steam supply mains should be pitched at ½ in./10 ft toward the main drips.
3. Pipe expansion compensation should be designed as described in the Pipe Expansion Compensation section earlier.
4. Shutoff valves should be designed for the mains serving each floor of a building.
5. Shutoff valves should be designed at all major branches in the piping system.

Condensate Load, lb/h	Drain Pipe to Trap,[2] in.	F&T Trap Size,[3,4] in.	Trap Discharge Pipe Size,[5] in.
100	¾	¾	1¼
250	1	1	1½
600	1¼	1¼	2½
1,200	1½	1½	3
2,500	2	2	4

[1]With modulating steam control valve.
[2]Same size as trap size. Pitch pipe at 1 in./10 ft.
[3]Incorporates a safety factor of 3 on condensate load.
[4]Based on Steam Heating Equipment Manufactures Association rating (built-in safety factor).
[5]Same size as dry, closed condensate return main size.
Note: F&T = float and thermostatic

TABLE 6-8   Low-Pressure (up to 15 psig) F&T Steam Trap Sizing (Steam Heating Equipment[1])

6. Steam main drips should be designed:
   a. At every 100 to 200 ft along horizontal mains.
   b. At every change in elevation.
      Exception: A drip is not required for undripped horizontal runouts to upfeed risers or equipment that is pitched toward the main (see Fig. 6-15).
   c. At the base of every downfeed riser.
   d. At the base of every dripped upfeed riser. In this case, the horizontal runout to the riser should be pitched toward the riser.
   e. At the end of every main.
7. Undripped horizontal runouts should be pitched at ½ in./ft toward the main and should be sized for condensate flowing against the steam flow.
8. Undripped horizontal runouts longer than 8 ft should be one pipe size larger than required if a minimum pitch of ½ in./ft toward the main cannot be obtained.
9. Takeoffs to upfeed risers and equipment should be made at a 45° angle above the horizontal centerline of the main pipe.
10. Takeoffs to downfeed risers should be made at a 45° angle below the horizontal centerline of the main pipe.
11. Upfeed risers should be sized for condensate flowing against the steam flow.
12. Downfeed risers should be sized the same as horizontal steam mains where the condensate flow does not inhibit the steam flow.
13. All reductions in pipe size should be made with eccentric reducers installed so they are flat on the bottom. The use of a normal concentric reducer creates a pocket on the bottom of the pipe where condensate can collect and disrupt the flow of steam and condensate.
14. The minimum size for steam supply pipes is ¾ in.

Dry, Closed Condensate Return Piping   The following are some basic guidelines for designing dry, closed condensate return piping systems:

- Condensate return piping between the equipment outlet and the trap should be pitched at 1 in./10 ft.
- Condensate return piping between the traps and the condensate receiver should be pitched at ½ in./10 ft toward the condensate receiver.
- Pipe expansion compensation should be designed as described in the Pipe Expansion Compensation section earlier.
- Shutoff valves should be designed for the mains serving each floor of a building
- Shutoff valves should be designed at all major branches in the piping system.
- The minimum size for condensate return pipes is 1 in.
- All reductions in pipe size should be made with eccentric reducers installed so they are flat on the bottom. This is done for the same reason stated in the 2-psig Steam Supply Piping section earlier.
- Condensate should not be lifted for 2-psig steam systems.

Traps   The following are some basic guidelines for trapping condensate in 2-psig steam systems:

- A 6-in.-long, full pipe size dirt leg should be designed upstream of all traps to collect scale and dirt.

- Steam main drips should be sized for the warm-up load and the running load of the pipe.

- Steam main drips downstream of undripped horizontal runouts to upfeed risers and equipment should include the load of the undripped horizontal runout and upfeed riser in addition to the load of the steam main pipe from the last main drip upstream.

- The inlet of F&T traps should be mounted a minimum of 15 in. below the outlet connection of all pieces of steam utilization equipment (e.g., air system heating coils, terminal heating equipment, heat exchangers, humidifiers) for which the steam flow is modulated by a control valve. When the control valve closes, the steam within the equipment will cool and condense and the pressure can drop to as low as 0 psig, provided the equipment is equipped with a vacuum breaker.[13] If any back pressure exists in the condensate return main, the F&T trap will not open with 0 psig pressure on the inlet of the trap and the equipment can become flooded with condensate. This can result in freezing of the condensate and damage to the equipment if it is exposed to freezing temperatures, such as for a steam heating coil where the entering air temperature is below freezing. The 15 in. of elevation difference between the outlet connection on the equipment and the inlet of the F&T trap will provide 15 in. of water column (approximately 0.5 psi) of positive pressure (minus the approximately 0.25 psi pressure drop through the vacuum breaker) on the trap to facilitate its operation should the pressure within the equipment drop to 0 psig.

- The inlet of an F&T trap installed in a steam main drip does not need to be 15 in. below the steam main pipe because the pressure within the steam main pipe is a constant 2 psig minus the friction losses in the piping system (maximum 0.5 psi). Thus, the pressure on the inlet of the F&T trap (minimum 1.5 psig) will always be greater than the pressure in the condensate return pipe (maximum 0.5 psig). This minimum 1 psi differential pressure across the trap will result in positive drainage of condensate through the trap.

- Sizing low-pressure traps based on the Steam Heating Equipment Manufacturers Association (SHEMA) ratings does not require any additional safety factor.

Steam Main Drips   Figure 6-15 indicates the locations of steam main drips in a typical steam system. Figure 6-16 shows the piping, valves, and specialties associated with an end of main drip (other main drips are similar).

Condensate Recovery

Condensate recovery equipment consists of two main parts: the receiver and the transfer pumps. Proper sizing of the receiver is required to prevent loss of condensate due to overflow and also to provide proper run time for the transfer pump. The HVAC system

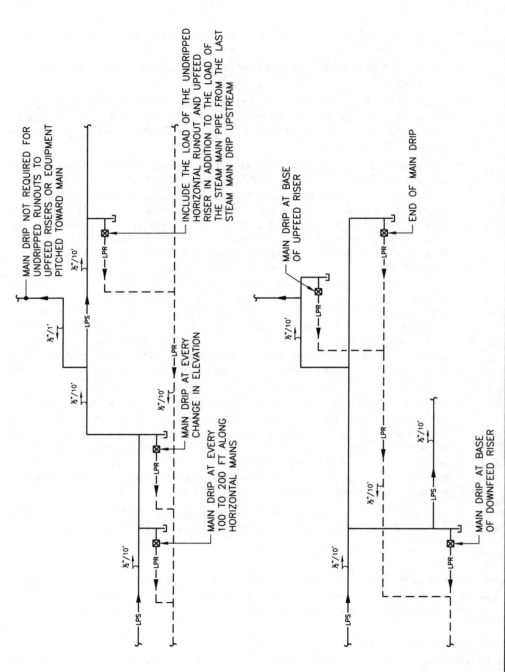

MAIN DRIP NOT REQUIRED FOR UNDRIPPED RUNOUTS TO UPFEED RISERS OR EQUIPMENT PITCHED TOWARD MAIN

INCLUDE THE LOAD OF THE UNDRIPPED HORIZONTAL RUNOUT AND UPFEED RISER IN ADDITION TO THE LOAD OF THE STEAM MAIN PIPE FROM THE LAST STEAM MAIN DRIP UPSTREAM

MAIN DRIP AT EVERY CHANGE IN ELEVATION

MAIN DRIP AT EVERY 100 TO 200 FT ALONG HORIZONTAL MAINS

MAIN DRIP AT BASE OF UPFEED RISER

END OF MAIN DRIP

MAIN DRIP AT BASE OF DOWNFEED RISER

Figure 6-15 Locations of steam main drips in a typical steam system.

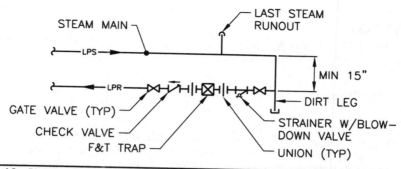

FIGURE 6-16   Piping, valves, and specialties associated with an end of main drip.

designer should contact a steam equipment manufacturer's representative for assistance in making a proper selection of the condensate recovery equipment. The following are some general sizing guidelines.

Condensate Transfer System   The condensate receiver should be sized for 1 minute of condensate storage at the design steam load. Steam condensate is generated at a rate of 1 gpm per 500 lb/h of steam flow. Thus, the condensate receiver should have 10 gal of storage for every 5,000 lb/h of design steam load.

The condensate transfer pump should be sized for two times the condensing rate. Thus, the condensate transfer pump should be sized for 20 gpm of condensate pumping capacity for every 5,000 lb/h of design steam load.

Boiler Feedwater System   The boiler feedwater receiver should be sized for 10 minutes of condensate storage at the design steam load. Thus, the boiler feedwater receiver should have 100 gal of storage for every 5,000 lb/h of design steam load.

The boiler feedwater pump should be sized for two times the condensing rate. Thus, the boiler feedwater pump should be sized for 20 gpm of condensate pumping capacity for every 5,000 lb/h of design steam load.

Refer to Chap. 4 for a further discussion of the boiler feedwater system.

Insulation

Insulation for hydronic and steam piping systems is typically rigid, molded fiberglass (also known as mineral-fiber) pipe insulation with a vapor barrier jacket. The insulation on pipe fittings is covered with a polyvinyl chloride (PVC) fitting covers. Typical insulation thicknesses are as follows:

- Chilled water supply and return piping—1½ in.
- Heating water supply and return piping—2 in.
- Low-pressure steam and condensate return piping—2 in.

Heat pump water piping systems are typically not insulated. Condenser water piping systems are also not insulated, unless the cooling tower is used for waterside economizer (refer to Chap. 4 for a brief discussion of waterside economizer operation). In this case, the condenser water piping would be insulated the same as chilled water piping.

Duct Systems

Ductwork is used to convey air between the air systems and the terminal equipment and spaces which these air systems serve.

Construction

Shapes

Common cross-sectional shapes for ductwork are rectangular, round, and flat oval. Rectangular ductwork is normally used for main ducts and runouts to air devices (air inlets and outlets) that have a rectangular neck. Round ductwork is normally used for runouts to variable air volume (VAV) terminal units or supply air devices, which commonly require round duct connections. Flat oval ductwork is not commonly used. However, when it is used, it is usually for portions of the duct systems that require a pressure classification exceeding 2 in. w.c.

Materials

Galvanized steel is the most common construction material for duct systems. The galvanized sheet steel should have a minimum galvanized coating of G60; however, G90 galvanized steel should be specified due to its greater corrosion resistance.[14] More costly materials such as stainless steel, aluminum, and carbon-steel have specific applications for duct systems. Stainless steel ductwork is often used for laboratory fume hood exhaust systems because of its corrosion resistance and the potential for corrosive chemical vapors to be conveyed through the duct system. Aluminum ductwork is not as resistant to corrosion from chemical vapors as stainless steel; however, it is suitable for use in damp environments and to convey moist air. For this reason, and because it is less costly than stainless steel, it is commonly used for HVAC systems serving indoor swimming pool areas (natatoriums) and exhaust systems serving commercial dishwashers and shower rooms. *NFPA Standard 96—Standard for Ventilation Control and Fire Protection of Commercial Cooking Operations* requires carbon-steel ductwork for exhaust systems serving kitchen exhaust hoods.

Duct systems are usually externally insulated. However, double-wall duct systems that consist of both outer and inner ducts separated by an insulating material are available. Sometimes the inner duct will be perforated for improved sound attenuation qualities. It is also common for portions of duct systems, especially those serving noise-sensitive areas, to be lined on the interior with 1 in. of sound attenuating/insulating material called duct liner.[15] Normally the length of ductwork that is lined on the interior is limited to the first 10 ft of supply and return air ductwork connected to air handling units and the first 5 ft of supply air ductwork downstream of VAV terminal units.

The final connections to supply air diffusers are often made with flexible round ductwork. Flexible ductwork consists of a wire-reinforced plastic inner duct surrounded by a plastic-faced layer of flexible fiberglass insulation. The use of flexible ductwork for the final connection to supply air devices simplifies the installation process and also provides a measure of sound attenuation for the supply air duct system. However, the length of flexible ductwork should be limited 8 ft because of the potential for excessive bending in the ductwork that could restrict the airflow through the duct.

Insulation

Duct systems that convey air that is cooler or warmer than room temperature are typically insulated with 1½ in. of foil-faced, flexible fiberglass blanket insulation. For duct systems conveying air that is cooler than room temperature, the foil facing should also be a vapor barrier to minimize the transfer of vapor from the environment through the insulation to the cold duct surface where it may condense. Insulation on ductwork installed in equipment rooms or other areas where it may be subject to damage is typically constructed of 1½-in. rigid fiberglass board insulation covered with a field-applied canvas jacket for increased durability. Insulation on supply and return air ductwork installed outdoors is typically constructed of 2-in. rigid fiberglass board insulation covered with a field-applied waterproof membrane.[16]

The supply and return air ductwork located within concealed spaces (e.g., above ceilings, within shafts) within commercial buildings is insulated to minimize the loss of heating and cooling energy from the conditioned air to the unconditioned spaces surrounding the ducts. One exception is that the return air ductwork installed within a ceiling space that is used as a return air plenum[17] is not required to be insulated because there is no temperature difference between the air within the ductwork and the surrounding environment. Outdoor air ductwork located anywhere within the building is also insulated to minimize the heat gains and losses from this ductwork to the indoor spaces and to prevent condensation that may occur on the duct surface at certain indoor and outdoor conditions. It is not necessary to insulate supply and return air ductwork that is installed within conditioned spaces of the building because, under normal conditions, the surface temperature of the supply air duct will be higher than the dew point temperature of the conditioned space and condensation will not form on the duct surface. Also, there is no temperature difference between the return air within the ductwork and the space temperature. Supply and return air ductwork installed outdoors, though it should be avoided wherever possible, will always be insulated to minimize the loss of heating and cooling energy from the conditioned air to the outdoor environment.

Pressure Classifications

Each portion of a duct system must be able to withstand the static pressure[18] of the air at that location exerted by the fan connected to the system. Thus, the strength of the ductwork is determined by the positive or negative static pressure that it must contain. The Sheet Metal and Air-Conditioning Contractors' National Association (SMACNA) publishes the *Duct Construction Standards—Metal and Flexible,* which defines the duct construction standards for various pressure classifications. The thickness (or gauge) of the sheet metal ductwork increases in proportion to its pressure classification and also increases in proportion to the ductwork cross-sectional dimensions. It may also be necessary for ductwork to be constructed with added reinforcement, such as steel angles mounted around the duct perimeter at specific intervals along the duct length, in order to obtain the required pressure classification.

The HVAC duct systems most commonly used in commercial buildings are low-pressure and medium-pressure duct systems. Low-pressure ductwork is commonly pressure classes that are 2 in. w.c. or less. Medium-pressure ductwork is commonly pressure classes that are 3 to 6 in. w.c.

The HVAC system designer must specify the duct pressure class required for all portions of the duct systems within the building based upon the estimated operating static pressure for each portion. This is typically written in the project specifications in a manner such as the following:

- Supply air ductwork between air handling units and VAV terminal units: 4 in. w.c. positive pressure class.

- Supply air ductwork downstream of VAV terminal units: 2 in. w.c. positive pressure class.

- Return and exhaust air ductwork: 2 in. w.c. negative pressure class.

Seams and Joints

Straight sections of rectangular ductwork are typically fabricated in a sheet metal shop in 4- or 5-ft-long sections. Each section is formed with the required cross-sectional dimensions from a flat sheet of duct material. This forming process results in a longitudinal seam in each section of ductwork, the construction of which depends upon the pressure class. The ends of each section are also prepared in the shop with the type of joint that is required by the pressure class. Refer to SMACNA's *Duct Construction Standards—Metal and Flexible* for details on the gauge of sheet metal, reinforcement, and types of seams and joints required to achieve a particular pressure classifications given a duct's cross-sectional dimensions.

Round ductwork can be formed in a manner similar to rectangular ductwork having a single longitudinal seam or, for increased strength with the same gauge of sheet metal, can be formed with a seam that spirals around the duct for the full length of the section. Flat oval ductwork[19] is formed in a similar manner as spiral-round ductwork.

All of the longitudinal seams, transverse joints, and duct wall penetrations (for takeoffs) in duct systems for commercial buildings should be sealed air tight. Sealing ductwork to this degree achieves a SMACNA Seal Class A. SMACNA Seal Class B omits sealing of duct wall penetrations, and Seal Class C further omits sealing of the longitudinal joints. In addition, duct systems for pressure classification greater than 2 in. w.c. should be leak-tested and proved air tight to standards established by SMACNA before the insulating systems are applied.

Duct Fittings

Duct fittings, such as transitions, elbows, tees, and branch fittings, are also shop-fabricated in accordance with the pressure class requirements of the ductwork. The most common fittings used for main rectangular ducts in commercial HVAC systems are concentric and eccentric transitions, radius elbows, square elbows with turning vanes, and tees constructed of back-to-back elbows. Fittings used for branch takeoffs include 45° entry rectangular fittings, bellmouth round or flat oval fittings, and flanged round or flat oval fittings. Figures 6-18 and 6-19 show the graphical representation of these commonly used duct fittings on a floor plan drawing. Other duct fittings that can be used are given in SMACNA's *Duct Construction Standards—Metal and Flexible*.

Duct Accessories

Flexible Duct Connectors   A flexible duct connector consists of a rubber-coated glass fabric that is about 4 in. in airway length with the same width and height dimensions as the ductwork to which it is connected. Flexible duct connectors should be used to connect ductwork to motor-driven equipment such as air handling units, fans, and fan-powered VAV terminal units. The purpose of flexible duct connectors is to prevent the vibrations that are generated within the motor-driven equipment from being transmitted through the duct system.

Dampers   Dampers perform various functions in HVAC duct systems. First, manual volume dampers are used to provide a means to balance the airflow to each branch within the duct system. Second, automatic dampers are used to modulate the airflow through certain sections of ductwork automatically. Finally, fire dampers, smoke dampers, and combination fire/smoke dampers are used to protect duct penetrations through fire-rated and/or smoke-tight building components, such as walls, floors, and shaft enclosures.

Manual Volume Dampers   The three types of manual volume dampers used most often in HVAC systems are butterfly dampers, opposed blade dampers, and parallel blade dampers. Butterfly dampers are commonly used in the flanged round branch takeoffs to supply air devices. Often, the flanged round takeoff fittings are equipped with an integral butterfly damper.

Opposed blade dampers are commonly used where mixing of two airstreams is required, such as in the mixing box for an air handling unit. Opposed blade dampers are also used where modulating control of the airflow through the damper is required. This is due to their linear performance. That is, there is a closer relationship between a change in the damper's position and a change in airflow for an opposed blade damper than for a parallel blade damper. Opposed blade dampers can be manually actuated, as is the case with balancing dampers installed within the duct system, and can also be automatically actuated, as is the case with the dampers in the mixing box for an air handling unit.

Parallel blade dampers can also be manually or automatically actuated. However, parallel blade dampers should not be used where mixing of two airstreams is required. The parallel blades can cause a stratification of the air within the downstream duct if the two airstreams are at significantly different temperatures. Parallel blade dampers also should not be used for modulating control of the airflow through the damper because they require a much higher pressure drop through the damper to exhibit the same linear response as opposed blade dampers. This higher pressure drop is less energy efficient and also generates noise. Thus, the only practical application for parallel blade dampers is in a two-position application where no mixing of two airstreams is required. An example of a parallel blade damper installation is in the open-end outdoor air or relief air duct within an equipment room.

The actuation of automatic dampers is most commonly performed with electric damper actuators. However, pneumatic damper actuators may also be used if the automatic temperature controls system within the building is pneumatic.

Backdraft Dampers   Backdraft dampers are similar to parallel blade manual volume dampers in that there are multiple blades within the damper that open and close. The difference is that the blades are weighted so that they allow airflow in one direction but not in the opposite direction. Thus, backdraft dampers are synonymous with check valves in a hydronic system. The weight of the blades can be adjusted to require more or less differential pressure across the damper to cause it to open.

Backdraft dampers are used in exhaust systems as a less costly option to motor-operated dampers to prevent outdoor air from infiltrating through the exhaust fan when it is off. They are also used in relief air systems to allow air to be relieved to the outdoors when there is a positive air pressure in the relief air system. It is common for the weighting of backdraft dampers used in relief air systems to be adjusted to require a differential pressure of 0.05 in. w.c. across the backdraft damper to open. The benefit of utilizing backdraft dampers in relief air systems is that they have a fairly constant pressure drop even with a variable airflow. This equates to a fairly constant building pressurization if the outdoor airflow to the building varies, such as is the case with the use of airside economizer for the air systems.

Fire Dampers, Smoke Dampers, and Combination Fire/Smoke Dampers   Some of the internal components of commercial buildings, such as walls, floors, and shaft enclosures, are required by the applicable building code to have a fire rating. That is, these components must be constructed as an assembly that is designed to resist the passage of fire for a certain period of time. These components are referred to as fire-resistance-rated assemblies, and their fire rating is given in terms of hours. In addition, some of the internal building components may be required to resist the passage of smoke. Smoke barriers, certain types of corridor enclosures, smoke partitions, and shaft enclosures are examples of smoke-tight building components. Both the fire-rated and smoke-tight components of a building are designed by the architect.

These components are important to the HVAC system designer because Sections 706 and 713 of the 2009 *International Building Code* and Section 607 of the 2009 *International Mechanical Code* define certain requirements for air transfer openings[20] and duct penetrations[21] in these components. Because these requirements are too numerous to list, we will discuss them only in general in this chapter. However, it is important that the HVAC system designer become thoroughly familiar with these code sections in order to properly apply the requirements to the HVAC system design. One notable exception to the requirements for fire, smoke, and combination fire/smoke dampers is that they are not to be installed in ducts where the closure of the damper(s) would interfere with the smoke control mode of operation of the HVAC system (if it is required to operate as a smoke control system).

Fire Dampers   A fire damper is a device that is mounted in an air transfer opening or duct penetration in a fire-resistance-rated assembly within a building. The purpose of a fire damper is to maintain the integrity of the fire-resistance-rated assembly by closing off the opening during a fire. Fire dampers consist of a spring-loaded shutter, the blades of which are folded like an accordion and are held by a retainer[22] and a fusible link. When the temperature exceeds the rating of the fusible link (usually 160°F), it melts, thereby releasing the retainer and causing the shutter to spring closed. The entire fire damper assembly, including its attachment to the building component and associated ductwork, if any, must comply with the requirements of UL 555. Fire dampers installed in air transfer openings or duct penetrations of fire-resistance-rated assemblies that have a fire rating that is less than 3 hours are required to have a 1½-hour fire rating. Fire dampers installed in air transfer openings or duct penetrations of fire-resistance-rated assemblies that have a fire rating that is 3 hours or more are required to have a 3-hour fire rating.

In general, fire dampers are required in air transfer openings in fire-resistance-rated assemblies that have a fire rating of 1 hour or more. Fire dampers are not required in duct penetrations of 1-hour fire-rated assemblies in fully sprinklered buildings. Fire dampers are required in duct penetrations of fire-resistance-rated assemblies that have a fire rating of 2 hours or more.

An access panel must be installed in ducts requiring fire dampers in order to provide a means by which the fire damper can be manually reset and a new fusible link installed if the damper closes.

Smoke Dampers   Smoke dampers are required in openings and duct penetrations of smoke-tight components in accordance with the applicable codes to prevent the passage of smoke through the openings or duct penetrations. Because smoke dampers are automatically actuated, they require an electrical connection. They are designed to close in response to a smoke detector which, for duct installations, must be installed in the duct within 5 ft of each smoke damper with no air outlets or inlets installed between the smoke detector and smoke damper. Installation of smoke dampers must comply with the requirements of UL 555S.

Smoke dampers must be labeled for use in dynamic systems if the HVAC system is required to operate during a fire as part of a smoke control system. Dynamic smoke dampers must be able to remain closed under the maximum differential static pressure that can exist across the damper while the fan is running. Because static smoke dampers are not able to remain closed while the fan is running, they are used in duct systems where the fan will be off during a fire. Smoke dampers must have a minimum leakage rating of Class II[23] and must have an elevated temperature rating of not less than 250°F. An access panel is normally installed in ducts requiring smoke dampers in order to provide a means by which the smoke damper can be inspected.

Combination Fire/Smoke Dampers   A combination fire/smoke damper is a UL 555S smoke damper that also has a fire rating and complies with the requirements of UL 555 for fire dampers. Thus, a combination fire/smoke damper requires an electrical connection and also closes in response to a smoke detector, similar to a smoke damper.

Combination fire/smoke dampers can be used wherever the applicable codes require the installation of both fire and smoke dampers. One notable location where both fire and smoke dampers are required is at air transfer openings and duct penetrations in shaft enclosures.

Sound Attenuators   Sound attenuators, also referred to as duct silencers, are used in ductwork distribution systems to attenuate (reduce) the noise generated by fans, VAV terminal units, and air turbulents within ducts. Dissipative silencers are the most common type of sound attenuator used for commercial buildings. They consist of perforated sheet metal baffles filled with sound-absorbing materials. Sound attenuators are normally 3 to 5 ft in airway length and are effective at air velocities up to about 2,000 fpm. Therefore, they can normally be selected with the same cross-sectional dimensions as the low-velocity or medium-velocity ductwork in which they are installed. The air pressure drop through sound attenuators needs to be included in the static pressure calculation for the duct system. A typical pressure drop through a 5-ft sound attenuator with an air velocity of 1,000 fpm is about 0.25 in. w.c.

Sound attenuators should be selected with the appropriate dynamic insertion loss[24] (DIL) in each octave band to achieve the desired room noise criteria (NC). A sound attenuator manufacturer's representative should be contacted for the selection of sound attenuators. Refer to Chap. 8 for a more detailed discussion of noise control measures.

Hangers

Refer to SMACNA's *Duct Construction Standards—Metal and Flexible* for details on the requirements for duct hangers, including the type and spacing of the hangers, which depend upon the duct size and pressure class.

General Design Guidelines

Ductwork Drawing Conventions

Ductwork is normally represented on floor plan drawings by single lines for ducts having widths up to 12 in. and for all ductwork downstream of VAV terminal units. Ducts having widths greater than 12 in. and all ductwork upstream of VAV terminal units are normally represented by double lines. For double-line ductwork drawings, the line work should represent the sheet metal dimensions, not the outside dimensions of the duct insulation. For the most part, double-line ductwork drawings present the ductwork as it actually appears. However, some symbols are used to show duct accessories, like dampers, in a schematic format.

Vertical portions of duct systems are denoted on the floor plan drawings by "duct up" and "duct down" symbols. These symbols can be understood by defining the viewing plane for a plan drawing as a horizontal plane located just below the underside of the deck that is above the ceiling of the floor being presented on the drawing. If the duct rises up through this plane, it is represented by a "duct up" symbol. If the duct drops down and does not rise up through this plane, it is represented by a "duct down" symbol. Furthermore, the rises and drops of the supply, return, and exhaust air duct systems are represented by different symbols. Refer to the sample legend and abbreviations in Chap. 10 for the supply, return, and exhaust air "duct up" and "duct down" symbols. These symbols should be shown in the legend for the drawing set.

Ductwork Layout

As with piping systems, duct systems should be laid out parallel and perpendicular to the building structure. This should be done utilizing the fewest number of bends and offsets in the vertical and horizontal planes necessary to connect the air systems to the terminal equipment and air devices in the spaces.

Some good design practices are as follows:

1. A duct size provides the cross-sectional dimensions of the duct.

2. Inch marks are normally omitted from the duct sizes shown on the drawings; thus, a 48" × 20" duct is notated as 48 × 20.

3. The first number in a duct size shown on a drawing indicates the cross-sectional dimension of the duct that is seen on the drawing. For example, a 48 × 20 duct shown on a floor plan drawing has a horizontal dimension of 48 in. and a vertical dimension of 20 in.

4. The first number in a duct size for a duct up or duct down indicates the cross-sectional dimension of the duct to which the dimension leader is pointing. For example, if a dimension leader points to the 20 in. side of a 48 × 20 duct rising up, the duct size would be notated as 20 × 48.

5. Even numbers are normally used for duct sizes, such as 10 × 12, 48 × 20, etc. Odd numbers for duct sizes, such as 11 × 13, are rarely used.

6. The aspect ratio (ratio of long dimension to short dimension) of ductwork should be kept at 4:1 or less. If it's not, reinforcement of the ductwork may be required to prevent the ductwork from "oil-canning" (producing that hollow metal sound that sheet metal makes when it is bent).

7. Ductwork should transition and offset gradually to minimize static pressure losses within the duct system.
   a. Transitions should have a maximum 30° combined angle wherever possible.
   b. Offsets in ductwork should be made with two 45° elbows instead of two 90° elbows.

8. Coordination of ductwork with other services above ceilings is a paramount concern. Since the ductwork is normally the largest component within a ceiling space, it should be located first and mounted as high as possible. The other services should be coordinated with the ductwork as required. Refer to the discussion of the vertical space above ceilings in Chap. 2.

9. A basic understanding of the terminology used in architectural and structural drawings is necessary to determine the space available below the building structure and above the finished ceiling for routing ductwork and other HVAC systems.
   a. Finished floor elevations[25] (FF EL) are shown on the architectural drawings and are indicated in terms of feet above sea level with two decimal places of precision (e.g., 261.00'). Sometimes a floor of a building is given a reference finished floor elevation of 0'-0" and the finished floor elevations of the other floors in the building are expressed in terms of that datum. For example, if the first floor in a building is given the reference finished floor elevation of 0'-0", the basement finished floor elevation would be indicated as a negative number with respect to the first floor datum (such as −15'-0") and the second floor finished floor elevation would be indicated as a positive number with respect to the first floor datum (such as +13'-0").
   b. The elevation of the underside of finished ceilings is given on the architectural reflected ceiling plans[26] in terms of the elevation above finished floor (AFF) (e.g., 9'-0" AFF).
   c. The elements on the reflected ceiling plan are drawn as if you were located above the ceiling looking down through the ceiling elements or, more descriptively, as if the floor were a mirror and you were viewing the reflection of the ceiling elements in the mirror.
   d. The top elevations of the structural steel members are shown on the structural drawings. The top of steel (TOS) elevations are indicated in terms of feet above sea level with two decimal places of precision (e.g., 260.50').
   e. Wide flange steel beams are identified with a capital W followed by two dimensions: the first is the height of the beam in inches and the second is the weight of the beam in pounds per linear foot. For example, a W8 × 31 steel beam is 8 in. high and weighs 31 lb/linear foot. The height of steel beams is the important information for the HVAC system designer.
   f. The first number in the designation of a structural steel open web joist (also called a bar joist) is its height. For example, a 14H4 bar joist is 14 in. high.

10. An allowance must be made for the usual 1½ in. of external insulation on ductwork and the appropriate thickness of insulation on other systems, such as piping, when coordinating ductwork with other services above ceilings. The space between systems above ceilings should usually be at least 1 in.

11. The height of branch duct takeoffs should be 2 in. less than the height of the main duct to which they are connected. This 2 in. dimension allows for connection of the 1-in. flange on the branch takeoff fitting to the main duct.

12. Ductwork with a height of up to 14 in. (sheet metal dimension) can normally be accommodated above ceilings without special coordination, provided a reasonable amount of ceiling space has been allowed by the architect. A 14-in.-high duct can handle a significant airflow at medium and low velocities and allows for a 12-in. branch takeoff to VAV terminal units or air devices. If main duct sizes require a height greater than 14 in. or if the ceiling space is unusually tight, special coordination with the architectural and structural components, as well as other systems above the ceilings, will be required. In some cases, it may be necessary to request that the architect lower the ceilings or design bulkheads in certain areas to accommodate large main ducts.

13. "U-turns" in ductwork should be avoided except in very rare cases where an unusual building feature must be accommodated. There will almost always be a more direct method of designing the ductwork distribution systems that does not utilize U-turns.

14. Both 90° elbows and 45° elbows for rectangular ductwork should either be radius elbows having a centerline radius equal to 1.5 times the duct width or square elbows equipped with turning vanes. Square elbows with turning vanes are often used where there is insufficient space to accommodate the airway length of radius elbows.

15. Radius elbows are always used for round ductwork.

16. It is common for the entering and leaving cross-sectional dimensions of duct elbows to be equal. However, it is possible for the entering and leaving width dimensions of a rectangular elbow to be unequal, but the entering and leaving height dimensions must remain equal. For example, a duct elbow may have an entering dimension (as shown on a floor plan drawing) of 24 × 16 and a leaving dimension of 20 × 16. Duct elbows having unequal entering and leaving duct widths are most often used in duct tees constructed of back-to-back duct elbows.

17. Duct tees consisting of one inlet and two outlets are normally constructed of back-to-back square elbows with turning vanes. This configuration is used to minimize the space requirement of the tee. However, back-to-back radius elbows can also be used for duct tees if space is available.

18. Duct tees consisting of two inlets and one outlet, called *bullhead tees*, should not be used. The two inlets should be separated by at least five duct diameters[27] of straight duct length in order to allow a uniform air velocity profile to develop between the two inlets.

19. A minimum of 6 in. clearance should be allowed between ductwork risers and other systems within shafts to accommodate ductwork flanges (if required) and insulation on the ductwork and other systems. A minimum clearance of 3 in.

should be allowed between the ductwork risers and the inside face of the shaft wall to accommodate the ductwork flanges and insulation. Refer to the sample shaft layout (Fig. 2-3) in Chap. 2.

20. Where it is necessary for a horizontal main duct to connect to a vertical duct riser within a shaft enclosure, most often a fire damper and smoke damper (or combination fire/smoke damper) will be required at the duct penetration of the shaft enclosure. An access door must be designed in the duct on the accessible side of the shaft enclosure and adequate space must be allowed below the access door in the duct for inspection and maintenance of the damper.

21. Manual volume dampers are required at all major branches in duct systems to provide a means for balancing the airflow of the HVAC system. However, manual volume dampers are not used in medium-velocity systems between air handling units and VAV terminal units.

22. Bellmouth round or flat oval fittings are normally used where the round runouts to VAV terminal units connect to the main ducts because bellmouth fittings (which have a radius corner at the connection to the main duct) have a lower static pressure drop than flanged fittings (which have a sharp corner at the connection to the main duct).

23. A minimum length of three duct diameters of straight duct should be designed for the inlet ductwork connected to VAV terminal units. This straight length of ductwork ensures a fairly uniform velocity profile at the inlet of the VAV terminal unit where the inlet velocity sensor is located and results in a more accurate airflow measurement.

24. Manual volume dampers are required at all duct runouts to air devices. Good practice is to design at least 5 ft of ductwork between manual volume dampers and air devices. This length of ductwork will help to attenuate any air noise that may be generated by the damper as it throttles the airflow.

25. Flanged round fittings with integral volume dampers are normally used where the round runouts to supply air diffusers connect to the main ducts.

26. Flexible round ductwork should be used above suspended ceilings or other accessible locations to make the final connections to supply air devices with round necks. The length of flexible ductwork should be limited to 8 ft.

27. Flexible ductwork should not be used above ceilings constructed of gypsum board or in other inaccessible locations. The connection of flexible ductwork to rigid ductwork (or the necks of air devices) is made with a nylon draw band; this band may become disconnected over time. Rigid ductwork, which is connected with sheet metal screws, makes a more permanent joint and should be used exclusively in inaccessible locations.

28. Air devices with integral volume dampers mounted in the neck of the device are not recommended unless these dampers are manually adjustable by the building occupants and are used for fine adjustments to the airflow only. If this is the case, manual volume dampers should also be installed at the connections of the duct runouts to the main ducts for use in performing the initial system air balance. Excessive air noise within the spaces may occur if the integral volume dampers mounted in the necks of the air devices are used for the initial system air balance.

29. Duct takeoffs to air devices are normally connected to the sides of main ducts. However, top and bottom connections to main ducts can also be utilized if necessary.

30. Typically a 12-in. "air cushion" is designed at the ends of main ducts serving air devices; that is, the last side takeoff to an air device is made 12 in. from the end of the duct. It is not customary to design a branch takeoff on the end of a main duct.

31. Elbows in close proximity to the discharge of fans or air handling units can create a condition called fan system effect, which results in an increased static pressure drop for that elbow. In order to minimize the fan system effect, a minimum straight duct length of three to five duct diameters should be designed between the fan discharge and the first elbow.[28] Also, the first elbow should turn the discharge air in the direction of the fan rotation. If the recommended length of straight duct cannot be accommodated, the duct elbow must still turn the discharge air in the direction of the fan rotation. However, a fan system effect factor must be applied to the normal pressure drop of the elbow to account for fan system effect.

32. A duct elbow within five duct diameters of the fan discharge that turns the discharge air in the opposite direction of the fan rotation should never be designed. This is referred to as "breaking the back" of the fan and results in an excessively high static pressure drop and noise in the duct system at that elbow. This configuration can be avoided by specifying a fan with an opposite rotation.

33. Elbows should be avoided at the inlet connection to centrifugal fans. Again, a minimum straight duct length of three to five duct diameters is recommended.

34. Ductwork should never penetrate egress stairwells. In fact, it is good practice to keep ductwork out of stairwells altogether.

35. Ductwork must not be installed above electrical panelboards or through electric or data closets. The vertical column above (and below) electrical panelboards is defined by the *National Electric Code* as dedicated electrical space and is reserved for the installation of electrical conduit and wiring only.

Figures 6-17 through 6-20 show the graphical representation of ductwork and commonly used duct fittings and accessories on a floor plan drawing.[29] Figure 6-21 shows the construction of a duct tee utilizing square elbows with turning vanes and radius elbows.

Penetrations of Building Components

All ductwork penetrations of interior walls require sleeves. A duct sleeve is a short section of galvanized sheet metal that is embedded in the wall. It is large enough to allow the duct to pass through without interruption to the duct insulation (if required). Sleeves for ducts penetrating nonfire-resistance-rated assemblies are used to isolate the ducts from the building components and to provide a finished appearance. Sleeves for duct penetrations that require fire, smoke, or combination fire/smoke dampers are an integral part of the UL-listed assembly. Duct penetrations through exterior walls and roofs must be waterproofed.

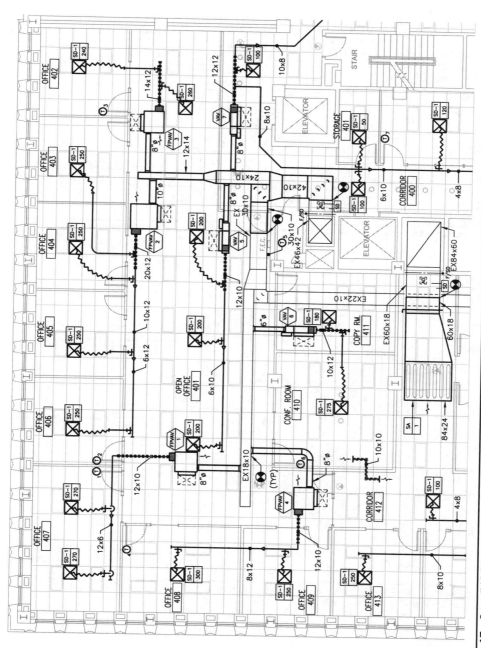

Figure 6-17 Overall HVAC plan.

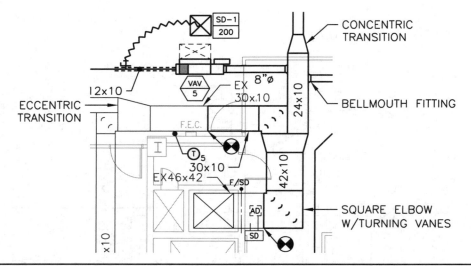

Figure 6-18 Enlarged area showing duct transitions, elbows, and bellmouth fittings.

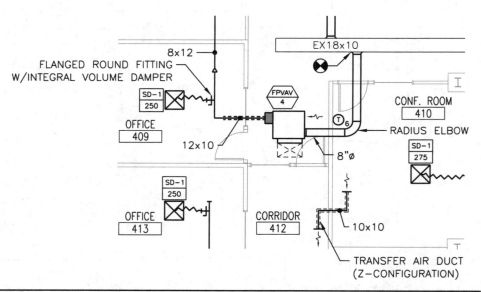

Figure 6-19 Enlarged area showing a radius elbow, flanged round fitting with integral volume damper, and transfer air duct.

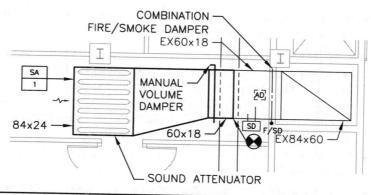

FIGURE 6-20 Enlarged area showing a sound attenuator, manual volume damper, and combination fire/smoke damper.

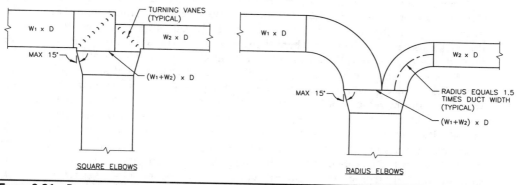

FIGURE 6-21 Duct tee detail.

Duct Sizing Criteria

Ductwork for commercial buildings is most commonly sized using the equal friction sizing method. This method maintains a constant static pressure drop per unit length throughout the entire duct system. The static regain method of sizing ductwork is another method that can be employed. However, this procedure is somewhat involved and is best suited to the use of ductwork design software for its application. We will not discuss the static regain method of sizing ductwork in this book.

When employing the equal friction method of sizing ductwork, it is first necessary to establish the static pressure drop per unit length criterion that will be used [expressed in terms of inches of water column (in. w.c.) per 100 feet of ductwork] and the maximum air velocity [expressed in terms of feet per minute (fpm)] within the ductwork that will be allowed. We will establish these criteria for both low-velocity duct systems and medium-velocity duct systems. We will define low-velocity duct systems as those having a maximum air velocity within the ductwork of 1,200 fpm. Medium-velocity duct systems are those having an air velocity within the ductwork that is between 1,200 and 2,500 fpm.

Low-Velocity Duct Systems   Low-velocity duct systems are typically utilized for the following air systems or portions of air systems:

- Any air system that does not have any terminal units or sound attenuators between the air handling unit and the air devices within the conditioned spaces. This typically applies to constant air volume systems.
- The portion of the duct system between the terminal units or sound attenuators and the air devices within the conditioned spaces for VAV or other air systems that utilize terminal units or sound attenuators.
- Return air duct systems between the air devices in the conditioned spaces and the air handling unit.
- Outdoor and relief air duct systems between the air handling unit and the outdoor air inlet or relief air outlet within the building envelope.

Low-velocity duct systems are used for these applications because the noise generated by the air turbulents within the ductwork for medium-velocity duct systems will be sufficiently loud enough to create objectionable noise within the conditioned spaces if there is no sound attenuating device, such as a VAV terminal unit or sound attenuator, inserted between these air turbulents and the air devices. In other words, medium-velocity duct systems can only be used where there is a VAV terminal unit or sound attenuator located between the conditioned spaces and the medium-velocity duct system.

A common sizing criterion for low-velocity duct systems that are installed above the ceilings of occupied areas is 0.08 in. w.c. static pressure drop per 100 ft of ductwork with a maximum velocity of 1,000 fpm.[30] For ductwork installed in unoccupied areas or within shafts, the same static pressure drop criterion should be used, but the maximum air velocity can be increased to 1,200 fpm. Table 6-9 lists the airflow capacities for even-sized ducts using the 0.08 in. w.c. static pressure drop per 100 ft, with a maximum velocity of 1,000 fpm criteria. Note that the static pressure drop criterion is used for airflows up to 1,534 cfm. For airflows exceeding this value, the maximum velocity criterion is used. Duct sizing calculators and software programs are available to assist in sizing ductwork if different criterion is desired.

Medium-Velocity Duct Systems (Upstream of VAV Terminal Units)   For VAV systems and any other system utilizing terminal units or sound attenuators between the air handling units and the air devices within the conditioned spaces, the ductwork between air handling unit and the terminal units or sound attenuators is normally sized for medium-velocity airflow. The sizing criterion typically used for medium-velocity ductwork is 0.30 in. w.c. static pressure drop per 100 ft of ductwork, with a maximum air velocity of 2,000 fpm. This sizing criterion allows for smaller duct sizes than the typical low-velocity criterion of 0.08 in. w.c. static pressure drop per 100 ft, with a maximum air velocity of 1,000 fpm. However, the result is that the static pressure within the medium-velocity ductwork typically exceeds the low-pressure ductwork classification limit of 2 in. w.c. Therefore, the supply air ductwork upstream of VAV terminal units is normally constructed to medium-pressure duct construction standards (3 to 6 in. w.c.). Table 6-10 lists the airflow capacities for even-sized ducts using the 0.30 in. w.c. static pressure drop per 100 ft, with a maximum velocity of 2,000 fpm criterion. Note that the static pressure drop criterion is used for airflows up to 3,086 cfm. For airflows exceeding this value, the maximum-velocity criterion is used. Duct sizing calculators and software programs are available to assist in sizing ductwork if different criterion is desired.

LOW VELOCITY DUCTWORK: 0.08 IN. S.P. DROP PER 100 FT; 1,000 FPM MAXIMUM VELOCITY

(DOWNSTREAM OF TERMINAL UNITS)

	4	6	8	10	12	14	16	18	20	22	24	26	28	30	32	34	36	38	40	42	44	46	48
4	43	72	103	134	166	198	230	262	294	327	359	391	424	456	489	521	554	586	619	651	683	716	748
6	72	125	181	239	299	360	421	483	545	607	669	732	795	857	920	983	1,046	1,109	1,172	1,235	1,298	1,361	1,424
8	103	181	266	356	448	543	639	736	833	932	1,031	1,130	1,229	1,329	1,429	1,530	1,630	1,725	1,801	1,875	1,949	2,022	2,094
10	134	239	356	479	608	740	875	1,012	1,150	1,290	1,430	1,572	1,713	1,820	1,925	2,029	2,131	2,232	2,332	2,430	2,528	2,624	2,719
12	166	299	448	608	775	948	1,125	1,306	1,489	1,674	1,823	1,961	2,097	2,230	2,362	2,492	2,620	2,747	2,872	2,995	3,118	3,238	3,358
14	198	360	543	740	948	1,164	1,386	1,613	1,811	1,982	2,151	2,317	2,481	2,642	2,801	2,958	3,113	3,266	3,417	3,566	3,714	3,861	4,005
16	230	421	639	875	1,125	1,386	1,656	1,875	2,079	2,280	2,477	2,672	2,864	3,053	3,240	3,425	3,607	3,787	3,965	4,141	4,315	4,488	4,658
18	262	483	736	1,012	1,306	1,613	1,875	2,112	2,345	2,575	2,801	3,025	3,245	3,463	3,678	3,891	4,101	4,309	4,514	4,717	4,918	5,118	5,315
20	294	545	833	1,150	1,489	1,811	2,079	2,345	2,607	2,866	3,122	3,375	3,624	3,871	4,115	4,356	4,594	4,830	5,063	5,294	5,522	5,749	5,973
22	327	607	932	1,290	1,674	1,982	2,280	2,575	2,866	3,155	3,440	3,722	4,000	4,276	4,549	4,818	5,085	5,349	5,611	5,870	6,126	6,380	6,632
24	359	669	1,031	1,430	1,823	2,151	2,477	2,801	3,122	3,440	3,754	4,065	4,373	4,678	4,980	5,279	5,574	5,867	6,157	6,444	6,729	7,011	7,291
26	391	732	1,130	1,572	1,961	2,317	2,672	3,025	3,375	3,722	4,065	4,406	4,743	5,077	5,408	5,736	6,061	6,382	6,701	7,017	7,330	7,640	7,948
28	424	795	1,229	1,713	2,097	2,481	2,864	3,245	3,624	4,000	4,373	4,743	5,110	5,473	5,833	6,190	6,544	6,895	7,242	7,587	7,929	8,268	8,604
30	456	857	1,329	1,820	2,230	2,642	3,053	3,463	3,871	4,276	4,678	5,077	5,473	5,866	6,255	6,642	7,025	7,404	7,781	8,155	8,525	8,893	9,258
32	489	920	1,429	1,925	2,362	2,801	3,240	3,678	4,115	4,549	4,980	5,408	5,833	6,255	6,674	7,090	7,502	7,911	8,317	8,720	9,119	9,516	9,910
34	521	983	1,530	2,029	2,492	2,958	3,425	3,891	4,356	4,818	5,279	5,736	6,190	6,642	7,090	7,535	7,976	8,414	8,850	9,281	9,710	10,136	10,559
36	554	1,046	1,630	2,131	2,620	3,113	3,607	4,101	4,594	5,085	5,574	6,061	6,544	7,025	7,502	7,976	8,447	8,915	9,379	9,840	10,298	10,753	11,205
38	586	1,109	1,725	2,232	2,747	3,266	3,787	4,309	4,830	5,349	5,867	6,382	6,895	7,404	7,911	8,414	8,915	9,412	9,905	10,396	10,883	11,367	11,848
40	619	1,172	1,801	2,332	2,872	3,417	3,965	4,514	5,063	5,611	6,157	6,701	7,242	7,781	8,317	8,850	9,379	9,905	10,428	10,948	11,465	11,978	12,488
42	651	1,235	1,875	2,430	2,995	3,566	4,141	4,717	5,294	5,870	6,444	7,017	7,587	8,155	8,720	9,281	9,840	10,396	10,948	11,497	12,043	12,586	13,125
44	683	1,298	1,949	2,528	3,118	3,714	4,315	4,918	5,522	6,126	6,729	7,330	7,929	8,525	9,119	9,710	10,298	10,883	11,465	12,043	12,618	13,190	13,759
46	716	1,361	2,022	2,624	3,238	3,861	4,488	5,118	5,749	6,380	7,011	7,640	8,268	8,893	9,516	10,136	10,753	11,367	11,978	12,586	13,190	13,792	14,390
48	748	1,424	2,094	2,719	3,358	4,005	4,658	5,315	5,973	6,632	7,291	7,948	8,604	9,258	9,910	10,559	11,205	11,848	12,488	13,125	13,759	14,390	15,017
50	781	1,487	2,165	2,814	3,476	4,149	4,827	5,510	6,195	6,882	7,568	8,253	8,938	9,620	10,300	10,978	11,653	12,326	12,995	13,661	14,324	14,984	15,641
52	813	1,550	2,236	2,907	3,593	4,290	4,995	5,704	6,416	7,129	7,843	8,556	9,269	9,980	10,688	11,395	12,099	12,800	13,499	14,194	14,887	15,576	16,262
54	846	1,613	2,306	2,999	3,709	4,431	5,161	5,896	6,634	7,374	8,116	8,857	9,597	10,336	11,074	11,809	12,542	13,272	14,000	14,724	15,446	16,164	16,879
56	878	1,676	2,375	3,091	3,824	4,570	5,325	6,086	6,851	7,618	8,386	9,155	9,923	10,691	11,456	12,220	12,982	13,741	14,497	15,251	16,002	16,749	17,494
58	910	1,732	2,443	3,182	3,938	4,709	5,488	6,275	7,065	7,859	8,655	9,451	10,247	11,042	11,836	12,629	13,419	14,207	14,992	15,775	16,554	17,331	18,105
60	943	1,779	2,511	3,272	4,051	4,846	5,650	6,462	7,279	8,099	8,921	9,745	10,568	11,392	12,214	13,034	13,853	14,670	15,484	16,295	17,104	17,910	18,713
62	975	1,826	2,579	3,361	4,164	4,981	5,810	6,647	7,490	8,337	9,186	10,036	10,888	11,738	12,589	13,437	14,285	15,130	15,973	16,813	17,651	18,486	19,318
64	1,007	1,873	2,646	3,449	4,275	5,116	5,970	6,832	7,700	8,573	9,448	10,326	11,204	12,083	12,961	13,838	14,714	15,587	16,459	17,328	18,195	19,059	19,920
66	1,040	1,919	2,712	3,537	4,385	5,250	6,128	7,014	7,908	8,807	9,709	10,614	11,519	12,425	13,331	14,236	15,140	16,042	16,942	17,840	18,735	19,628	20,518
68	1,072	1,965	2,778	3,624	4,495	5,383	6,284	7,196	8,115	9,040	9,968	10,899	11,832	12,766	13,699	14,632	15,564	16,494	17,423	18,349	19,273	20,195	21,114
70	1,104	2,010	2,843	3,711	4,603	5,515	6,440	7,376	8,321	9,271	10,226	11,183	12,143	13,104	14,064	15,025	15,985	16,944	17,900	18,856	19,808	20,759	21,707
72	1,137	2,056	2,908	3,796	4,711	5,645	6,595	7,555	8,525	9,500	10,481	11,465	12,452	13,439	14,428	15,416	16,404	17,391	18,376	19,359	20,341	21,320	22,297
74	1,169	2,100	2,972	3,882	4,818	5,775	6,748	7,731	8,727	9,729	10,735	11,745	12,759	13,773	14,789	15,805	16,820	17,835	18,848	19,860	20,870	21,878	22,884
76	1,201	2,145	3,036	3,966	4,925	5,904	6,901	7,910	8,929	9,955	10,987	12,024	13,064	14,105	15,148	16,191	17,235	18,277	19,319	20,359	21,397	22,434	23,468
78	1,234	2,189	3,100	4,050	5,030	6,033	7,052	8,085	9,129	10,180	11,238	12,301	13,367	14,435	15,505	16,576	17,646	18,717	19,786	20,855	21,921	22,986	24,049
80	1,266	2,233	3,163	4,134	5,135	6,160	7,203	8,260	9,327	10,404	11,487	12,576	13,668	14,763	15,860	16,958	18,056	19,154	20,252	21,348	22,443	23,536	24,628
82	1,298	2,277	3,226	4,217	5,240	6,286	7,352	8,433	9,525	10,627	11,735	12,849	13,968	15,089	16,213	17,338	18,464	19,589	20,715	21,839	22,962	24,084	25,204
84	1,331	2,320	3,288	4,299	5,343	6,412	7,501	8,605	9,722	10,848	11,982	13,121	14,266	15,414	16,564	17,716	18,869	20,022	21,175	22,327	23,479	24,629	25,777
86	1,363	2,363	3,350	4,381	5,446	6,537	7,649	8,776	9,917	11,068	12,227	13,392	14,562	15,737	16,913	18,092	19,272	20,453	21,633	22,814	23,993	25,171	26,347
88	1,395	2,406	3,411	4,463	5,549	6,662	7,796	8,947	10,111	11,286	12,470	13,661	14,857	16,057	17,261	18,467	19,674	20,881	22,090	23,297	24,505	25,711	26,915
90	1,427	2,449	3,473	4,544	5,651	6,785	7,942	9,116	10,304	11,504	12,712	13,929	15,150	16,377	17,607	18,839	20,073	21,308	22,543	23,779	25,014	26,248	27,481
92	1,460	2,491	3,534	4,624	5,752	6,908	8,087	9,284	10,496	11,720	12,953	14,195	15,442	16,694	17,950	19,209	20,470	21,732	22,995	24,258	25,521	26,783	28,044
94	1,492	2,533	3,594	4,704	5,853	7,030	8,231	9,452	10,687	11,935	13,193	14,459	15,732	17,010	18,293	19,578	20,866	22,155	23,445	24,735	26,026	27,315	28,604
96	1,524	2,575	3,654	4,784	5,953	7,152	8,375	9,618	10,877	12,149	13,432	14,723	16,021	17,325	18,633	19,945	21,259	22,575	23,892	25,210	26,528	27,846	29,162

ROUND DUCTS

	4	6	8	10	12	14	16	18	20	22	24	26	28	30	32	34	36	38	40	42	44	46	48
	34	99	210	379	613	921	1,309	1,767	2,182	2,640	3,142	3,687	4,276	4,909	5,585	6,305	7,069	7,876	8,727	9,621	10,559	11,541	12,566

DUCTS HANDLING AIR QUANTITIES GREATER THAN 1,534 CFM ARE SIZED USING THE MAXIMUM VELOCITY CRITERIA

DUCTS HAVING AN ASPECT RATIO GREATER THAN 4:1 REQUIRE ADDITIONAL REINFORCEMENT OR HEAVIER GAUGE METAL

NOTE: DUCTS HANDLING AIR QUANTITIES GREATER THAN 1,534 CFM ARE SIZED USING THE MAXIMUM VELOCITY CRITERIA

TABLE 6-9 Airflow Capacities for Even-Sized Ducts Using 0.08 in. w.c. Static Pressure Drop per 100 ft, 1,000 fpm Maximum Velocity

MEDIUM VELOCITY DUCTWORK: 0.30 IN. S.P. DROP PER 100 FT; 2,000 FPM MAXIMUM VELOCITY (UPSTREAM OF TERMINAL UNITS)

	4	6	8	10	12	14	16	18	20	22	24	26	28	30	32	34	36	38	40	42	44	46	48
4	85	143	204	267	330	393	457	521	586	650	714	779	844	908	973	1,037	1,102	1,166	1,231	1,295	1,360	1,425	1,489
6	143	248	360	476	595	716	838	960	1,084	1,208	1,332	1,456	1,581	1,706	1,831	1,956	2,081	2,207	2,332	2,457	2,583	2,708	2,834
8	204	360	530	708	892	1,080	1,271	1,464	1,658	1,854	2,051	2,248	2,446	2,645	2,844	3,044	3,243	3,443	3,601	3,750	3,898	4,043	4,188
10	267	476	708	954	1,210	1,473	1,741	2,013	2,289	2,567	2,846	3,128	3,410	3,639	3,850	4,058	4,262	4,464	4,664	4,861	5,055	5,248	5,439
12	330	595	892	1,210	1,543	1,887	2,239	2,598	2,962	3,331	3,645	3,921	4,193	4,461	4,724	4,984	5,241	5,494	5,744	5,991	6,235	6,477	6,716
14	393	716	1,080	1,473	1,887	2,317	2,759	3,210	3,621	3,964	4,302	4,634	4,962	5,284	5,602	5,916	6,226	6,532	6,834	7,133	7,429	7,721	8,010
16	457	838	1,271	1,741	2,239	2,759	3,295	3,751	4,158	4,560	4,955	5,344	5,728	6,107	6,481	6,849	7,214	7,574	7,930	8,282	8,630	8,975	9,317
18	521	960	1,464	2,013	2,598	3,210	3,751	4,224	4,690	5,149	5,602	6,050	6,491	6,927	7,357	7,782	8,202	8,617	9,028	9,435	9,837	10,235	10,630
20	586	1,084	1,658	2,289	2,962	3,621	4,158	4,690	5,214	5,732	6,244	6,749	7,249	7,742	8,229	8,711	9,188	9,659	10,126	10,588	11,045	11,498	11,946
22	650	1,208	1,854	2,567	3,331	3,964	4,560	5,149	5,732	6,300	6,880	7,443	8,001	8,552	9,097	9,637	10,170	10,699	11,222	11,739	12,252	12,760	13,264
24	714	1,332	2,051	2,846	3,645	4,302	4,955	5,602	6,244	6,880	7,508	8,131	8,747	9,356	9,960	10,557	11,148	11,734	12,314	12,888	13,458	14,022	14,581
26	779	1,456	2,248	3,128	3,921	4,634	5,344	6,050	6,749	7,443	8,131	8,812	9,487	10,155	10,816	11,472	12,121	12,764	13,402	14,034	14,660	15,281	15,896
28	844	1,581	2,446	3,410	4,193	4,962	5,728	6,491	7,249	8,001	8,747	9,487	10,221	10,947	11,667	12,381	13,088	13,790	14,485	15,174	15,858	16,536	17,208
30	908	1,706	2,645	3,639	4,461	5,284	6,107	6,927	7,742	8,552	9,356	10,155	10,947	11,732	12,511	13,283	14,049	14,809	15,562	16,310	17,051	17,786	18,516
32	973	1,831	2,844	3,850	4,724	5,602	6,481	7,357	8,229	9,097	9,960	10,816	11,667	12,511	13,348	14,179	15,004	15,822	16,634	17,439	18,239	19,032	19,819
34	1,037	1,956	3,044	4,058	4,984	5,916	6,849	7,782	8,711	9,637	10,557	11,472	12,381	13,283	14,179	15,069	15,952	16,829	17,699	18,563	19,421	20,272	21,117
36	1,102	2,081	3,243	4,262	5,241	6,226	7,214	8,202	9,188	10,170	11,148	12,121	13,088	14,049	15,004	15,952	16,894	17,829	18,758	19,681	20,596	21,506	22,410
38	1,166	2,207	3,443	4,464	5,494	6,532	7,574	8,617	9,659	10,699	11,734	12,764	13,790	14,809	15,822	16,829	17,829	18,823	19,811	20,792	21,766	22,734	23,696
40	1,231	2,332	3,601	4,664	5,744	6,834	7,930	9,028	10,126	11,222	12,314	13,402	14,485	15,562	16,634	17,699	18,758	19,811	20,857	21,896	22,930	23,956	24,976
42	1,295	2,457	3,750	4,861	5,991	7,133	8,282	9,435	10,588	11,739	12,888	14,034	15,174	16,310	17,439	18,563	19,681	20,792	21,896	22,995	24,086	25,172	26,250
44	1,360	2,583	3,898	5,055	6,235	7,429	8,630	9,837	11,045	12,252	13,458	14,660	15,858	17,051	18,239	19,421	20,596	21,766	22,930	24,086	25,237	26,381	27,518
46	1,425	2,708	4,043	5,248	6,477	7,721	8,975	10,235	11,498	12,760	14,022	15,281	16,536	17,786	19,032	20,272	21,506	22,734	23,956	25,172	26,381	27,583	28,779
48	1,489	2,834	4,188	5,439	6,716	8,010	9,317	10,630	11,946	13,264	14,581	15,896	17,208	18,516	19,819	21,117	22,410	23,696	24,976	26,250	27,518	28,779	30,034
50	1,554	2,959	4,330	5,627	6,952	8,297	9,655	11,020	12,391	13,763	15,136	16,507	17,875	19,240	20,601	21,957	23,307	24,652	25,990	27,323	28,649	29,969	31,282
52	1,618	3,084	4,471	5,814	7,187	8,581	9,989	11,407	12,831	14,258	15,686	17,113	18,537	19,959	21,377	22,790	24,198	25,601	26,998	28,389	29,773	31,152	32,524
54	1,682	3,210	4,611	5,999	7,419	8,862	10,321	11,791	13,268	14,749	16,231	17,714	19,194	20,673	22,147	23,616	25,084	26,544	27,999	29,448	30,891	32,328	33,759
56	1,747	3,335	4,750	6,182	7,649	9,141	10,650	12,172	13,701	15,236	16,773	18,310	19,847	21,381	22,913	24,440	25,963	27,482	28,995	30,502	32,003	33,499	34,988
58	1,811	3,461	4,887	6,363	7,877	9,411	10,976	12,549	14,131	15,719	17,310	18,902	20,494	22,085	23,673	25,257	26,838	28,413	29,984	31,549	33,109	34,662	36,210
60	1,876	3,558	5,023	6,543	8,103	9,691	11,300	12,923	14,557	16,198	17,843	19,490	21,137	22,783	24,427	26,069	27,706	29,339	30,968	32,591	34,208	35,820	37,426
62	1,940	3,652	5,158	6,722	8,327	9,963	11,621	13,295	14,980	16,673	18,372	20,073	21,775	23,477	25,177	26,875	28,569	30,260	31,945	33,626	35,302	36,972	38,636
64	2,005	3,745	5,291	6,898	8,549	10,232	11,939	13,663	15,400	17,145	18,897	20,652	22,409	24,166	25,922	27,676	29,427	31,174	32,918	34,656	36,389	38,117	39,839
66	2,069	3,838	5,424	7,074	8,770	10,500	12,255	14,029	15,817	17,614	19,419	21,227	23,039	24,851	26,662	28,472	30,280	32,084	33,884	35,680	37,471	39,257	41,037
68	2,133	3,930	5,556	7,248	8,989	10,765	12,569	14,392	16,230	18,080	19,937	21,799	23,664	25,531	27,398	29,264	31,127	32,988	34,845	36,698	38,547	40,390	42,228
70	2,198	4,021	5,686	7,421	9,206	11,029	12,880	14,752	16,641	18,542	20,451	22,366	24,286	26,207	28,129	30,050	31,970	33,887	35,801	37,711	39,617	41,518	43,414
72	2,262	4,111	5,816	7,593	9,422	11,291	13,189	15,110	17,049	19,001	20,962	22,930	24,903	26,879	28,856	30,832	32,808	34,781	36,752	38,719	40,681	42,640	44,593
74	2,326	4,201	5,945	7,763	9,636	11,551	13,496	15,466	17,454	19,457	21,470	23,491	25,517	27,547	29,578	31,610	33,641	35,670	37,697	39,721	41,741	43,756	45,767
76	2,391	4,290	6,073	7,933	9,849	11,809	13,801	15,819	17,857	19,910	21,975	24,048	26,127	28,210	30,296	32,383	34,469	36,554	38,637	40,718	42,794	44,867	46,935
78	2,455	4,378	6,200	8,101	10,060	12,065	14,104	16,170	18,257	20,361	22,476	24,602	26,734	28,870	31,010	33,151	35,293	37,434	39,573	41,709	43,843	45,973	48,098
80	2,519	4,466	6,326	8,268	10,270	12,320	14,405	16,519	18,655	20,808	22,975	25,152	27,336	29,526	31,720	33,916	36,113	38,308	40,503	42,696	44,886	47,073	49,255
82	2,583	4,554	6,451	8,434	10,479	12,573	14,704	16,866	19,050	21,253	23,470	25,699	27,936	30,179	32,426	34,676	36,927	39,179	41,429	43,678	45,924	48,167	50,407
84	2,648	4,640	6,576	8,599	10,686	12,824	15,002	17,210	19,443	21,695	23,963	26,243	28,532	30,828	33,128	35,432	37,738	40,044	42,350	44,655	46,957	49,257	51,553
86	2,712	4,727	6,700	8,763	10,893	13,075	15,297	17,553	19,834	22,135	24,453	26,784	29,125	31,473	33,827	36,185	38,545	40,906	43,267	45,627	47,986	50,342	52,695
88	2,776	4,813	6,823	8,926	11,097	13,323	15,591	17,893	20,222	22,572	24,940	27,322	29,714	32,115	34,522	36,933	39,347	41,763	44,179	46,595	49,009	51,421	53,831
90	2,840	4,898	6,945	9,088	11,301	13,570	15,883	18,232	20,608	23,007	25,425	27,857	30,301	32,754	35,214	37,679	40,146	42,616	45,087	47,558	50,028	52,496	54,962
92	2,904	4,983	7,067	9,249	11,504	13,816	16,174	18,568	20,992	23,440	25,907	28,389	30,884	33,389	35,901	38,419	40,940	43,465	45,990	48,516	51,042	53,566	56,087
94	2,969	5,067	7,188	9,409	11,705	14,061	16,463	18,903	21,374	23,870	26,386	28,919	31,465	34,021	36,585	39,156	41,731	44,309	46,890	49,470	52,051	54,631	57,209
96	3,033	5,151	7,309	9,569	11,905	14,304	16,760	19,236	21,754	24,298	26,863	29,443	32,042	34,650	37,266	39,890	42,518	45,150	47,785	50,420	53,056	55,691	58,325
	4	6	8	10	12	14	16	18	20	22	24	26	28	30	32	34	36	38	40	42	44	46	48

ROUND DUCTS

4	6	8	10	12	14	16	18	20	22	24	26	28	30	32	34	36	38	40	42	44	46	48
67	196	419	754	1,220	1,832	2,605	3,534	4,363	5,280	6,283	7,374	8,552	9,817	11,170	12,610	14,137	15,752	17,453	19,242	21,118	23,082	25,133

NOTE: DUCTS HANDLING AIR QUANTITIES GREATER THAN 3,086 CFM ARE SIZED USING THE MAXIMUM VELOCITY CRITERIA

(shaded) DUCTS HAVING AN ASPECT RATIO GREATER THAN 4:1 REQUIRE ADDITIONAL REINFORCEMENT OR HEAVIER GAUGE METAL

TABLE 6-10 Airflow Capacities for Even-Sized Ducts Using 0.30 in. w.c. Static Pressure Drop per 100 ft, 2,000 fpm Maximum Velocity

One very important thing to understand when designing medium-velocity duct systems is that the static pressure losses for fittings where the air velocity is 2,000 fpm are four times the static pressure losses for the same fittings where the air velocity is 1,000 fpm. The higher static pressure losses for both the straight duct and the duct fittings in medium-velocity duct systems must be included in the static pressure calculation for the duct system.

Air Devices

Air devices, which include diffusers, registers, and grilles, are the components at the ends of the duct distribution system through which air is either delivered to the spaces within the building (air outlets) or removed from the spaces within the building (air inlets). Air devices consist of a neck, to which the ductwork is connected, and the face, which is the visible portion of the air device. There are many different types of air devices, some of which have very specialized applications, such as for clean rooms and laboratories. In this section, we will focus on the air devices commonly used in commercial buildings and discuss some of the basic criteria used to make proper selections and layout the air devices within the occupied spaces. Manufacturers' product catalogs should be consulted for more specific information about a particular type of air device or application and also for engineering guidelines that cover the topic of air distribution in much more detail than we will discuss in this book.

The selection and layout of air devices within a space that will coordinate with the interior design of the architect incorporate both form (appearance) and function (performance). Air devices not only provide a finished appearance to the air distribution system but also contribute to the overall comfort of the occupants within the space. Two factors that significantly influence the level of comfort perceived by the occupants are the uniformity of the space air temperature and the airspeed in the occupied zone.[31] The number and locations of the air devices are as important to achieving a uniform space temperature and airspeed in the occupied zone as the selection of the air devices themselves. The layout of the air devices within a space must consider factors such as localized space heat gains or losses, location of occupants, obstructions to airflow, and the height of the space. An improperly designed air distribution system can result in excessive noise, hot or cold spots within the occupied zone, drafty or stagnant conditions within the occupied zone, stratification (warmer air near the top of the room and cooler air near the bottom of the room), and potentially added cost to the project.

First, we will discuss the distinctions between a diffuser, a register, and a grille. A diffuser is a ceiling air outlet that is usually designed for a high induction of room air into the supply airstream.[32] The induction of room air into the supply airstream aids in the mixing of the room air and contributes to a uniform space temperature and airspeed. A grille can be either an air outlet or an air inlet and is often a wall-mounted device. Grilles generally do not have the same level of room air induction as diffusers. Grilles often have a louvered appearance where the face of the grille consists of parallel blades. Double-deflection grilles have a second set of blades behind the visible blades that are oriented perpendicular to the visible blades. As a result, double-deflection grilles distribute air both horizontally and vertically. A register is a grille that has an adjustable volume damper mounted within the neck of the grille. Registers are not commonly used for commercial projects because the system air balance is affected by adjustments made to the dampers in the registers and because of the excessive air noise that can result in a space if the damper is closed significantly. Registers should only be used if it is necessary for the occupants to have manual control of the airflow (which is not common). In this case, the dampers should be equipped with a lever adjustment (not a screwdriver adjustment) and should be used for fine adjustments to the airflow only. Manual volume dampers should also be installed at the connections of the duct runouts to the main ducts for use in performing the initial system air balance.

Once the type of air device for a particular space has been identified, two main criteria must be considered in order to make an appropriate selection: noise and throw criteria. Noise criteria (NC) is discussed in Chap. 8 and is a measure of the noisiness of a space. The NC levels given in the manufacturer's product data for air devices represent the room NC that results from the noise generated by the air device and are based on an assumed sound absorption of the space. Air devices for occupied spaces in commercial buildings should be selected with an NC level that is less than 30.

Throw values only apply to supply air devices and are normally given for terminal (air) velocities of 150, 100, and 50 fpm. A terminal velocity of 50 fpm in the occupied zone is desirable for occupant comfort. Thus, the 50-fpm throw value should be used to make an appropriate air device selection; that is, the distance from the air device to the occupants should be approximately equal to the 50-fpm throw value.

Air devices should be identified on the floor plan drawings with a unique designation given to each different type of air device. For example, all air devices that have the same configuration and construction, that is, the same manufacturer and model number, should be given the same designation, such as SD-1 (supply diffuser, type 1) or RG-2 (return grille, type 2). The design airflow should also be included in the tag for each air device shown on the floor plan drawings.[33] Also, for air devices that are the same type, but have varying neck sizes or lengths, which may be the case for some types of linear slot diffusers, the neck size (or length) would also need to be included in the tag. In short, air device tags require either two or three lines of information: type, design airflow, and neck size (or length). Refer to the sample legend (Fig. 10-2) in Chap. 10 for sample air device tags. Also refer to Fig. 6-17 which shows air device tags on a floor plan drawing.

The types, sizes, airflow ranges, and physical characteristics of air devices should be given in a diffuser, register, and grille schedule. Figure 6-22 is a schedule that lists some common types of diffusers and grilles and associated airflows, which result in an NC level that is less than 30.

The following are some basic guidelines for designing air distribution systems:

- The layout of ceiling air devices needs to be coordinated with the locations of lighting fixtures, sprinkler heads, and other systems installed in the ceiling.

- Ceiling air devices should be laid out on the floor plan drawings in a pattern that is similar to the pattern of the lighting fixtures.

- For a typical open office plan with a 9 ft ceiling height, 24 in. × 24 in. ceiling supply air diffusers serving interior zones normally supply between 150 and 250 cfm (depending upon the load) and are located on approximately 16-ft centers.

- The quantity of 24 in. × 24 in. ceiling supply air diffusers should normally be based on a supply airflow of about 250 cfm per diffuser. It is not recommended that 24 in. × 24 in. ceiling supply air diffusers supply more than about 400 cfm.

DIFFUSER, REGISTER, AND GRILLE SCHEDULE

DESIG.	SERVICE	BORDER	CFM RANGE	NECK SIZE	DESCRIPTION	BASIS OF DESIGN	NOTES
SD-1	SUPPLY	LAY-IN	0 - 100 101 - 175 176 - 275 276 - 400 401 - 535	6"Ø 8"Ø 10"Ø 12"Ø 14"Ø	24"x24" LOUVERED FACE CEILING SUPPLY AIR DIFFUSER, 4-WAY BLOW, STEEL CONSTRUCTION, WHITE FINISH.	TITUS TMS, BORDER TYPE 3.	1
SG-1	SUPPLY	SURFACE MOUNT	NOTED ON PLANS		35° DOUBLE-DEFLECTION SUPPLY AIR GRILLE, 3/4" BLADE SPACING, FRONT BLADES PARALLEL TO LONG DIMENSION, STEEL CONSTRUCTION, WHITE FINISH.	TITUS 300RL, BORDER TYPE 1.	1
LS-1	SUPPLY	FLUSH	0 - 80	6" OVAL	4 FT LINEAR SLOT DIFFUSER, (1)-1" SLOT, END BORDERS, ALUMINUM CONSTRUCTION, WHITE FINISH.	TITUS ML-39, BORDER TYPE 3, WITH MPI-39 INSULATED PLENUM.	1
LS-2	SUPPLY	FLUSH	81 - 160	8" OVAL	4 FT LINEAR SLOT DIFFUSER, (2)-1" SLOTS, END BORDERS, ALUMINUM CONSTRUCTION, WHITE FINISH.	TITUS ML-39, BORDER TYPE 3, WITH MPI-39 INSULATED PLENUM.	1
LS-3	SUPPLY	FLUSH	161 - 240	10" OVAL	4 FT LINEAR SLOT DIFFUSER, (3)-1" SLOTS, END BORDERS, ALUMINUM CONSTRUCTION, WHITE FINISH.	TITUS ML-39, BORDER TYPE 3, WITH MPI-39 INSULATED PLENUM.	1
LS-4	SUPPLY	FLUSH	241 - 320	12" OVAL	4 FT LINEAR SLOT DIFFUSER, (4)-1" SLOTS, END BORDERS, ALUMINUM CONSTRUCTION, WHITE FINISH.	TITUS ML-39, BORDER TYPE 3, WITH MPI-39 INSULATED PLENUM.	1
RG-1	RETURN	SURFACE MOUNT	0 - 100 101 - 200 201 - 300 301 - 400 401 - 500 501 - 700 701 - 800 801 - 1,000 1,001 - 1,300	6"x6" 8"x8" 10"x10" 12"x12" 14"x14" 16"x16" 18"x18" 20"x20" 22"x22"	35° FIXED-BLADE RETURN AIR GRILLE, 3/4" BLADE SPACING, BLADES PARALLEL TO LONG DIMENSION, STEEL CONSTRUCTION, WHITE FINISH.	TITUS 350RL, BORDER TYPE 1.	1
RG-2	RETURN	SURFACE MOUNT	NOTED ON PLANS		35° FIXED-BLADE RETURN AIR GRILLE, 3/4" BLADE SPACING, BLADES PARALLEL TO LONG DIMENSION, STEEL CONSTRUCTION, WHITE FINISH.	TITUS 350RL, BORDER TYPE 1.	1
RG-3	RETURN	LAY-IN	0 - 1,000	22"x22"	24"x24" PERFORATED FACE CEILING RETURN AIR GRILLE, STEEL CONSTRUCTION, WHITE FINISH.	TITUS PAR, BORDER TYPE 3.	

NOTES:
1. RUNOUT DUCT SIZE SHALL BE SAME AS NECK SIZE.

Figure 6-22 Diffuser, register, and grille schedule.

- Linear slot diffusers perform well to condition the localized heat gains and losses from windows. If linear slot diffusers are designed for this purpose, a two-slot linear slot diffuser would typically be designed in the ceiling within 2 ft of the window. The airflow of the exterior slot would be adjusted in a downward direction to "wash" the window with supply airflow, and the airflow of the interior slot would be adjusted in a horizontal direction to supply air toward the interior of the space.

- Multiple levels of supply air diffusers should be designed for spaces that are two or more stories high. Long-throw diffusers may be required to distribute the supply air down to the occupied zone.

- Air inlets should be designed near the floor for spaces that are two or more stories high to ensure air movement from the top to the bottom of the space. This will reduce the stratification of the air within the space and also prevent *short-circuiting* of the supply airflow to the air inlets. This can occur if the air inlets are mounted high within the space along with the air outlets.

- Air inlets should be located near sources of heat or odors in order to capture heat and odors at their source.

- Air inlets and air outlets should be separated sufficiently to prevent *short-circuiting* of the air.

- The architect should review the air device selections to ensure that they coordinate with the interior design.

- The color of air devices is normally white. However, custom colors can be specified if requested by the architect.

- The construction of air devices is normally painted steel. However, painted or anodized aluminum should be specified for damp environments, such as for dishwasher exhaust and shower exhaust grilles.

- Air devices are available with different types of borders for installation in different types of ceiling, wall, floor, or window sill construction. For example, an air device mounted within a suspended ceiling will have one type of border and an air devices installed within a gypsum board ceiling will have a different type of border. The type of border for each air device should be given in the diffuser, register, and grille schedule.

- Some types of diffusers are available with different discharge patterns. For example, square ceiling diffusers are normally specified with a 4-way discharge pattern which means air will be discharged from the diffuser evenly in all four directions. However, some situations, such as a diffuser located in the corner of a room or in a narrow corridor, may require a diffuser with a 2-way discharge pattern in order to prevent air from blowing directly onto a wall or other obstruction. The manufacturer's product data should be consulted to determine the discharge patterns that are available for each type of diffuser.

Coordination With Other Disciplines

The architect is responsible for identifying the locations and fire ratings of all fire-resistance-rated assemblies and smoke barriers within the building. This enables the HVAC system designer to design the fire, smoke, or combination fire/smoke dampers that may be required in the duct penetrations and/or transfer air openings in these assemblies.

In addition to locating the fire-resistance-rated assemblies and smoke barriers designed by the architect, it is important for the HVAC system designer to locate all full-height walls designed by the architect, that is, the walls that are solid from the floor to the underside of the deck above. If a ceiling return air plenum is used, transfer air ducts will need to be designed in these walls above the ceiling to allow return air to migrate from all of the spaces back to the main open-end return air duct.

The architect should review the layout of all air devices, particularly those that will be mounted in the ceiling, to ensure that the layout coordinates with the work of all disciplines.

The space above the ceiling must be coordinated with the work of all other disciplines so that reasonable ceiling heights can be established by the architect. If the ceiling heights proposed by the architect in the usable areas do not allow sufficient space for the HVAC systems, the HVAC system designer should consider routing portions of the piping and ductwork systems in alternate locations, such as above the ceilings of corridors and storage areas, where the ceilings can usually be lowered. If these locations do not afford an acceptable route for the HVAC systems, sections should be developed for particular areas of concern, identifying the maximum allowable ceiling heights. The sections will provide the architect with the information needed to revise the ceiling heights, design bulkheads, or perhaps reconfigure some rooms to suit the needs of the HVAC systems for space above the ceilings.

As discussed in Chap. 2, it is necessary for the HVAC system designer to coordinate the sizes and locations of all openings in the floors and exterior walls with the architect and structural engineer.

Finally, the electrical engineer will need to know the locations of all smoke and combination fire/smoke dampers in the building because it is common for these dampers to utilize 120V/1Ø electrical power. Furthermore, the electrical power for these dampers will need to be connected to a back-up (emergency) power distribution system so that power is provided to these dampers during a fire.

Endnotes

1. Btu per hour (Btu/h) is frequently represented as Btuh.
2. Standard air is defined as air at standard temperature and pressure and can be used without volumetric correction for typical HVAC applications at elevations below 1,000 ft above sea level.
3. 1 MBH in English units is equal to 1,000 Btuh. The prefix "M" in the English system represents thousands of units, which is not to be confused with the prefix "M" in the metric system, which represents millions of units.
4. EDR (equivalent direct radiation), expressed in terms of square feet, is the amount of heating surface at 215°F (1 psig steam) that will dissipate 240 Btuh of heat when surrounded by still air at 70°F. The capacity of steam heating equipment is sometimes given in terms of EDR and sometimes given in terms of Btuh or MBH.
5. Condenser water piping is used in a generic sense to include piping systems operating at temperatures lower than 100°F, which includes heat pump water piping systems.

6. Manual air vents are recommended for locations where their operation will be infrequent, such as at the high points in the hydronic system distribution piping and the piping connections to HVAC equipment. Once the air is released from the system during start-up, the system and equipment air vents should not have to be opened unless a section of piping nearby or a piece of equipment has to be drained for maintenance or replacement. The seals in automatic air vents may dry out over time if they are not required to vent any air. If this occurs, they may leak or fail to function properly. Automatic air vents are recommended for locations where they will operate frequently, such as at the air separator and expansion tank for the system.

7. A butterfly valve with a memory stop can perform both shutoff and balancing duty for pipe sizes 2½ in. and larger.

8. A takeoff is defined as a branch pipe or duct connection to a main pipe or duct.

9. Lifting of condensate for higher-pressure steam systems is possible, but these systems are beyond the scope of this book.

10. The flash steam percentage for saturated condensate at 2 psig supply pressure and 0 psig return pressure is 0.7%. The flash steam percentage for saturated condensate at 15 psig supply pressure and 0 psig return pressure is 4.0%.

11. For more in-depth coverage of steam utilization for a variety of applications, refer to the engineering manuals published by the various steam equipment manufacturers.

12. The equivalent length of a piping system is equal to the length of the straight pipe sections to the farthest point in the system plus the pressure drop through the pipe fittings and valves, expressed in terms of equivalent lengths of straight pipe. It is common practice to allow 100% of the straight pipe length to account for the pressure drop through the pipe fittings and valves.

13. If no vacuum breaker is provided, the pressure within the equipment can drop below atmospheric pressure, resulting in a partial vacuum.

14. Galvanized coatings are described in terms of one hundredths of an ounce (oz.) of coating per square foot of sheet steel (total coating, both sides of the sheet). A G60 galvanized coating has 0.60 oz. of galvanized coating per square foot of sheet steel and a G90 coating has 0.90 oz. per square foot.

15. Duct liner should be specified with an anti-microbial agent to protect the coating of the duct liner from the potential growth of fungus and bacteria.

16. Outdoor supply and return air ductwork should be avoided whenever possible due to the potential for leakage in the waterproof membrane. If water penetrates the waterproof membrane, damage could occur to the ductwork and mold may grow within the insulation and possibly within the ductwork itself. Mold growth within the HVAC ductwork distribution system is a serious concern because it adversely affects the indoor air quality within the building and can cause health problems for the building occupants.

17. A plenum is a space formed by the interior components a building (walls, floors, ceilings) through which air is conveyed by an HVAC system. A shaft, the space above a ceiling, or the space below a raised floor can be a plenum. It is common for the space above the ceiling in commercial buildings to be used as a return air plenum. Air enters the plenum from the conditioned spaces through return air grilles mounted within the ceiling. From there it migrates through the ceiling plenum to a central open-end return air duct. From this point the air is transferred through the return air duct to the air handling unit. If the area served by the HVAC system is extensive, return air ductwork may be distributed through the ceiling

plenum to multiple open-end ducts in order to return air more evenly from the entire area served.

18. Static pressure is the pressure exerted transversely on walls of a duct system by the air within the duct system. Static pressure can be positive or negative and is independent of air velocity; that is, static pressure exerted on a duct system by a fan can exist even in the absence of airflow. The resistance of a duct system to airflow (friction loss) is proportional to the square of the air velocity through the duct system. Friction loss, commonly expressed in terms of in. w.c. static pressure drop per 100 ft of duct, reduces the static pressure (potential energy) of the air as it travels through the duct system.

19. Flat oval ductwork has similar performance characteristics as rectangular ductwork. The dimensions of flat oval ductwork are given in terms of the duct width (major dimension) and the diameter of the rounded sides connecting the flat portions of the duct (minor dimension).

20. An air transfer opening is simply an opening in a wall through which air is transferred during the normal operation of an HVAC system. Air transfer openings may consist of a framed opening in the wall or may be a short section of ductwork that has openings on both sides of the wall.

21. A duct penetration is defined as one in which the sheet metal ductwork is at least 26-gauge and is continuous from the air handling unit to the air devices.

22. The blades are retained either within the airstream (Type A) or within a pocket mounted outside of the airstream (Type B). Type B fire dampers have a lower resistance to airflow.

23. Smoke dampers are available with Class I or Class II leakage ratings. Class I smoke dampers are rated for a maximum of 8 cfm/ft^2 at 4.0 in. w.c. differential pressure. Class II smoke dampers are rated for a maximum of 20 cfm/ft^2 at 4.0 in. w.c. differential pressure.

24. Dynamic insertion loss is the difference in sound pressure levels measured within a space before and after the insertion of the sound attenuator. DIL of a sound attenuator decreases as the air velocity through the sound attenuator increases.

25. The finished floor elevation is the elevation to the top of the finished floor. For example, the finished floor elevation of a raised floor system in a computer room may be 12 to 18 in. higher than the top elevation of the floor slab that supports it.

26. The reflected ceiling plan is a drawing that shows the type, elevation, and components of the ceilings within a building. Ceiling components include the ceiling grid (for suspended ceilings), lights, air devices, sprinkler heads, exit signs, speakers, and similar components.

27. A duct diameter is the diameter of a round duct or the equivalent round duct diameter of a rectangular duct.

28. This straight length of duct allows a uniform air velocity profile to develop between the fan discharge and the elbow.

29. A ceiling return air plenum with return air-type lighting fixtures is used in this example. Thus, there is no return air ductwork routed to each space nor are there any return air devices shown on the plan.

30. A static pressure drop of 0.10 in. w.c. per 100 ft of ductwork can be used; this will result in slightly smaller ducts for airflows up to about 1,100 cfm. This sizing criterion will result in a slightly lower first cost for these portions of the duct systems but will also result in a slightly higher fan brake horsepower for the air handling unit serving these portions of the duct systems.

31. The occupied zone is from 0 to 6 ft above the floor.
32. One exception to this definition is a laminar flow diffuser designed for a minimal induction of room air into the supply airstream. Laminar flow diffusers are often used in laboratories that contain fume hoods but are also used in commercial kitchens as the makeup air outlet for the kitchen exhaust hood.
33. One exception is for return air grilles that are simply openings to a return plenum where there is no volume damper associated with the air device. In this case, these air devices would not be balanced by the testing, adjusting, and balancing contractor and would not require a design airflow in the tag.

CHAPTER 7
Terminal Equipment

Terminal equipment was introduced in Chap. 2 as the equipment that delivers the heating and/or cooling energy to the HVAC zones in response to the zone thermostats. The terminal equipment we will discuss in this chapter is commonly used in commercial buildings and includes finned-tube radiators, electric radiators, duct heating coils, unit heaters, cabinet unit heaters, fan-coil units, ductless split-system units, and variable air volume (VAV) terminal units.

Before we begin a further discussion of terminal equipment, it is appropriate to discuss the piping connections for terminal equipment coils in general. Many types of terminal equipment have heating coils, cooling coils, or both that require connection to the building hydronic system(s), steam system, or electrical system.

Connections for the various types of coils used in terminal equipment are as follows:

1. Hot water heating coils
   a. Heating water supply and return piping.

2. Steam heating coils
   a. Steam supply and condensate return piping.

3. Chilled water cooling coils
   a. Chilled water supply and return piping.
   b. Condensate drain pipe connection to the drain pan.

4. Electric coils
   a. Electrical power (common voltages are listed below)
      (1) Single-phase voltage (120V/1Ø, 208/230V/1Ø, or 277V/1Ø)
      (2) Three-phase voltage (208/230V/3Ø or 480V/3Ø)

5. Direct expansion (DX) refrigerant coils
   a. Refrigerant suction.
   b. Refrigerant liquid.
   c. Possibly hot gas piping.
   d. Condensate drain pipe connection to the drain pan.

Valves and specialties are required for shutoff, balancing, pressure and temperature measurement, and automatic temperature control. These are common for the hot water and steam heating coils (Figs. 7-1 through 7-3) and chilled water cooling coils used in terminal equipment. The actual type of control valve selected for the hydronic terminal equipment coil will depend upon the operation of the overall hydronic system. The piping connections for chilled water coils are similar to the piping connections for hot water heating coils.

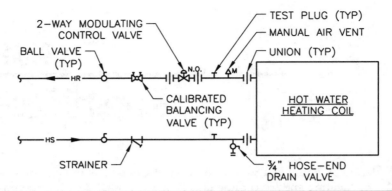

FIGURE 7-1   Hot water heating coil piping connections with 2-way valve control.

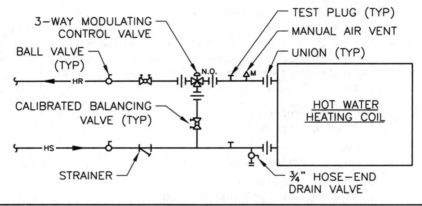

FIGURE 7-2   Hot water heating coil piping connections with 3-way valve control.

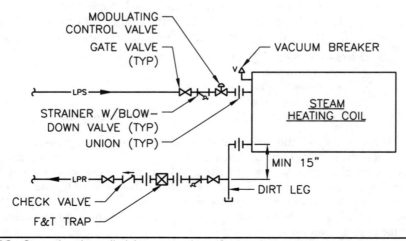

FIGURE 7-3   Steam heating coil piping connections (2-way valve control).

Heating-Only Equipment

In this section, we will discuss terminal equipment that provides only heating for the spaces served.

Finned-Tube Radiators

Purpose

Finned-tube radiators (FTRs) are designed for use in heating water or steam systems and consist of a copper tube with mechanically bonded aluminum fins. This finned-tube assembly is normally factory-mounted within a painted sheet metal enclosure or it can be field-installed within a custom enclosure. FTRs are normally used to supplement the heating capacity of the building HVAC systems rather than being the only source of heat for the spaces within the building. One area where FTRs are used extensively is below windows. In this location, FTRs warm the air at the bottom of the windows, which initiates natural convection currents of warm air rising upward. These warm air currents offset the cool air currents that flow downward near the cold window surface.

Physical Characteristics

There are various configurations of FTRs, including wall-mounted (Fig. 7-4) and pedestal-mounted units, which have different shapes, sizes, and heating capacities. The one thing that all FTRs have in common is the need for air to enter through the bottom of the FTR and exit through the top of the FTR in order to initiate the natural convection flow of warm air upward. It is common for the bottom of FTR enclosures to be open and the top of the enclosures to have some type of grille. FTRs can also be installed within a floor trough with a grille on top, but this configuration compromises their effectiveness.

Figure 7-4 Photograph of a wall-mounted FTR with manual damper, which controls the airflow through the FTR.

Connections

Connections for hot water FTRs are limited to the heating water supply and return piping. Connections for steam FTRs are limited to the steam supply and condensate return piping. FTRs do not require an electrical connection; however, automatic temperature control (ATC) connections are required for the control valve and thermostat if the FTR is so equipped. The hot water or steam piping connections as well as the shutoff and control valves are normally located within the FTR enclosure. Figure 7-20 in the Fan-Powered VAV Terminal Units section illustrates a floor-mounted hot water FTR (FTR-1) on a floor plan drawing.

Design Considerations

Capacity   The heating capacity of FTRs is dependent upon the quantity and length of the finned tube(s) and the density of fins on each finned tube (usually expressed in terms of fins per inch of tube length). The heating capacity of hot water FTRs is also dependent upon the entering water temperature, whereas the heating capacity of steam FTRs is normally given for 1 psig (215°F) steam. Steam pressures in excess of 2 psig are not recommended for FTRs because the surface temperature of the enclosure would be too hot.

Control   The heating output of hot water FTRs can be controlled in a number of ways. The most common method is to vary the heating water flow through each FTR, or group of FTRs, with an automatic control valve that responds to a space thermostat. Another method is for the occupant to manually vary the airflow through the finned tube by means of a manual damper installed within the FTR enclosure. Individual control valves and thermostats for the FTRs are not required for this configuration. A third method is to reset the temperature of the heating water supplied to a group of FTRs based on the outdoor temperature. This may be done for all of the FTRs serving a particular building exposure, such as the north exposure, or it can be done for all the FTRs in the building. Reset of the heating water temperature is performed at a central location for this configuration. Thus, individual control valves and thermostats for the FTRs are not required.

The heating output of steam FTRs can be controlled manually through a two-position (opened/closed) or modulating radiator supply valve. The heating output can also be controlled automatically through a 2-way control valve in response to a space thermostat. Steam FTRs are not used very often in commercial buildings because the heating output is difficult to control.

Selection   In general, FTRs are used below windows in colder climates (winter design outdoor temperature less than 15°F) where the windows are more than 5 ft high. The capacity of the FTR should be equal to the design heat loss through the window.

Because FTRs are visible components of the HVAC system, the project architect will usually want to review and comment on the FTR selections proposed by the HVAC system designer.

Installation   FTRs are mounted to walls with brackets that support the finned tube separately from the enclosure. This allows the enclosure to be removed for cleaning and inspection. Floor-mounted FTRs are typically supported by pedestals, again with the finned tube supported separately from the enclosure. FTRs are available in various standard lengths. If longer sections of FTR are required, multiple FTR sections can be joined together.

If an FTR is to be mounted within a trough, the trough should be at least 9 in. wide and the FTR should be mounted on the interior side of the trough. This will allow the cool air downdrafts near the window surface to enter the trough on the exterior side, thus creating a circular motion of the air currents as the warm air rises up through the FTR.

Electric Radiators

Electric radiators perform the same function as hot water and steam FTRs, the only difference is that electricity is the source of heating energy. The heating capacity of electric radiators is dependent upon the electrical power input of the electric heating element given in terms of watts per linear foot of radiator length or total radiator watts. The heating output of electric radiators is normally controlled by a contactor (heavy-duty relay) for on/off control in response to a unit-mounted or space thermostat.

Electric radiators only require an electrical connection. The thermostat is usually specified to be furnished with the radiator for stand-alone operation [no interface with the building automation system (BAS)]. The thermostat will be either a line voltage[1] thermostat or a low-voltage thermostat. If a low-voltage thermostat is specified, the control voltage (usually 24V ac) is derived from a unit-mounted, step-down control transformer that must also be specified to be furnished with the electric radiator.

The representation of an electric radiator on a floor plan drawing is the same as that of an FTR without the piping. The thermostat controlling the radiator should be shown on the radiator if it is unit-mounted or on the wall if it is a space thermostat.

Duct Heating Coils

Purpose

Duct heating coils are heating coils that are mounted within the duct distribution systems to provide supplemental heat, either for zone heating or for zone humidity control.

Physical Characteristics

Duct heating coils can utilize heating water, steam, or electricity as the source of heating energy. Hot water and steam heating coils will typically utilize flanges for connection to the ductwork.

Electric heating coils can utilize either flanges or be of the slip-in configuration. A slip-in electric heating coil is designed to be inserted through a hole in the side of the ductwork. Although both open coil and finned tubular heating elements are available for electric heating coils, the less costly open coil-type elements are used predominantly for HVAC applications.

Connections

In addition to the ductwork connections required for all duct heating coils, hot water coils require heating water supply and return piping connections, and steam coils require steam supply and condensate return piping connections. Electric duct heating coils require an electrical power connection that is usually terminated within a terminal box that is furnished as an integral part of the electric coil. ATC connections are required for the control valve (hot water and steam coils) and the thermostat. Figure 7-5 illustrates an electric duct heating coil (EHC-1) on a floor plan drawing.

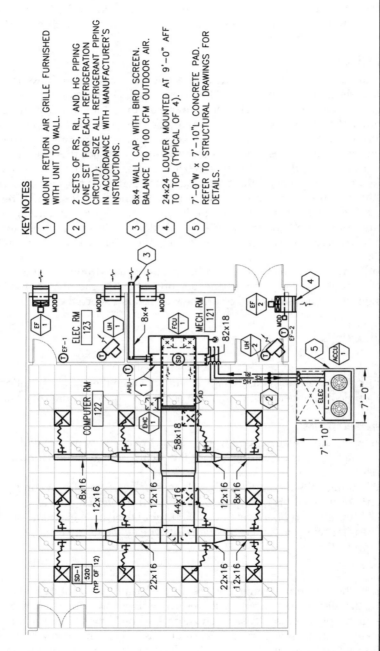

KEY NOTES

1. MOUNT RETURN AIR GRILLE FURNISHED WITH UNIT TO WALL.

2. 2 SETS OF RS, RL, AND HG PIPING (ONE SET FOR EACH REFRIGERATION CIRCUIT). SIZE ALL REFRIGERANT PIPING IN ACCORDANCE WITH MANUFACTURER'S INSTRUCTIONS.

3. 8x4 WALL CAP WITH BIRD SCREEN. BALANCE TO 100 CFM OUTDOOR AIR.

4. 24x24 LOUVER MOUNTED AT 9'-0" AFF TO TOP (TYPICAL OF 4).

5. 7'-0"W x 7'-10"L CONCRETE PAD. REFER TO STRUCTURAL DRAWINGS FOR DETAILS.

Figure 7-5 Floor plan representation of a vertical split-system fan-coil unit with electric duct heating coil.

250

Design Considerations

Capacity   The heating capacity of hot water and steam duct heating coils is dependent upon the face area, number of rows, and number of fins per inch of the heating coil. The heating capacity of hot water duct heating coils is also dependent upon the entering water temperature. The heating capacity of steam duct heating coils is dependent upon the steam pressure. The heating capacity of electric duct heating coils is dependent upon the electrical power input of the electric resistance element given in terms of kilowatts (kW).

Control   The heating output of hot water and steam duct heating coils is controlled by an automatic control valve that modulates the water or steam flow through the coil in response to either a duct-mounted thermostat or a space thermostat. The thermostat and control valve may be connected to the BAS for central monitoring and control.

An electric duct heating coil is also controlled either by a duct-mounted thermostat or a space thermostat, but its capacity is controlled either in steps by a step controller[2] or in a modulating manner by a silicon-controlled rectifier (SCR) controller.[3] Magnetic contactors are the standard offering for step controllers, but these can be noisy. For quiet operation of step controllers, mercury contactors should be specified; however, these are more costly than magnetic contactors. Also, the number of steps of capacity control that are available is limited by the manufacturer. More steps of capacity control are available as the capacity of the coil increases. In general, the number of steps of capacity control should be based on an approximate 5 to 10°F air temperature rise per step. Thus, if an electric duct heating coil is designed to raise the air temperature 20°F, two to four steps of capacity control should be specified. If more control is required than is available with a step controller, an SCR controller should be specified, keeping in mind this will add to the cost of the project. The thermostat may be connected to the BAS; however, the electric heating coil controller is normally not connected to the BAS. If the operational status of the electric duct heater needs to be monitored through the BAS, a temperature sensor connected to the BAS can be installed downstream of the coil to measure the leaving air temperature. If the entering air temperature is already known, it can be determined whether the coil is energized or not. If the entering air temperature is not known, a separate temperature sensor will have to be installed upstream of the coil to report the entering air temperature to the BAS.

Selection   Hot water and steam duct heating coils are usually selected for a face velocity of 500 to 1,000 fpm. However, electric duct heating coils must be selected to ensure that a minimum air velocity through the coil is always present to prevent overheating of the electrical elements and nuisance tripping of the thermal cutouts. The minimum required air velocity varies depending upon the power density of electric heater expressed in terms of kilowatts per square foot of cross-sectional duct area. The manufacturer's product data should be consulted to determine the minimum airflow required for each electric duct heating coil.

Installation   In addition to the connections listed above, the following are some important installation considerations that should be addressed in the design of duct heating coils:

- Access doors should be installed both upstream and downstream of duct heating coils for cleaning and inspection of the coils.

- Four duct diameters of straight duct should be designed upstream of electric duct heating coils to ensure a uniform air velocity distribution across the heating element.

- Transitions upstream and downstream of electric heating coils, if required, should be gradual, with a maximum combined angle of 30°.

- Slip-in duct heaters are normally suitable for installation in ducts lined on the interior with up to 1 in. of duct liner. Since 1 in. is a standard thickness for duct liner, slip-in duct heaters can be used for most commercial applications unless there is another reason why a flanged connection would be desired.

- A door-interlocked disconnect switch should be specified for the door on the terminal box of electric duct heaters. This will disconnect power to the electric duct heater whenever the terminal box door is opened. Specifying this option will avoid the need for a field-installed electrical disconnect switch. This should be coordinated with the electrical engineer so that a redundant disconnect switch is not designed.

- An airflow switch should be specified for electric duct heaters that will prevent operation of electric duct heater when there is no airflow.

- A minimum of 18 in. clearance should be allowed in front of the terminal box for the door swing and for maintenance access.

Unit Heaters

Purpose

Unit heaters are normally used in unoccupied areas, such as equipment rooms, to keep these areas from freezing in the winter.

Physical Characteristics

Unit heaters consist of a fan and a heating coil contained within a single enclosure. Small-capacity unit heaters utilize a propeller fan and large-capacity unit heaters utilize a centrifugal fan. The heating coil can be hot water, steam, electric, or gas.[4] Gas-fired unit heaters are available with natural draft and forced draft burners. Hot water, steam, and electric unit heaters are available in a horizontal-blow or vertical-blow (downflow) configuration. Gas-fired unit heaters are available only in a horizontal-blow configuration.

Connections

All unit heaters require an electrical connection for the fan. Hot water unit heaters require heating water supply and return piping connections. Steam unit heaters require steam supply and condensate return piping connections. Gas unit heaters require gas and vent connections. If an electric coil is used, the unit heater will normally require a single-point electrical connection for both the fan and the electric heating coil. Unit heaters typically do not require an ATC connection because they are normally controlled by a unit-mounted thermostat.

Hot water unit heaters can be used in 2-pipe hydronic systems provided a strap-on aquastat is utilized to sense the pipe temperature and cause the control valve to close completely whenever the pipe temperature is below 90°F. This will keep chilled water from circulating through the unit heater during the cooling mode of the 2-pipe system, preventing any damage to the unit heater that could occur due to condensation. Figure 7-5 in the Duct Heating Coils section illustrates electric unit heaters (UH-1 and UH-2) on a floor plan drawing.

Design Considerations

Capacity   Unit heaters are available in standard-size fan/coil combinations, which are determined by the manufacturer. The capacity of unit heaters is given in terms of the heating output capacity for hot water, steam, and electric unit heaters and in terms of the gas burner input rating for gas-fired unit heaters.[5] Standard-size hot water and steam unit heaters range from 8,000 to more than 300,000 Btuh output. Standard-size electric unit heaters range from 3 kW (10,239 Btuh) to 100 kW (341,300 Btuh). Standard-size gas-fired unit heaters range from 25,000 to 400,000 Btuh input rating. Unit heaters with a capacity greater than 30,000 Btuh are rarely used for commercial building projects. Unit heaters in this size range are typically used for warehouses or other large open spaces.

Control   Unit heaters are typically controlled by unit-mounted thermostats, although space thermostats are available as an option. Unit heaters typically operate with on/off control; that is, the unit runs when there is a call for heat and shuts off when no heating is required. For hot water and steam unit heaters, the control valve is a two-position (opened/closed) control valve. For electric unit heaters, the electric heating coil has a single contactor that fully energizes or de-energizes the coil. For gas unit heaters, the gas burner operates at high fire or it is off.

Selection   Unit heaters are typically sized to maintain the unoccupied areas that they serve at approximately 60°F during the winter. In order to select the appropriate unit heater(s) to serve an area, it is necessary to estimate the heat losses through the building envelope as well as the heating load associated with any outdoor air that may need to be conditioned. Sources of outdoor air that must be conditioned include infiltration from frequently opened doors and combustion air for equipment rooms containing fuel-fired equipment that uses room air for combustion.

Once the heating load is estimated, the number and capacity of unit heater(s) required to meet the space needs can be determined. However, an exact match of unit heater capacity to heating load is not necessary as long as the capacity exceeds the load. For commercial building projects, it is common to utilize one or two different sized unit heaters that can be applied to all of the unoccupied spaces within the building. If more heating capacity is required for a particular space than is available through one unit heater, multiple unit heaters are designed for the space. Common sizes for hot water unit heaters are 8,000 Btuh and 24,000 Btuh. The smallest size for a steam water heater is about 18,000 Btuh and the smallest size for a gas-fired unit heater is about 25,000 Btuh input rating.

Another factor to consider is the throw of supply air from the unit heater. An 8,000-Btuh unit heater has a throw of about 20 ft and a 24,000-Btuh unit heater has a throw of about 30 ft. This provides the basic guideline that a separate unit heater should be designed for about every 400 to 900 ft² of space.

Hot water heating coil capacities are typically given in the manufacturer's product data for 200°F entering water temperature and 60°F entering air temperature. Correction factors must be applied to the capacities if different entering water or entering air temperatures are used. The heating water flow rate through the heating coil is determined by the heating water temperature drop (Δt) through the coil, which is commonly 20°F. The heating water Δt must be consistent with the design parameters of the overall heating water system.

Steam heating coil capacities are typically given in the manufacturer's product data based on 2 psig steam (219°F saturation temperature) and 60°F entering air temperature. Correction factors must be applied to the capacities if a different steam pressure or entering air temperature is used.

Electric heating coil capacities are given in the manufacturer's product data in terms of kilowatts. The heating capacity of electric heating coils is independent of the entering air temperature.

Installation   Unit heaters are normally suspended from the building structure. Unit heaters should be positioned so that they blow warm air along exterior walls and in the direction of doors and combustion air openings within equipment rooms. This will help to condition the heat losses through the building envelope and the outdoor air that is drawn in through these openings. This is particularly important for locations where the outdoor air may be drawn across pipes that can freeze. For larger areas where multiple unit heaters are used, the unit heaters should be mounted near the perimeter walls of the room and should blow in such a manner as to create a circular flow of air within the room.

Gas-fired unit heaters typically use the room air for combustion. Section 9.3.2.1 of *NFPA Standard 54—National Fuel Gas Code* requires 50 ft³ of room volume per 1,000 Btuh of gas input. If the unit heater has a natural draft burner, a draft hood is required at the connection of the vent to the unit heater and the vent will have to terminate at least 5 ft higher than the unit heater draft hood (*NFPA Standard 54*, Section 12.7.2). If the unit heater has a forced draft burner, the vent can terminate through a wall or roof. All vent terminations must be in compliance with the requirements of *NFPA Standard 54*, Chap. 12.

Cabinet Unit Heaters

Purpose
Cabinet unit heaters are most frequently used near exterior doors to heat the outdoor air that infiltrates through these doors. However, they can also be used to condition areas that are heated only, such as locker rooms.

Physical Characteristics
Cabinet unit heaters are similar to unit heaters in that they consist of a fan and a heating coil that are contained within a single enclosure. The heating coil can be hot water, steam, or electric. Cabinet unit heaters are available in surface-mounted and recessed configurations for both wall-mounted and ceiling-mounted arrangements. The supply and return air openings are typically stamped into the front of the painted sheet metal enclosure.

Connections
Connections for hot water, steam, and electric cabinet unit heaters are the same as those for unit heaters (refer to the Unit Heaters section earlier). The hot water or steam piping connections as well as the shutoff and control valves are normally located within cabinet

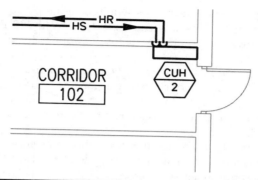

FIGURE 7-6   Floor plan representation of wall-mounted hot water cabinet unit heater.

unit heater enclosure. Figure 7-6 is a graphical representation of a wall-mounted hot water cabinet unit heater on a floor plan drawing. Figure 7-20 in the Fan-Powered VAV Terminal Units section is a graphical representation of a ceiling-mounted hot water cabinet unit heater (CUH-1) on a floor plan drawing.

Design Considerations

Capacity   Cabinet unit heaters are available in standard-size fan/coil combinations, which are determined by the manufacturer. The capacity of cabinet unit heaters is given in terms of the heating output capacity for hot water, steam, and electric unit heaters. Standard-size hot water and steam cabinet unit heaters range from around 12,000 to 80,000 Btuh output. Standard-size electric cabinet unit heaters range from 750 W (2,560 Btuh) to 24 kW (81,912 Btuh). Cabinet unit heaters with a capacity greater than 30,000 Btuh are rarely used for commercial building projects.

Control   The control of hot water, steam, and electric cabinet unit heaters is the same as that of unit heaters (refer to the Unit Heaters section earlier).

Selection   Cabinet unit heaters serving exterior doors are typically sized to heat the outdoor air infiltration through these doors to 70°F. In addition, if a cabinet unit heater is installed within a vestibule, it should also be sized to condition the heat losses through the envelope of the vestibule.

Outdoor air infiltration through exterior doors should be estimated according to the procedure described in Chap.16 of the 2009 *ASHRAE Handbook—Fundamentals*. For an exterior door that serves approximately 75 people per hour, the outdoor air infiltration can be estimated at about 50 cfm per 3 ft × 7 ft door, based on a 0.10 in. w.c. pressure difference across the door. Thus, for a typical 50 ft² vestibule with two 3 ft × 7 ft doors, the cabinet unit heater should have a capacity of at least 15,000 Btuh based on a 0°F winter design temperature (7,500 Btuh envelope heat loss[6] plus 7,500 Btuh outdoor air heating load). For an exterior door consisting of two 3 ft × 7 ft doors without a vestibule, the cabinet unit heater should have a capacity of at least 7,500 Btuh (outdoor air heating load only). For an exterior door consisting of one 3 ft × 7 ft door without a vestibule, the terminal heating equipment should have a capacity of at least 3,750 Btuh (outdoor air heating load only). However, this heating load is much less than the lowest capacity available for a hot water or steam cabinet unit heater (approximately 12,000 Btuh).

Therefore, a hot water or steam FTR would be a more appropriate choice of terminal heating equipment for this application. A 1,500 W (5,120 Btuh) electric cabinet unit heater would, however, be appropriate for this application if electricity is used for heating in the building.

For cabinet unit heaters serving exterior doors, an exact match of cabinet unit heater capacity to heating load is not necessary as long as the capacity exceeds the load by no more than about 50% of the load. For commercial building projects, it is common to utilize one to three different-sized cabinet unit heaters that can be applied to the various exterior doors of the building.

Cabinet unit heaters serving other areas should be sized for the heat losses through the building envelope as well as the heating load associated with any outdoor air that may need to be conditioned.

Hot water heating coil capacities are typically given in the manufacturer's product data for 200°F entering water temperature and 60°F entering air temperature. Correction factors must be applied to the capacities if different entering water or entering air temperatures are used. The heating water flow rate through the heating coil is determined by the heating water Δt through the coil, which is commonly 20°F. The heating water Δt must be consistent with the design parameters of the overall heating water system.

Steam heating coil capacities are typically given in the manufacturer's product data based on 2 psig steam (219°F saturation temperature) and 60°F entering air temperature. Correction factors must be applied to the capacities if a different steam pressure or entering air temperature is used.

Electric heating coil capacities are given in the manufacturer's product data in terms of kilowatts. The heating capacity of electric heating coils is independent of the entering air temperature.

Installation   The configuration of the cabinet unit heaters, that is, whether they are wall- or ceiling-mounted and whether they are surface-mounted to or recessed within the walls or ceilings, needs to be coordinated with the design of the architectural components in which they are installed. Recessed cabinet unit heaters require about 10 in. clear inside depth of the wall or ceiling in which they are installed. Surface-mounted cabinet unit heaters will protrude approximately 10 in. into the space. It is necessary to coordinate the protrusion of cabinet unit heaters within the spaces they serve with door swings and any clearances that must be maintained in accordance with the Americans with Disabilities Act (ADA). Also, the location of ceiling-mounted cabinet unit heaters must be coordinated with the ceiling grid for suspended ceilings and with the ceiling lighting fixture locations.

Heating and Cooling Equipment

In this section, we will discuss terminal equipment that provides heating and cooling for the spaces served.

Fan-Coil Units

Purpose
Fan-coil units are normally small, single-zone, constant air volume air handling units (AHUs), commonly used in 2-pipe and 4-pipe systems and also in conjunction with

split-system air-cooled condensing units or split-system air-cooled heat pump units. Fan-coil units are often used to serve areas less than 2,000 ft² where little or no outdoor air ventilation is required or where outdoor air ventilation is provided by a separate system. However, they can also be used to serve larger areas as long as the HVAC loads are within the capabilities of the standard equipment offered by the manufacturer.

Physical Characteristics

Fan-coil units are available in various horizontal and vertical (upflow and downflow) configurations. Horizontal and vertical fan-coil units are normally installed in concealed locations, such as above ceilings or within equipment rooms. Cabinet configurations are also available for below-the-window installations.

Because fan-coil units are packaged pieces of equipment, they are only available in the standard configurations and fan/coil combinations offered by the manufacturer. A common configuration includes a filter, heating coil, cooling coil, and supply fan. Heating coils can be hot water, steam, or electric. Cooling coils can be chilled water or DX refrigerant. Supply fans are typically centrifugal. Return fans are not an option for fan-coil units.

Connections

All fan-coil units require an electrical connection for the supply fan. The supply fan motors for small fan-coil units (2,000 cfm or less) are typically fractional horsepower (less than 1 hp) utilizing single-phase voltage (120V/1Ø, 208/230V/1Ø, or 277V/1Ø). The supply fan motors for larger fan-coil units (2,000 to 12,000 cfm) will normally require three-phase voltage (208/230V/3Ø or 480V/3Ø) because the motor sizes are much larger (1 to 10 hp).

Depending upon the installation, supply and return air ductwork connections may be required. Flexible duct connectors should be used for duct connections to fan-coil units. However, fan-coil units are also available with supply and return air grilles that are integral to the unit cabinet; thus they do not require any ductwork connections. Such is the case with cabinet units for exposed mounting below windows and also for horizontal or vertical units that are installed either within the conditioned space or recessed above a ceiling or within a wall.

If outdoor air ventilation is supplied by the fan-coil unit, the outdoor air duct normally connects to the return duct as opposed to connecting to a mixing box on the unit, as is the case with a modular central station AHU. The outdoor air duct will normally be equipped with a two-position (opened/closed) motor-operated damper that opens when the unit is running and closes when the unit is off.

Fan-coil units with hot water and/or chilled water coils require heating and/or chilled water supply and return piping connections. Connections for DX refrigerant coils include the refrigerant suction, liquid, and possibly hot gas piping. Chilled water, dual-temperature water, and DX refrigerant cooling coils also require a condensate drain pipe connection to the drain pan. Steam coils require steam supply and condensate return piping connections. Electric coils utilize the same voltage as the supply fan (single-phase voltage for small fan-coil units, three-phase voltage for larger fan-coil units). If an electric coil is used, the fan-coil unit should be specified with a single-point electrical connection as an optional accessory. Otherwise, the fan-coil unit will require separate electrical connections for the supply fan and electric heating coil.

Figure 7-5 in the Duct Heating Coils section is a graphical representation of a large (12-ton), vertical split-system fan-coil unit on a floor plan drawing. The unit serves a computer room with high space sensible heat gains. Therefore, the airflow is high on a cubic foot per minute (cfm) per square foot basis, but the outdoor ventilation rate is a low percentage of the supply airflow because the occupant density is very low. Figure 7-7 is a graphical representation of cabinet fan-coil units on a floor plan drawing; Fig. 7-8 is a photograph of a cabinet fan-coil unit in its installed condition.

In the configuration shown in Fig. 7-7, space would need to be made available within the exterior wall on the interior side of the wall insulation for routing the dual-temperature water piping from the ceiling space to the fan-coil unit mounted below the window. Outdoor air for the office would have to be provided by a separate system. The condensate drain pipe is often connected to a vertical condensate drain riser that is shared by two fan-coil units on each floor, as shown in Fig. 7-7. The capacities and configurations of fan-coil units are often scheduled as typical "types" rather than as

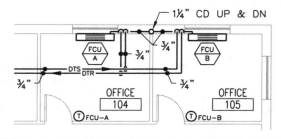

FIGURE 7-7 Floor plan representation of cabinet fan-coil units.

FIGURE 7-8 Photograph of a cabinet fan-coil unit.

individual pieces of equipment because the same unit configuration and capacity may be utilized in multiple locations for a project. For example, there may be several Type "A" fan-coil units in a project, all of which have the same configuration and capacity. This is done to simplify the scheduling of the unit capacities and configurations and also to make the contractor's job of pricing, ordering, and installing the units easier.

Design Considerations

Capacity The capacity of a fan-coil unit is often given in terms of its cooling capacity (in tons) or its supply airflow (in cfm). Fan-coil unit cooling capacities are usually 5 tons or less with supply airflows of 2,000 cfm or less. However, fan-coil units as large as 25 tons with supply airflows up to 10,000 cfm are also available, usually as a less costly option to modular central station AHUs. Fan-coil units larger than 5 tons are similar to modular central station AHUs, with the exception that fan-coil units are available only with the standard components and capacities offered by the manufacturer. Thus, there is less flexibility in selecting the components of a large fan-coil unit compared to the options available with a modular central station AHU.

Another limitation of fan-coil units is the external static pressure[7] that the supply fan is capable of developing to overcome the static pressure loss of the supply and return air duct system. Small fan-coil units (2,000 cfm or less) are limited to duct systems having a static pressure loss of about 0.50 in. w.c. or less. Larger fan-coil units (2,000 to 10,000 cfm) are capable of serving duct systems with a static pressure loss of 1.00 in. w.c. or less for the standard-size motor. Typically, a larger motor and drive[8] are available as an option to serve duct systems with a static pressure loss of up to 2.00 in. w.c.

Control Fan-coil units connected to duct systems are typically controlled by space thermostats. Cabinet units and units without ductwork connections that are installed in accessible locations are usually controlled by unit-mounted thermostats. Both space thermostats and unit-mounted thermostats may also be connected to the BAS, depending upon the requirements of the project. The unit fans are typically two-speed or three-speed and may run continuously or intermittently when heating or cooling is required. The control valves for the hot water or steam heating and/or chilled water cooling coils are normally two-position (opened/closed) since modulating control valves are not required for capacities up to 5 tons. Electric heaters for fan-coil units less than 5 tons are normally single stage (on/off). Normally, fan-coil units are available with all required ATC components as an optional accessory.

Fan-coil units larger than 5 tons are normally equipped with modulating control valves. Step controllers are used for electric heaters in fan-coil units larger than 5 tons.

Selection

HVAC Load As with any air system, fan-coil units should be selected to meet the heating, ventilating, and air-conditioning needs of the areas that they serve. However, because the capabilities of fan-coil units are limited by the standard fan/coil combinations offered by the manufacturers, they may not be suitable for areas that require a high percentage of outdoor air or have significant internal latent heat gains, such as people or cooking appliances.

In order to determine the suitability of a fan-coil unit for a particular application, HVAC load calculations must first be performed for the areas served to determine the sensible and total cooling load[9] and the heating load that will be imposed upon the fan-coil unit cooling and heating coils. The calculations must include the outdoor airflow that is conditioned by the fan-coil unit, if any. The cooling supply airflow should be based on a cooling supply air temperature of 55 to 58°F and a space temperature of 75°F for occupied areas and 85°F for unoccupied areas that require cooling.[10] The output of the HVAC load calculations will show the calculated space relative humidity (RH), which should be between 40 and 60% RH (preferably closer to 50% RH). The HVAC load calculations will also identify the conditions of the air entering and leaving the cooling coil, the total and sensible cooling load on the cooling coil, and the heating load on the heating coil.

Fan Selection   In order to properly select a fan-coil unit, the unit must first be able to supply the calculated airflow required for cooling against the static pressure losses of the supply and return air duct systems, if any. In the manufacturer's product data, supply fan airflows are given for each unit size, fan speed, total number of coil rows (cooling coil rows plus heating coil rows), and the external static pressure of the supply and return air duct systems.

This introduces the term *unit size*, which is commonly used to describe terminal equipment such as fan-coil units and VAV terminal units. Although different manufacturers may use different conventions, it is common for fan-coil unit sizes to correspond to the nominal supply airflow capability. For example, a unit size 06 may supply 600 cfm of airflow at a certain fan speed, number of coil rows, and external static pressure determined by the manufacturer. The capacities of the cooling and heating coils available for the various unit sizes are based on this nominal airflow. Correction factors must be applied to the coil capacities for airflows that differ from the nominal values.

Cooling Coil Selection   If chilled water is used, the proper cooling coil must be selected based on the entering chilled water temperature and design temperature rise (Δt) of the chilled water through the coil. Typically, the entering chilled water temperature is 45°F and the design Δt is 10°F. However, this is determined by the design parameters of the overall chilled water system. If a DX refrigerant cooling coil is used, the proper saturated suction temperature[11] must be identified, which is usually 45°F.

The cooling coil entering air conditions from the calculations must be matched with the closest entering air conditions supplied by the manufacturer in the product data. Three common entering air conditions supplied in manufacturers' product data are 75°F dry bulb (db)/63°F wet bulb (wb), 80°F db/67°F wb, and 85°F db/71°F wb. The 75°F db/63°F wb entering air condition represents a 75°F/52% RH space with no outdoor air ventilation (occupied space, 100% recirculating air). The 80°F db/67°F wb entering air condition represents a 78°F/53% RH space with approximately 10 to 12% outdoor air ventilation (occupied space, outdoor air ventilation for a minimum number of occupants). The 85°F db/71°F wb entering air condition represents an 85°F/50% RH space with no outdoor air ventilation (unoccupied space, 100% recirculating air).

The fan-coil unit manufacturer's product data will also provide the sensible and total cooling capacity for the cooling coils that are available for each unit size based on the entering air dry bulb/wet bulb temperatures listed above, entering water temperatures, and chilled water temperature differentials through the coil (Δt). For chilled water coils, performance data are provided for 40°F, 45°F, and 50°F entering water temperatures. An entering water temperature of 45°F and a 10°F Δt temperature rise is common. If a greater degree of dehumidification is required, a 40°F entering water temperature should be utilized, keeping in mind that the chilled water plant must be designed to supply this chilled water temperature. Similar performance data are provided for DX refrigerant coils for 35°F, 40°F, 45°F, and 50°F saturated suction temperatures, where a 45°F saturated suction temperature is the most common. Greater dehumidification capacity can be achieved with a lower saturated suction temperature, but the refrigeration system must be designed accordingly. Cooling coils for fan-coil units are normally available in two-row, three-row, and four-row configurations, although a six-row configuration may be available for larger units. The latent cooling capacity of cooling coils increases with an increase in the number of rows.

The sensible cooling capacity of the fan-coil unit cooling coil must exceed the sensible cooling load of the areas served in order to be capable of maintaining the design space temperature setpoint. Furthermore, the cooling coil's ratio of latent capacity to total capacity must be equal to, or greater than, the ratio of latent load to total load of the areas served (including the outdoor air ventilation load) to ensure that the cooling coil has sufficient ability to dehumidify the air flowing through the coil. This "latent heat ratio" is commonly expressed as a sensible heat ratio (SHR), or q_s/q_t, where q_s is the sensible heat flow (in Btuh) and q_t is the total heat flow (in Btuh). Although "latent heat ratio" it is not a common term, it is equal to 1 minus the SHR. Therefore, if the *latent heat ratio* of the cooling coil must be equal to, or greater than, the latent heat ratio of the load, it follows that the SHR of the cooling coil must be equal to, or less than, the SHR of the load in order to provide adequate dehumidification of the supply airflow.

If the SHR of the coil is greater than the SHR of load, the cooling coil will not have sufficient dehumidification capacity to maintain the design space relative humidity determined by the HVAC load calculations. In this case, the HVAC system designer must determine if an increase in the space relative humidity is acceptable or if a different cooling coil should be selected. More cooling coil rows, an increase in chilled water flow (with a corresponding decrease in chilled water Δt), and a lower entering chilled water temperature will all provide greater dehumidification capabilities for the cooling coil. Remember, all of these factors must be consistent with the design parameters of the overall chilled water system.

Heating Coil Selection Heating coils are much easier to select than cooling coils because air warming is strictly a sensible heating process; that is, there is no latent heat component. Therefore, the heating coil selected for the fan-coil unit must simply have a greater heating capacity than the heating load of the areas served, including the heating load of the outdoor airflow, if any. The heating coil capacity should be selected as close as possible to the heating load in order to prevent drastic swings in the space temperature when the heating coil is energized and de-energized. Sufficient heating capacity is typically available from a one-row or two-row hot water or steam heating coil for most commercial applications.

Hot water heating coil capacities are typically given in the manufacturer's product data for 180°F entering water temperature and 70°F entering air temperature. Correction factors must be applied to the capacities if different entering water or entering air temperatures are used. The heating water flow rate through the heating coil is determined by the heating water Δt through the coil, which is commonly 20°F. The heating water Δt must be consistent with the design parameters of the overall heating water system.

Steam heating coil capacities are typically given in the manufacturer's product data based on 2 psig steam (219°F saturation temperature) and 70°F entering air temperature. Correction factors must be applied to the capacities if a different steam pressure or entering air temperature is used.

Electric heating coil capacities are given in the manufacturer's product data in terms of kilowatts. The heating capacity of electric heating coils is independent of the entering air temperature.

General Guidelines   The following are some general guidelines that should be followed when selecting fan-coil units:

- Fan-coil units should not be used where the outdoor airflow requirement for occupant ventilation or exhaust air makeup exceeds 10 to 12% of the supply airflow.

- Fan-coil units are suitable for serving the heating and cooling needs for areas such as offices and conference rooms where the occupant density ranges between 10 and 25 people per 1,000 net occupiable square feet.

- Fan-coil units should not be used to serve the heating and cooling needs of areas where the occupant density exceeds 25 people per 1,000 net occupiable square feet because the units typically will not have the latent cooling capacity required to maintain an acceptable space relative humidity for these occupant-dense areas.

- A two-row or three-row cooling coil is normally acceptable for fan-coil units that do not condition any outdoor air.

- A four-row cooling coil is normally required if the fan-coil unit must condition any outdoor air.

- Space thermostats provide better control of space temperature than unit-mounted thermostats.

- An outdoor air duct connection with or without a manual damper is normally available as an option for cabinet units mounted on exterior walls. However, this option is not recommended because there is the potential for freezing conditions within the wall box housing and subsequent freezing of the water coil(s) should the heating coil in the fan-coil unit fail to function properly.

- Fan-coil units with capacities that are 5 tons or less are only capable of utilizing 1-in.-thick air filters. Larger fan-coil units with capacities up to 25 tons are only capable of utilizing 2-in.-thick air filters. A filtration efficiency of minimum efficiency reporting value (MERV) 8 is possible with 1-in. and 2-in. filters, which is standard for commercial buildings. However, if superior air filtration is required, such as is required to achieve the Leadership in Energy and Environmental Design (LEED) Indoor Environmental Quality (EQ) Credit 5, fan-coil units cannot be used because they are incapable of obtaining the MERV 13 air filtration efficiency that is required.

- The heating and sensible cooling capacities of the coils in fan-coil units should be selected as close as possible to the heating and sensible cooling loads of the areas served. Oversizing the heating and cooling coils is not recommended because it results in poor control of space temperature, reduced dehumidification capability,[12] and potential short-cycling of the refrigeration system compressor(s) for units utilizing DX refrigerant cooling coils.

- A cooler supply air temperature utilizing less airflow results in a lower space relative humidity than a warmer supply air temperature utilizing more airflow. Therefore, a supply air temperature closer to 55°F (rather than 58°F) is recommended for spaces having occupant densities that are at the higher end of the acceptable range.

- If a split-system is used for cooling, there is a limit to the separation distance between the AHU and the outdoor air-cooled condensing unit for the standard-sized refrigerant piping. This separation distance is typically in the range of 75 to 100 ft. Also, the allowable separation distance between the two units may be decreased if the condensing unit is installed at a higher elevation than the AHU. The manufacturer's representative should be consulted if the separation distance between the two units exceeds 75 ft because special sizing of the refrigerant lines may be required.

- The lowest outdoor temperature in which the outdoor air-cooled condensing units and heat pump units can operate in the cooling mode is approximately 23°F. If cooling operation of the outdoor units is required for temperatures below 23°F, variable speed condenser fans must be specified; this will enable low ambient cooling operation down to −20°F outdoor temperature.

Installation   The following are some installation and maintenance requirements that need to be respected in the design of HVAC systems utilizing fan-coil units:

- Fan-coil units suspended from the building structure should be suspended with neoprene vibration isolation hangers.

- Sufficient clearance around fan-coil units should be provided for piping and electrical connections, filter replacement, access to removable panels in the unit, and coil removal.

- The locations of fan-coil units installed above ceilings and their associated maintenance clearances need to be coordinated with the locations of lighting fixtures, sprinkler heads, and other systems installed in the ceiling.

- Sufficient space must be allowed, especially above ceilings, for the required ¼ in./ft pitch required for the condensate drain piping from the fan-coil units to the points of discharge to the building storm water system.

- Sufficient clear space between the ceilings and building structure must be allowed for fan-coil units installed above ceilings.

- Ceiling access panels (minimum 24 in. × 24 in.) are required for maintenance access to fan-coil units installed above gypsum board ceilings.

Ductless Split-System Units

Purpose
Ductless split-system air-conditioning and heat pump units are often used where it is difficult or impossible to route ductwork to the area served; where heating water, steam, or chilled water is unavailable or undesirable for use; and/or where a stand-alone unit

separate from the central HVAC system is desired. Applications for ductless split-system units frequently include elevator machine rooms, data closets, and other equipment rooms that require cooling year-round, 24 hours per day. Although some models are equipped with a connection for an outdoor air duct, most units do not provide any outdoor air ventilation to the spaces they serve.

Physical Characteristics

As the name implies, ductless split-system units do not have any supply or return air ductwork. They consist of an indoor fan-coil unit with integral supply and return air grilles and an outdoor air-cooled condensing unit or heat pump unit, depending upon the configuration. The indoor unit is equipped with a DX refrigerant coil and possibly an electric heating coil.

The indoor unit is either surface wall-mounted within the conditioned space or recessed within the ceiling. The outdoor air-cooled condensing unit or heat pump unit is usually of a compact design, which allows it to be installed in tight spaces outdoors.

Ductless split-system units are available in cooling capacities of up to approximately 3½ tons. Configurations include cooling-only, cooling with electric heat, heat pump, and heat pump with backup electric heat.

Certain models are also capable of serving multiple indoor units from one outdoor unit. The manufacturer's product data should be consulted for specific products that are available for each application.

Connections

Connections to ductless split-system units include the refrigerant suction, liquid, and possibly hot gas piping between the indoor and outdoor units, condensate drain pipe connection to the drain pan, and electrical connections to the indoor and outdoor units. Because the indoor units are wall-mounted below the ceiling or recessed within the ceiling, it is common for there to be insufficient space for the pitch of the condensate drain piping. Therefore, a small condensate pump is usually installed adjacent to the indoor unit to receive condensate from the cooling coil drain pan and pump it to the point of discharge to the building storm water system.

Ductless split systems typically utilize 208/230V/1Ø electrical power. If a condensate pump is required, it should be specified to utilize 120V/1Ø power and be furnished with a cord and plug to serve as the disconnecting means. In this case, the electrical engineer would design a ground fault circuit interrupter (GFCI) receptacle near the condensate pump as its source of electrical power.

Figures 7-9 and 7-10 illustrate the indoor and outdoor components of ductless split-system units on a floor plan drawing and a connection detail for a ductless split-system unit, respectively. Figures 7-11 and 7-12 are photographs of these units in their installed conditions.

Design Considerations

The following are some design considerations that apply to most ductless split-system units:

- Ductless split-system units should be selected with heating and sensible cooling capacities that are greater than the heating and sensible cooling loads of the areas served.

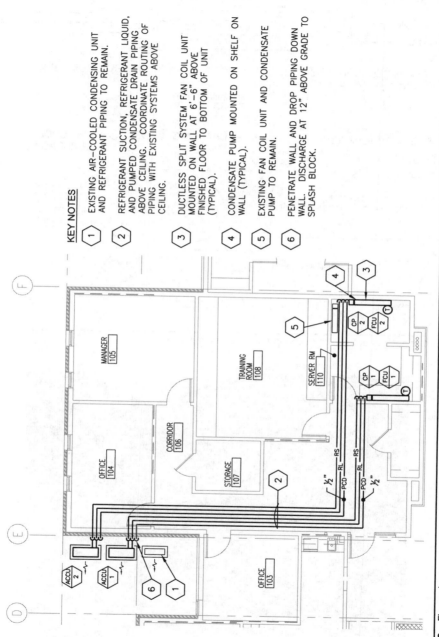

KEY NOTES

1. EXISTING AIR-COOLED CONDENSING UNIT AND REFRIGERANT PIPING TO REMAIN.

2. REFRIGERANT SUCTION, REFRIGERANT LIQUID, AND PUMPED CONDENSATE DRAIN PIPING ABOVE CEILING. COORDINATE ROUTING OF PIPING WITH EXISTING SYSTEMS ABOVE CEILING.

3. DUCTLESS SPLIT SYSTEM FAN COIL UNIT MOUNTED ON WALL AT 6'-6" ABOVE FINISHED FLOOR TO BOTTOM OF UNIT (TYPICAL).

4. CONDENSATE PUMP MOUNTED ON SHELF ON WALL (TYPICAL).

5. EXISTING FAN COIL UNIT AND CONDENSATE PUMP TO REMAIN.

6. PENETRATE WALL AND DROP PIPING DOWN WALL. DISCHARGE AT 12" ABOVE GRADE TO SPLASH BLOCK.

FIGURE 7-9 Floor plan representation of ductless split-system units.

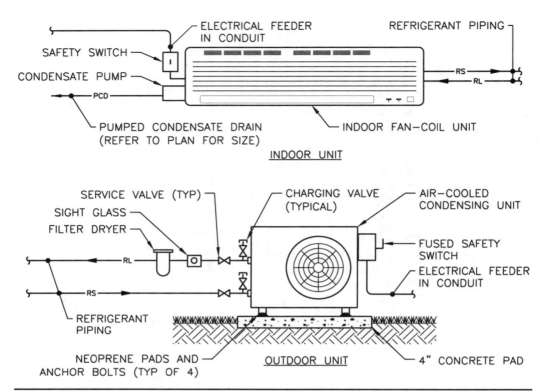

ELECTRICAL FEEDER
IN CONDUIT

SAFETY SWITCH

CONDENSATE PUMP

REFRIGERANT PIPING

RS

RL

PCD

PUMPED CONDENSATE DRAIN
(REFER TO PLAN FOR SIZE)

INDOOR FAN–COIL UNIT

INDOOR UNIT

SERVICE VALVE (TYP)

SIGHT GLASS

FILTER DRYER

CHARGING VALVE
(TYPICAL)

AIR–COOLED
CONDENSING UNIT

FUSED SAFETY
SWITCH

ELECTRICAL FEEDER
IN CONDUIT

RL

RS

REFRIGERANT
PIPING

NEOPRENE PADS AND
ANCHOR BOLTS (TYP OF 4)

OUTDOOR UNIT

4" CONCRETE PAD

FIGURE 7-10   Connection detail of a ductless split-system unit.

FIGURE 7-11   Photograph of a ductless split-system indoor unit.

FIGURE 7-12   Photograph of a ductless split-system outdoor unit.

- Ductless split-system units typically have a high sensible heat ratio; thus they are not suitable for serving areas with high internal latent heat gains, such as occupant-dense areas.

- There is no flexibility in the coil selection for the indoor units as there is with fan-coil units.

- Ductless split-system units utilize 1-in.-thick air filters; thus they cannot be used to serve areas that require a filtration efficiency higher than MERV 8.

- Ductless split-system units are controlled by either wall-mounted space thermostats or by wireless handheld thermostats.

- The lowest outdoor temperature in which the outdoor air-cooled condensing units and heat pump units can operate in the cooling mode is approximately 23°F. If cooling operation of the outdoor units is required for temperatures below 23°F, the optional wind baffle accessory must be specified to enable low ambient cooling operation down to 0°F outdoor temperature.

- The heating capacity of heat pump units decreases with a decrease in outdoor temperature. The rated heating capacities for heat pump units are normally given for an outdoor temperature of 17°F. If the design winter temperature is less than 17°F, the manufacturer's representative should be consulted to determine the heating capacity at the design winter temperature and see if it is adequate for the application.

- If the heating capacity of a heat pump unit is insufficient at the design winter temperature or for an air-conditioning unit that requires heating capabilities, an electric heater should be specified for the indoor unit.

- Sufficient clear space must be provided around outdoor units for free airflow. The manufacturer's product data should be consulted to determine the minimum separation distances required between adjacent units, from units to nearby walls, and for the proper orientation of discharge air with respect to adjacent units and walls.

- Outdoor units should be securely attached to rigid bases with anchor bolts and neoprene vibration isolation pads.

- As with all split-system units, there is a limit to the separation distance between the indoor and outdoor units for the standard-sized refrigerant piping. The manufacturer's representative should be consulted if the separation distance between the two units exceeds 75 ft because special sizing of the refrigerant lines may be required.

- The locations of ceiling-recessed units need to be coordinated with the locations of lighting fixtures, sprinkler heads, and other systems installed in the ceiling.

- Indoor units should be mounted at least 6 ft 6 in. above the floor to the bottom of the unit so that the units do not become a head-bumping hazard.

Variable Air Volume Terminal Units

In Chap. 5 we discussed, in general, the operation of VAV terminal units within VAV air systems. In this chapter we will look more closely at the two types of VAV terminal units most commonly used in commercial buildings: single-duct and fan-powered VAV terminal units. We will first discuss the similarities of single-duct and fan-powered VAV terminal units and then look at the features that distinguish them. We will also discuss the two types of fan-powered VAV terminal units: series fan-powered and parallel fan-powered VAV terminal units.

Similarities of Single-Duct and Fan-Powered VAV Terminal Units

Purpose
Both single-duct and fan-powered VAV terminal units have the same purpose, to provide heating, cooling, and ventilation to the temperature zone (group of spaces controlled by a single zone temperature sensor) that each VAV terminal unit serves.

Physical Characteristics

Primary Air Inlet   Single-duct and fan-powered VAV terminal units have one (typically round) primary air[13] inlet.

Primary Air Damper and Controller   The primary airflow for single-duct and fan-powered VAV terminal units is modulated by an automatic damper mounted within the primary air inlet. The primary air damper actuator is typically electric and is controlled by a direct digital control (DDC) system controller mounted on the VAV terminal unit.[14] The primary air damper actuator and controller are normally housed within a terminal box mounted on the outside of the VAV terminal unit.

Inlet Velocity Sensor   The primary air inlet is equipped with a velocity sensor consisting of multiple pitot tubes[15] that constantly measure the total pressure and static pressure of the primary airflow at multiple locations across the primary air inlet. This type of sensor is called a multipoint averaging sensor because the sensor averages all of the total pressure measurements and all of the static pressure measurements of the pitot tubes. The velocity pressure, which is the difference between the average total pressure and average static pressure, is read at the controller mounted on the VAV terminal unit. The controller is connected to the inlet velocity sensor by two tubes, one for the average total pressure and one for the average static pressure measured by the inlet velocity sensor. From the velocity pressure, the controller is able to continuously calculate the primary airflow (in cfm).[16] The inlet velocity sensor ensures that the airflow called for by the space temperature sensor is what is actually delivered by the VAV terminal unit (regardless of the static pressure within the primary air duct system) and also ensures that the primary airflow delivered by the VAV terminal unit remains within the preset minimum and maximum primary airflows for the VAV terminal unit.[17]

Outlet   Single-duct and fan-powered VAV terminal units have one rectangular outlet to which the downstream ductwork is connected.

Heating Coil   A hot water or electric heating coil can be specified as an accessory to a single-duct or fan-powered VAV terminal unit if heating is required by the zone.

Terminal Box   The primary air damper actuator and the unit controller for both single-duct and fan-powered VAV terminal units are normally housed within a terminal box mounted on the side of the terminal unit. Figure 7-13 shows the components within the

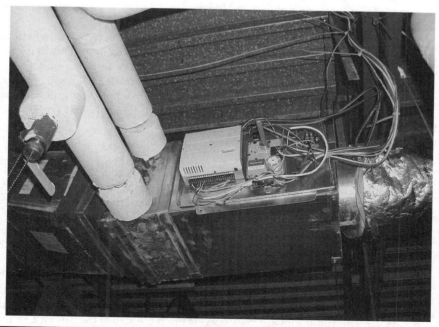

Figure 7-13   Photograph of the terminal box for a VAV terminal unit.

terminal box of a single-duct VAV terminal unit with a hot water heating coil. The damper actuator is mounted on the primary air damper shaft and is located on the right-hand side of the terminal box in Fig. 7-13 and the unit controller is located on the left-hand side of the terminal box.

Liner   Typically, VAV terminal unit casings are lined on the interior with coated, dual-density fiberglass insulation. Other liners with a less porous surface can be specified to reduce the accumulation of dirt; these include fiber-free (foam) liner, foil-faced liner, and a double-wall casing with a galvanized steel inner wall. Specification of these smooth liners slightly reduces the sound attenuation qualities of the VAV terminal unit.

Connections

Both single-duct and fan-powered VAV terminal units require duct connections to the inlet and outlet of the terminal units. The primary air duct system (upstream of VAV terminal units) is usually a medium-velocity duct system (between 1,200 and 2,500 fpm air velocity) having a medium-pressure classification (3 in. w.c. to 6 in. w.c.). The duct distribution system downstream of VAV terminal units will be a low-velocity duct system (maximum 1,000 fpm air velocity) having a low-pressure classification (2 in. w.c. or less). If a hot water heating coil is designed, it will require heating water supply and return piping connections. If an electric heating coil is designed, it will require an electrical power connection that is usually terminated within the terminal box, which is furnished as an integral part of the VAV terminal unit. All VAV terminal units that are controlled by a DDC system require low-voltage (24V ac) control power.

A minimum length of three duct diameters of straight duct should be designed for the inlet ductwork connected to VAV terminal units. This straight length of ductwork ensures a fairly uniform velocity profile across the inlet velocity sensor, increasing the accuracy of the primary airflow measurement.

It is common to design the first 5 ft of supply air ductwork downstream of VAV terminal units to be lined on the interior with 1 in. of duct liner. This helps to attenuate some of the discharge noise that is generated by the VAV terminal unit.

One important distinction between VAV terminal units and many of the other types of terminal equipment that provide cooling is that VAV terminal units do not require a condensate drain connection because they do not contain a cooling coil.

Design Considerations

Primary Airflow Capabilities   All primary air dampers can shut off the primary airflow completely. However, complete shutoff of primary airflow is not recommended if the VAV terminal unit serves a zone that has a heating load (refer to the VAV Terminal Units section in Chap. 5). If the zone has a heating load, the primary air damper should be specified to have a minimum airflow position in order for the VAV terminal unit to provide ventilation to the zone during times of low cooling load and during the heating mode of operation.

The inlet velocity sensor requires a minimum air velocity of about 400 fpm through the inlet in order to accurately measure the primary airflow. Therefore, all VAV terminal units have a minimum primary airflow below which the terminal unit cannot accurately deliver primary air. This minimum airflow depends upon the size of the primary air inlet.

The maximum primary airflow that can be delivered by single-duct and fan-powered VAV terminal units is limited by the acceptable radiated and discharge noise criteria (NC) levels for the terminal units. The noise that is generated by a VAV terminal unit increases as the air velocity through the primary air inlet increases and also increases as the differential static pressure between the inlet and outlet of the terminal unit increases. Most VAV terminal unit manufacturers test their equipment in accordance with the *Air-Conditioning, Heating, and Refrigeration Institute (AHRI) Standard 880-98* to obtain the sound power levels for their terminal units at various primary airflows and various differential static pressures between the inlet and outlet of the terminal unit. Table 7-1 summarizes the minimum allowable airflows for different-sized VAV terminal unit inlets (based on the minimum airflow required by the inlet velocity sensor) and the maximum recommended airflows (based on the AHRI-rated airflows with a 1.5 in. w.c. differential static pressure between the inlet and outlet of the terminal unit). These maximum airflows result in radiated and discharge NC levels for the terminal units of about 30, which is an acceptable NC for spaces within commercial buildings.

Capacity   Once the heating and cooling loads of the zones within the building have been calculated, the heating and cooling capacity of the VAV terminal units serving these zones can be determined. The cooling capacity (or maximum primary airflow) of a VAV terminal unit must be equal to or greater than the cooling supply airflow calculated for the zone. The zone cooling supply airflow is based on the zone sensible cooling load, the design primary air temperature (which is normally 55°F), and the zone cooling setpoint (which is normally 75°F).

The heating capacity of a VAV terminal unit must be equal to the zone heat losses through the building envelope, plus the heating load of any outdoor air infiltration to the zone, plus the heat required to raise the heating primary airflow from the primary air temperature (which is normally 55°F) to the zone heating setpoint (which is normally 70°F). Thus, the VAV terminal unit heating coil not only needs to meet the heating load of the zone served by the terminal unit, it also needs to heat the heating primary airflow up to room temperature. Once the load on the heating coil is known, the heating supply airflow must be calculated. The heating supply airflow should be calculated based on a supply air temperature that is 15°F warmer than the zone heating setpoint,

Inlet Size, in.	Minimum Airflow, cfm	Maximum Airflow,[1] cfm
6	80	400
8	145	700
10	230	1,100
12	325	1,600
14	450	2,100
16	580	2,800

[1]Maximum airflows are the Air-Conditioning, Heating, and Refrigeration Institute–rated airflows (based on a 1.5 in. w.c. differential static pressure between the inlet and outlet of the terminal unit).

TABLE 7-1   Primary Airflow Capabilities for Single-Duct and Fan-Powered VAV Terminal Units

which equates to 85°F for a 70°F zone heating setpoint. If the heating supply air temperature is greater than 20°F warmer than the room temperature, the air within the room will stratify. That is, the warmer air will rise to the top of the room and the cooler air will remain near the bottom of the room, resulting in an uncomfortable condition for the occupants. Refer to the Selection sections of the Single-Duct VAV Terminal Units and Fan-Powered VAV Terminal Units sections later for sample calculations of the heating and cooling supply airflows and the capacities of the heating coils for single-duct and fan-powered VAV terminal units.

Standard hot water heating coils for VAV terminal units are one-row and two-row coils. The heating capacity of the hot water heating coil depends upon the number of rows, entering water temperature, water flow, and entering air temperature. Standard-sized electric heating coils are available for both single-duct and fan-powered VAV terminal units depending upon the unit size. The heating capacity of the electric heating coil is independent of the entering air temperature.

The minimum primary airflow designed for VAV terminal units that do not shut off is commonly 25% of the maximum airflow. This normally provides adequate outdoor air ventilation for commercial buildings. Thus, based on the information presented in Table 7-1, any size VAV terminal unit can be selected to deliver its maximum allowable airflow and still be capable of throttling the primary airflow to 25% of the maximum airflow.

Control   VAV terminal units are either pressure independent or pressure dependent. Pressure-independent VAV terminal units utilize an inlet velocity sensor and are capable of maintaining the required airflow delivered by the units regardless of the pressure fluctuations within the primary air duct system between the AHU and the VAV terminal units. Because pressure-independent VAV terminal units are used in almost all VAV air systems for commercial buildings, we will limit our discussion in this chapter to pressure-independent units. Pressure-dependent VAV terminal units do not have a means of compensating for pressure fluctuations within the primary air duct system and are, therefore, dependent upon the system pressure for the airflow delivered by the units.

Each VAV terminal unit is equipped with a DDC system controller that receives input from the inlet velocity sensor and the space temperature sensor. The controller sends output to the primary air damper actuator to position the damper to deliver the airflow required to maintain the cooling setpoint of the zone temperature sensor. If the VAV terminal unit is equipped with a heating coil, the controller will also send output to the control valve actuator (hot water heating coil) or electric heating coil controller to control the output of the heating coil to maintain the heating setpoint of the zone temperature sensor. The controller on fan-powered VAV terminal units also sends output to the fan to turn it on and off.

The DDC system controller on each VAV terminal unit requires low-voltage (24V ac) power for operation. The 24V ac control power for single-duct VAV terminal units is normally fed from a control transformer mounted in a main ATC panel called a network DDC panel (refer to Chap. 9 for a discussion of DDC systems). This eliminates the need for separate line voltage electrical power connections to all of the single-duct VAV terminal units. For fan-powered VAV terminal units and single-duct VAV terminal units with electric heat, the low-voltage control power is normally derived from the line voltage feed to the terminal unit through a control transformer mounted in the terminal unit terminal box.

Selection In the cooling mode of operation, single-duct and fan-powered VAV terminal units function basically the same. Consequently, the selection of the inlet size for both single-duct and fan-powered VAV terminal units is based solely on the zone cooling supply airflow requirement (refer to the Primary Airflow Capabilities section earlier).

The electric heating coils for VAV terminal units must be selected to ensure that a minimum air velocity through the coil is always present to prevent overheating of the electrical elements and nuisance tripping of the thermal cutouts. The manufacturer's product data should be consulted to determine the minimum airflow required for the electric heating coil.

The supply fan in the AHU must have sufficient external static pressure to overcome the static pressure losses associated with the primary air duct system between the AHU and the farthest VAV terminal unit. Furthermore, for single-duct VAV terminal units and parallel fan-powered terminal units, the AHU supply fan must also overcome the static pressure losses through the farthest VAV terminal unit and the duct distribution system downstream of the terminal unit. For series fan-powered VAV terminal units, the terminal unit fan overcomes the static pressure losses through the terminal unit and the duct distribution system downstream of the terminal unit.

Installation Duct strap[18] can be used to suspend both single-duct and fan-powered VAV terminal units from the building structure. However, if greater vibration isolation is desired, neoprene vibration isolation hangers can be used to suspend fan-powered VAV terminal units from the building structure.

A clear space of at least 18 in. should be provided in front of the terminal box for both single-duct and fan-powered VAV terminal units to allow room for the door swing and for maintenance access. This access space should be coordinated so that it is not above lighting fixtures. Access to the fan within fan-powered VAV terminal units is usually through the bottom panel of the unit casing. Therefore, it is necessary to provide clear access to the bottom of all fan-powered VAV terminal units; that is, they should not be located above lights or any other services that would restrict this access. Ceiling access panels (minimum 24 in. × 24 in.) are required for maintenance access to all VAV terminal units installed above gypsum board ceilings.

Single-Duct VAV Terminal Units

Purpose

Single-duct VAV terminal units are used for zones that require cooling only and where the zone heating supply airflow requirement (at 85°F) is close to the minimum primary airflow required for ventilation. Single-duct VAV terminal units are not equipped with fans to recirculate air and are, therefore, not recommended for zones that have a heating supply airflow requirement at 85°F, which is significantly higher than the minimum primary airflow required for ventilation. Fan-powered VAV terminal units are recommended in this case, as is discussed in the Fan-Powered VAV Terminal Units section later.

Single-duct VAV terminal units can also be used to serve zones that require a constant primary airflow, such as where the zone has continuous exhaust airflow. In this case, the terminal units would be considered constant air volume (CAV) terminal units and the minimum and maximum primary airflows would be the same. For CAV terminal units, the primary air damper only modulates to compensate for pressure fluctuations within the primary air duct system. A heating coil is required for CAV terminal units, whether zones have a heating load or not, to prevent overcooling. The output of the heating coil would be modulated to maintain the setpoint of the zone temperature sensor.

Physical Characteristics

The height of single-duct VAV terminal units ranges from 8 to 18 in., the width ranges from 12 to 24 in., and the length is usually about 20 in. (Fig. 7-14). The rectangular outlet size for single-duct VAV terminal units is approximately the same size as the terminal unit casing.

Connections

Figure 7-15 illustrates the connections associated with a single-duct VAV terminal unit with an electric heating coil. The connections for a single-duct VAV terminal unit with a hot water heating coil would be similar, but the hot water heating coil would be

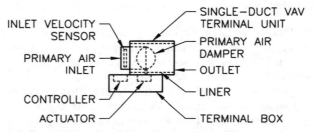

Figure 7-14   Diagram of a single-duct VAV terminal unit.

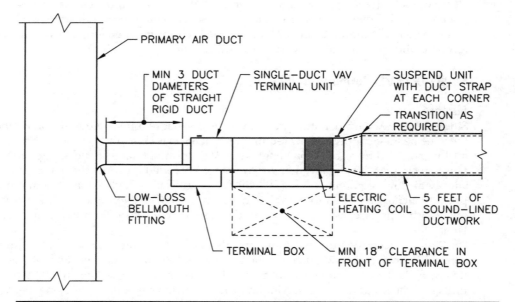

Figure 7-15   Single-duct VAV terminal unit with electric heating coil connection detail.

mounted directly on the outlet of the terminal unit. The heating water piping connections (depending upon the type of control valve) would be similar to those shown in Figs. 7-1 and 7-2. If the VAV terminal unit has no heating coil, the lined discharge ductwork would connect directly to the outlet of the terminal unit.

Design Considerations

Control   The normal sequence of operation for a single-duct VAV terminal unit is for the primary air damper to modulate closed as the space temperature drops below the cooling setpoint of the zone temperature sensor and to modulate open as the space temperature rises above the cooling setpoint of the zone temperature sensor. The primary airflow will modulate between the preset cooling minimum and maximum airflows. If the VAV terminal unit is equipped with a heating coil, the unit will change its mode of operation from cooling to heating mode once the primary air damper reaches its cooling minimum airflow setting and the zone temperature drops below the heating setpoint. In the heating mode of operation, the primary air damper will position to deliver the heating airflow (which may be the same as, or higher than, the cooling minimum airflow) and the output of the heating coil will be modulated as required to maintain the heating setpoint of the zone temperature sensor. Upon a rise in space temperature above the heating setpoint of the zone temperature sensor, the reverse will occur.

Selection   The following are sample calculations for the heating and cooling supply airflows and the capacity of the heating coil for a single-duct VAV terminal unit:

- Zone cooling load: 15,120 Btuh
- Zone heating load: 4,860 Btuh
- Cooling setpoint: 75°F
- Primary air temperature: 55°F
- Heating setpoint: 70°F
- Heating supply air temperature: 85°F
- Cooling supply airflow = (15,120 Btuh)/[(1.08)(75°F − 55°F)] = 700 cfm
- Minimum primary airflow required for ventilation = (700 cfm)(25%) = 175 cfm
- Minimum heating supply airflow[19] = (4,860 Btuh)/[(1.08)(85°F − 70°F)] = 300 cfm
- Heating coil capacity[20] = (4,860 Btuh) + (1.08)(300 cfm)(70°F − 55°F) = 9,720 Btuh

Based on the cooling supply airflow of 700 cfm, a single-duct VAV terminal unit with an 8-in.-diameter primary air inlet should be selected to serve the zone in this example, realizing that the terminal unit is sized for its maximum airflow capability. If any additional zone cooling load is anticipated in the future, a VAV terminal unit with a 10-in.-diameter primary air inlet should be selected to provide some spare capacity for the future.

This example shows how the recommended heating supply airflow of 300 cfm at 85°F is 125 cfm greater than 25% of the maximum cooling supply airflow, which is 175 cfm. What this means is that this VAV terminal unit should have two minimum primary airflows: the cooling minimum airflow of 175 cfm and the heating minimum airflow of 300 cfm. If the heating minimum airflow were allowed to be the cooling minimum airflow of 175 cfm, the heating supply air temperature would have to be 95.7°F in order to provide the 4,860 Btuh of heating required for the zone [4,860 Btuh = (1.08)(175 cfm)(95.7°F − 70°F)].

Since the heating minimum primary airflow is greater than 25% of the maximum primary airflow, the heating coil must heat an additional 125 cfm of primary air than what is required by the zone for ventilation. It would require less energy for the VAV terminal unit to heat 125 cfm of 70°F return air to 85°F than it would to heat 125 cfm of primary air at 55°F to 85°F. It is for this reason that fan-powered VAV terminal units were developed. Fan-powered VAV terminal units have the capability to reduce the primary airflow to the minimum required by the zone for ventilation and utilize return air for the additional airflow required for heating. Refer to the discussion of fan-powered VAV terminal units below for the fan-powered VAV terminal unit selection procedure.

Fan-Powered VAV Terminal Units

Purpose

Fan-powered VAV terminal units should be designed for zones that have a significantly higher heating supply airflow requirement (at an 85°F supply air temperature) than the minimum primary airflow required by the zone for ventilation. Fan-powered VAV terminal units are normally designed for perimeter zones for this reason. Fan-powered VAV terminal units are also most suitable for use in air systems that utilize a ceiling return air plenum. The ceiling return air plenum provides a ready source of return air for the terminal unit fan to draw through the induced air inlet. Fan-powered VAV terminal units can also be used in air systems that do not utilize a ceiling return air plenum. However, a lined return air duct must be connected from the induced air inlet to return air grilles in all of the spaces served by the terminal unit.

Physical Characteristics

Fan-powered VAV terminal units have all of the same components that single-duct VAV terminal units have with the addition of a fan and an induced air inlet. The fan can be oriented within the terminal unit such that it is in series with the primary airflow or parallel to the primary airflow. Thus, the two types of fan-powered VAV terminal units are called series fan-powered and parallel fan-powered VAV terminal units. Figures 7-16 and 7-17 illustrate the relationship of the fan to the primary airflow for these two types of fan-powered VAV terminal units.

Because of the fan that is installed within the terminal unit, fan-powered VAV terminal units are significantly larger than single-duct VAV terminal units. The height of fan-powered VAV terminal units ranges from 17 to 20 in., although low-profile units that are approximately 10 in. high are available. The width ranges from 36 to 48 in., and the length ranges from 41 to 47 in. The rectangular outlet size for fan-powered VAV terminal units ranges from 14 to 17 in. wide by 11 to 14 in. high. Because of their size, fan-powered VAV terminal units require close coordination with ductwork and other services above the ceiling.

Fan-powered VAV terminal units will almost always be equipped with a heating coil. If no heating coil is required in the terminal unit, a single-duct VAV terminal unit would be a more cost-effective choice to serve the cooling needs of the zone. Hot water

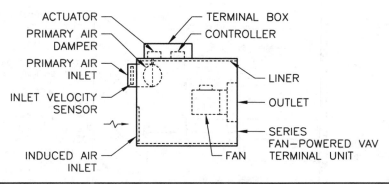

FIGURE **7-16** Diagram of a series fan-powered VAV terminal unit.

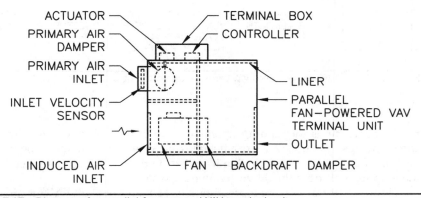

FIGURE **7-17** Diagram of a parallel fan-powered VAV terminal unit.

or electric heating coils are mounted on the outlet connection of series fan-powered VAV terminal units. For parallel fan-powered VAV terminal units, the hot water coil is normally mounted on the induced air inlet and the electric heating coil is mounted on the outlet connection of the unit.

A filter rack on the induced air inlet is usually available as an optional accessory and is recommended to keep the heating coil clean. However, filter replacement will become an ongoing maintenance item for the building owner.

Terminal Box Fan-powered VAV terminal units with an electric heating coil normally have a larger terminal box to allow room for the terminal block (where the electrical power is connected) and the electric heating coil controller to be installed within the terminal box.

Connections
Figure 7-18 illustrates the connections associated with a parallel fan-powered VAV terminal unit with a hot water heating coil. The connections for a series fan-powered VAV terminal unit with a hot water heating coil would be similar, but the hot water heating coil would be mounted on the outlet of the terminal unit. The heating water

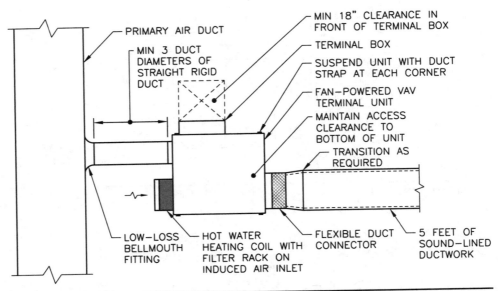

FIGURE 7-18   Parallel fan-powered VAV terminal unit with hot water heating coil connection detail.

piping connections (depending upon the type of control valve) are similar to those shown in Figs. 7-1 and 7-2. Also, a flexible duct connector is required on the outlet connection of fan-powered terminal units to isolate the vibrations generated by the fan in the terminal unit from the downstream duct system.

Figures 7-19 and 7-20 are representations of a fan-powered VAV terminal with a hot water heating coil along with the associated ductwork and piping shown on separate floor plans. Also shown on the HVAC piping plan is a ceiling-mounted cabinet unit heater and floor-mounted finned-tube radiators.

Design Considerations

Control   The operation of the primary air damper for both types of fan-powered VAV terminal units is the same as the cooling sequence of operation for the primary air damper in single-duct VAV terminal units. The fan in series fan-powered VAV terminal units runs continuously during the occupied mode of operation. The fan in parallel fan-powered VAV terminal units only runs when heating is required by the zone temperature sensor. During heating operation, the output of the heating coil is modulated as required to maintain the heating setpoint of the zone temperature sensor. Therefore, the only difference in the operation of series and parallel fan-powered terminal units is in the operation of the fan. Series fan-powered VAV terminal units deliver a constant supply airflow to the zone at a variable temperature for both cooling and heating operation. Parallel fan-powered VAV terminal units deliver a variable supply airflow to the zone at a constant temperature for cooling operation and a constant supply airflow to the zone at a variable temperature during heating operation.

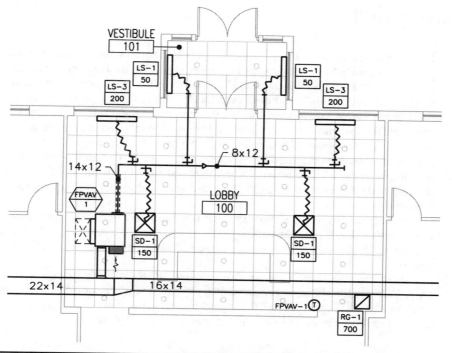

FIGURE 7-19 HVAC duct plan.

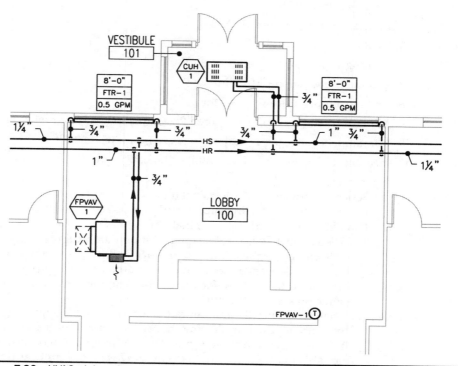

FIGURE 7-20 HVAC piping plan.

Selection   The fan in a series fan-powered VAV terminal unit must be capable of supplying the maximum cooling supply airflow since the fan is in series with the primary air. The fan in a parallel fan-powered VAV terminal unit only needs to be sized to deliver the heating supply airflow (based on an 85°F supply air temperature) minus the minimum primary airflow required for ventilation since the fan is in parallel with the primary air. The manufacturer's product data must be consulted to determine the minimum terminal unit fan airflow for parallel fan-powered VAV terminal units because it may be greater than what is required based on an 85°F supply air temperature. The minimum terminal unit fan airflow is usually in the range 300 to 400 cfm, but this depends upon the size of the fan and the external static pressure of the duct system downstream of the terminal unit.

The fan in both series and parallel fan-powered VAV terminal units must be capable of delivering the design fan airflow at the external static pressure required to overcome the losses in the downstream duct system, heating coil, and induced air inlet filter (if any).

Using the same zone heating and cooling loads for the example given in the Single-Duct VAV Terminal Units section earlier, the following are sample calculations for the heating and cooling supply airflows and the capacity of the heating coil and terminal unit fan for both series and parallel fan-powered VAV terminal units:

1. Zone cooling load: 15,120 Btuh
2. Zone heating load: 4,860 Btuh
3. Cooling setpoint: 75°F
4. Primary air temperature: 55°F
5. Heating setpoint: 70°F
6. Heating supply air temperature: 85°F
7. External static pressure: 0.25 in. w.c.
8. Cooling supply airflow = 700 cfm
9. Minimum primary airflow required for ventilation = (700 cfm)(25%) = 175 cfm
10. Minimum heating supply airflow[21] = (4,860 Btuh)/[(1.08)(85°F – 70°F)] = 300 cfm
11. Heating coil capacity[22] = (4,860 Btuh) + (1.08)(175 cfm)(70°F – 55°F) = 7,695 Btuh
12. Fan airflow
    a. Series fan-powered VAV terminal unit
       (1) Fan airflow = cooling supply airflow = 700 cfm
    b. Parallel fan-powered VAV terminal unit
       (1) The fan airflow is ideally equal to minimum heating supply airflow of 300 cfm minus the minimum primary airflow of 175 cfm, which equals 125 cfm. However, based on the manufacturer's product data, the minimum fan airflow for the terminal unit at 0.25 in. w.c. of external static pressure is 300 cfm.
       (2) Therefore, since the minimum fan airflow at the design external static pressure is equal to 300 cfm, the terminal unit will deliver the sum of the minimum fan airflow plus the minimum primary airflow, which equals a total of (300 cfm + 175 cfm) = 475 cfm of supply airflow during heating operation.

13. Heating supply air temperature
   a. Series fan-powered VAV terminal unit
      (1) Heating supply air temperature = {(zone heating load)/[(1.08)(heating supply airflow)]} + heating setpoint
      (2) Heating supply air temperature = {(4,860 Btuh)/[(1.08)(700 cfm)]} + 70°F = 76.4°F
   b. Parallel fan-powered VAV terminal unit
      (1) Heating supply air temperature = {(zone heating load)/[(1.08)(heating supply airflow)]} + heating setpoint
      (2) Heating supply air temperature = {(4,860 Btuh)/[(1.08)(475 cfm)]} + 70°F = 79.5°F

The primary air inlet size for both the series and parallel fan-powered VAV terminal units would be the same as for the single-duct VAV terminal unit illustrated in the single-duct VAV terminal unit example above. The fan for the series fan-powered VAV terminal unit would be selected to deliver 700 cfm of air at 0.25 in. w.c. external static pressure. The fan for the parallel fan-powered VAV terminal unit would be selected to deliver 300 cfm of air at 0.25 in. w.c. external static pressure.

Comparison

The following are some of the advantages and disadvantages of single-duct, series fan-powered, and parallel fan-powered VAV terminal units.

Single-Duct VAV Terminal Units

Advantages

- Most energy-efficient choice because there is no fan energy associated with the terminal unit.
- Lowest first cost.
- Lowest maintenance cost because there is no air filter or fan that requires maintenance.

Disadvantages

- Not suitable for use when the heating airflow is significantly higher than the minimum primary airflow required for ventilation.
- Mixing of room air is decreased as the primary airflow is reduced, which occurs during times of low cooling load and during heating operation.

Series Fan-Powered VAV Terminal Units

Advantages

- Constant supply airflow:
  - Consistent noise from the terminal unit.
  - Effective mixing of room air at all times.

Disadvantages

- Least energy-efficient because the fan runs continuously during occupied operation.
- Lowest supply air temperature during heating operation, which may feel drafty to occupants.

Parallel Fan-Powered VAV Terminal Units

Advantages

- More energy-efficient than series fan-powered VAV terminal units:
  - The fan runs only during the heating mode of operation.
  - A smaller fan is required.
- Warmer supply air temperature than series fan-powered VAV terminal units during heating operation, which will feel more comfortable to occupants.

Disadvantages

- Variable supply airflow:
  - Variable noise from the terminal unit.
  - Variable mixing of room air.
- Mixing of room air is decreased as the primary airflow is reduced, which occurs during times of low cooling load.

Coordination With Other Disciplines

Coordination with other disciplines includes the following:

- The architect should review the selections of all terminal equipment that will be visible in the occupied spaces to ensure that they coordinate with the interior design.
- The electrical requirements of the terminal equipment should be communicated to the electrical engineer through the equipment list and a highlighted set of HVAC plans, as discussed in Chap. 2.

Endnotes

1. Line voltage is a term that refers to the operating voltage of the equipment.
2. A step controller is a series of contactors that close individually to energize the electric heater element in stages. For example, a three-step controller energizes the electric heater in one-third capacity increments.
3. An SCR controller is a solid-state device that uses a pulsed signal to control the percentage of time the electric heater is energized (it does not control the electrical power input). SCR controllers are capable of modulating the output of the electric heater from 0 to 100%.

4. Unit heaters can also burn fuel oil, but fuel-burning unit heaters are not common in commercial buildings.

5. Gas-fired unit heaters typically have an 80% thermal efficiency; thus, the gross output is typically 80% of the burner input rating.

6. The envelope heat loss is the sum of the heat losses through the walls, windows, doors, roof, and floor slab of the vestibule.

7. External static pressure refers to the static pressure losses external to the unit. Static pressure losses associated with the components of a fan-coil or air handling unit, such as the filter and coils, are called internal static pressure losses. The supply fan must develop a total static pressure sufficient to overcome both the external and internal static pressure losses.

8. Drive refers to the pulleys on the motor and fan for belt-driven fans. A larger motor pulley and/or smaller fan pulley increases the fan speed and, as a result, causes the fan to develop a greater static pressure. However, this also requires more power (brake horsepower) from the motor.

9. The total cooling load is equal to the sensible cooling load plus the latent cooling load. The sensible component of the cooling load (in Btuh) is met by the cooling coil through a decrease in the dry bulb temperature of the air flowing through the cooling coil. The latent component of the cooling load (also in Btuh) is met by the cooling coil through the condensation of the water vapor within the air flowing through the cooling coil. Thus, latent cooling of moist air is not related to a change in dry bulb temperature of the air; rather, it is related to a reduction in the moisture content of the air.

10. The cooling supply airflow is strictly a function of the supply air temperature, the space sensible heat gains, and the space temperature. The cooling supply airflow required to maintain the design space temperature setpoint is independent of the amount of outdoor air ventilation that is mixed with return air upstream of the cooling coil.

11. The saturated suction temperature is the temperature at which the refrigerant evaporates within the cooling coil. This is the temperature of the cooling coil.

12. The average coil temperature of an oversized cooling coil will be warmer than the average coil temperature of a properly sized cooling coil. Less moisture is condensed on a warmer surface; thus the dehumidification capability of an oversized cooling coil is reduced.

13. Primary air is the air that is delivered to the VAV terminal units by the AHU. Typically, the primary air temperature is 55°F so that cooling can always be provided through the VAV terminal units to the temperature zones, if required.

14. A pneumatic damper actuator may also be used with a pneumatic control system, although this is less common.

15. A pitot tube measures total air pressure and static air pressure. The difference between the total pressure and the static pressure is the velocity pressure.

16. Airflow can be calculated if both the velocity pressure of the airflow through a duct and the cross-sectional area of the duct are known.

17. The maximum and minimum primary airflows for the VAV terminal units designed for a project are given in the VAV terminal unit equipment schedule on the construction drawings. These values are preset at the factory before the units are shipped to the contractor and are also verified by the testing, adjusting, and balancing contractor after installation (during start-up).

18. Duct strap is a light gauge strip of sheet metal about 1 in. wide that is cut to length and is used for suspending ductwork in accordance with SMACNA guidelines. It is also used to suspend small pieces of HVAC equipment such as VAV terminal units and fans.

19. Based on an 85°F heating supply air temperature.

20. Based on the heating supply airflow of 300 cfm.

21. Based on an 85°F heating supply air temperature.

22. Based on the minimum primary airflow required for ventilation of 175 cfm.

CHAPTER 8

Noise and Vibration Control

Although sound and vibration are complex subjects, a basic understanding of the fundamental methods of controlling noise (unwanted sound) and vibration is necessary to avoid what can be the most objectionable qualities of a poorly designed HVAC system.

Noise Control

Sound Power and Sound Pressure

Sound power and sound pressure are two related, but different, terms that are used in the study of sound. *Sound power* is the intensity of a source, such as a chiller, fan, or cooling tower. It cannot be measured directly but must be calculated through measurements of sound pressure in a controlled environment. Sound power for HVAC equipment is commonly determined through measurements of sound pressure conducted in accordance with the Air-Conditioning, Heating, and Refrigeration Institute (AHRI) standards. *Sound pressure* is measureable, diminishes in proportion to the square of the distance from the source, and is affected by intervening obstacles. Because sound pressure varies with the distance from the source, sound pressure levels will always be given in terms of the distance from the source.

Sound power and sound pressure are both expressed in terms of decibels (dB), a unit based on a reference sound power of 10^{-12} W,[1] which is defined as having a sound power level of 0 dB. Sound power is represented by the following equation, which converts sound power from watts to decibels:

$$L_W = 10 \log W + 120$$

where

L_W = sound power (dB)
W = sound power (W)

A sound power level of 1 W is equivalent to 120 dB,[2] a sound power level of 2 W is equivalent to 123 dB,[3] and a sound power level of 10 W is equivalent to 130 dB.[4] Thus, we can see that an increase of 3 dB in sound power level is a two-fold increase in sound power level, and an increase of 10 dB in sound power level is a ten-fold increase in sound power level.

Some common sources and their associated sound power levels are as follows:

Source	Sound Power, dB
Soft whisper	30
Conversational speech	70
Shouting voice	90
Loud radio	110
Small aircraft engine	120
Loud rock band	130
Afterburning jet engine	170

Tonal Qualities of Sound

Sounds have different tonal qualities; that is, they have predominantly low-, mid-, or high-frequency tones. The tonal quality of a sound can be defined by its sound spectrum. In order to describe the sound spectrum of a source, sound power levels and sound pressure levels are often reported in terms of decibels in each of eight octave bands having center frequencies of 63, 125, 250, 500, 1,000, 2,000, 4,000, and 8,000 Hz. By analyzing the sound spectrum, the tonal quality of the sound can be determined. The characteristics of sound in the various octave bands can be described as follows:

Octave Band	Characteristic
Below 63 Hz	Throb
63 to 125 Hz	Rumble
125 to 500 Hz	Roar, hum, and buzz
500 to 2,000 Hz	Whine and whistle
2,000 to 8,000 Hz	Hiss

Balanced and Unbalanced Sound

Sound can also be defined as being balanced or unbalanced. Unbalanced sound has strong tonal components; that is, the sound spectrum has peaks in certain octave bands. Balanced sound conforms to established octave band spectra, such as noise criteria (NC) curves, which were established to rate the noisiness of indoor spaces (refer to the discussion of noise criteria below). Generally, unbalanced sound is more objectionable than balanced sound and may require noise control measures.

A-Weighted Sound Pressure Level

Sounds may also be rated by means of the A scale, which approximates the response of the human ear to sound. The A-scale (or A-weighted) sound pressure level is expressed in terms of dBA. To obtain the A-weighted sound pressure level of a source, a weighting factor is applied to the sound pressure level measured in each octave band to adjust the spectrum to the A-scale.[5] The adjusted sound pressure levels in adjacent octave bands are then summed through logarithmic addition to arrive at the overall A-weighted sound pressure level. A-weighted sound pressure levels are a good indicator of the perceived loudness of a source but give no indication of the tonal quality or spectral balance of the sound.

To provide a frame of reference for the perceived loudness of sources, the following are some common sources and their associated A-weighted sound pressure levels:

Source	Sound Pressure, dBA
Soft whisper at 5 ft	34
Conversational speech at 5 ft	50
Loud radio in an average room	70
Vacuum cleaner at arm's length	74
Jackhammer at 50 ft	88
Home lawn mower at 5 ft	95
Chain saw at arm's length	116
Threshold of pain	120
Immediate damage to unprotected ears	130

Table 8-1 provides an example of the sound pressure level spectrum and A-weighted sound pressure level as they might appear in a manufacturer's product data for a particular piece of HVAC equipment.

Outdoor Noise

Noise generated by outdoor HVAC equipment can be a concern because there may be local or state regulations that establish maximum values of noise at the property lines. These values will depend upon the location of the building (residential, commercial, or industrial district, etc.) and the time of day. Also, the building owner may want to keep certain areas outside the building free from HVAC equipment noise.

It is important to understand that noise in any octave band is additive to the ambient noise in the same octave band. For example, if the ambient sound pressure level in an octave band is 60 dB and a piece of HVAC equipment has a sound pressure level of 60 dB in the same octave band, the resulting sound pressure level in that octave band will be doubled (i.e., it will be measured at 63 dB). However, if the HVAC equipment has a sound pressure level that is 10 dB (or more) lower than the ambient sound pressure level in the same octave band, the ambient sound pressure level in that octave band will be virtually unaffected and will, for this example, remain at 60 dB. This is important to keep in mind when considering the effect that HVAC equipment noise will have on ambient noise both outdoors and indoors.

Octave Band Number (Center Frequency)									Overall A-weighted[1]
1 (63 Hz)	2 (125 Hz)	3 (250 Hz)	4 (500 Hz)	5 (1,000 Hz)	6 (2,000 Hz)	7 (4,000 Hz)	8 (8,000 Hz)		
66	68	65	65	62	56	56	55		67

[1]30 ft from side of unit.

TABLE 8-1 Sample HVAC Equipment Sound Pressure Data, dB

Indoor Noise

Maximum Noise Level
For equipment rooms, the noise level should be kept below 85 dBA. If it isn't, the Occupational Safety and Health Administration (OSHA) will require hearing conservation measures to protect employees who could be exposed to this noise for an 8-hour day.

Noise Criteria Rating
For other indoor locations, such as offices, conference rooms, corridors, and similar areas, the noise should be kept below the accepted indoor NC level for the space. The NC method of rating indoor noise is the most widely used method, although other methods such as the room criteria (RC) and balanced noise criteria (NCB) are also used, particularly to identify interference with speech and imbalances in the noise spectrum. Table 8-2 and Fig. 8-1 illustrate the NC curves.

Noise Criteria	Octave Band Number (Center Frequency)							
	1 (63 Hz)	2 (125 Hz)	3 (250 Hz)	4 (500 Hz)	5 (1,000 Hz)	6 (2,000 Hz)	7 (4,000 Hz)	8 (8,000 Hz)
20	51	41	33	26	22	19	17	16
30	57	48	41	35	31	29	28	27
40	64	57	51	45	41	39	38	37
50	71	64	59	54	51	49	48	47
60	77	71	67	63	61	59	58	57
70	83	79	75	72	71	70	69	68

TABLE 8-2 NC Curve Tabulations, dB

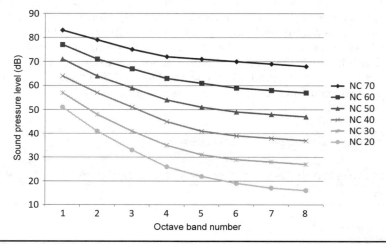

FIGURE 8-1 NC curves.

The sound pressure level in the octave band having the highest NC level determines the NC level of the space. For example, a space with a sound pressure level of 41 dB in the 1,000-Hz octave band would have an NC level of 40. Noise criteria design guidelines for various commercial spaces are as follows:

Space	Design NC level
Conference room	30
Private office	35
Open office	30
Corridor	45

Discharge and Radiated Sound Power Levels

For airside HVAC equipment, it is common for the intensity of the noise generated by the equipment to be given in terms of the discharge sound power and radiated sound power levels.

The discharge sound power is the noise generated by the equipment that will be transmitted through the air within the ductwork to the occupied spaces. The discharge sound pressure that will be perceived in the areas served by the equipment will be transmitted through the air inlets and outlets located in these areas. Discharge sound pressure diminishes in proportion to the square of the distance from the source and is also affected by intervening obstacles within the ductwork distribution system, such as sound attenuators and variable air volume (VAV) terminal units.

The radiated sound power is the noise that is generated by vibration of the equipment walls. The radiated sound power is transmitted to the space in which the equipment is installed and can also be transmitted to adjacent spaces through vibration of the building walls and floors and the ductwork that is connected to the equipment. Radiated sound pressure also diminishes in proportion to the square of the distance from the source and is affected by intervening obstacles such as sound attenuating materials installed on the equipment, building walls, floors, and ductwork.

Noise Control Measures

In order to properly attenuate the noise generated by HVAC equipment to acceptable levels, the sound power level spectrum of the source and the desired NC level for the space must first be defined. Then it is necessary to identify the sound paths from the source to the building occupants and apply insertion losses for the intervening obstacles. Insertion losses of the intervening obstacles are described as such because when they are "inserted" between the source and the receiver (building occupants), they attenuate the sound pressure levels in each octave band to varying degrees depending upon the nature of the obstacle. Once the insertion losses for all intervening obstacles along a sound path have been applied, the resultant sound spectrum is what will be perceived by the building occupants. The resultant sound spectrum is then compared to the acceptable NC level for the space to determine whether noise control measures need to be applied or not and, if so, to what level they are required in each of the eight octave bands.

In general, dense materials have good sound attenuation qualities for low-frequency noise and less dense materials have good sound attenuation qualities for high-frequency noise. For example, an acoustical consultant may specify a lead blanket to be installed below the floor of a penthouse equipment room to attenuate the radiated noise of a centrifugal chiller (which generates predominantly low-frequency noise). On the other hand, partition walls filled with fiberglass batt insulation are effective in attenuating the radiated noise of speech (which has sound in the mid- to high-frequency range).

The following are some generally accepted practices for reducing the HVAC noise that will be perceived by building occupants:

- Locate equipment rooms as far away from noise-sensitive areas as possible.

- Coordinate with the project architect to ensure that the walls of equipment rooms are constructed of concrete masonry units.

- Locate outdoor HVAC equipment, such as chillers, cooling towers, and air-cooled condensing units, away from noise-sensitive areas such as building entrances and operable windows.

- Locate rooftop HVAC equipment, especially compressorized HVAC equipment, as far away from noise-sensitive areas on the top floor of the building as possible.

- In general, a larger fan wheel rotating at a lower speed generates less noise than a smaller fan wheel rotating at a higher speed. Therefore, if two fans meet the design requirements for airflow and static pressure, the larger of the two fans will normally generate less noise.

- Design sound-lined ductwork for the first 10 ft of supply and return air ductwork connected to air handling units.

- Design sound attenuators[6] in the supply and/or return air ductwork connected to large air handling units or air handling units that serve noise-sensitive spaces. A sound attenuator manufacturer's representative should be contacted for the selection of sound attenuators. The representative will need the sound power level spectrum of the equipment, duct sizes, airflows, and the intervening obstacles along the sound paths in order to select sound attenuators with the appropriate dynamic insertion loss[7] (DIL) in each octave band to achieve the desired room NC level.

- The size of the opening for open-end return air ductwork should be based on an air velocity of 500 fpm through the opening wherever possible. A maximum air velocity of 750 fpm is acceptable if a 500 fpm air velocity cannot be achieved. Transition the open-end duct to the return air duct main with a maximum combined angle of 30°. Design sound-lined ductwork for the open-end return air ductwork transition and the first 5 ft of return air main ductwork.

- Design open-end return air ductwork to air handling units to have at least one elbow between the unit and the open end of the ductwork.

- Design sound-lined ductwork for the first 5 ft of supply air ductwork downstream of VAV terminal units, especially fan-powered VAV terminal units.

- Locate VAV terminal units above corridors wherever possible and not above private offices or conference rooms. This is especially true for fan-powered VAV terminal units, which have a higher radiated sound power than single-duct VAV terminal units.

- Design transfer air ducts to be sound-lined and have a Z-configuration to eliminate a straight line sound path between the openings on each end of the transfer air duct.

- Design flexible duct connectors for all ductwork connections to equipment with rotating elements, such as fans or compressors.

- Design flexible pipe connectors for all piping connections to equipment with rotating elements, such as pumps, cooling towers, and chillers.

- Design sleeves for all duct penetrations of walls. The sleeves should be at least 1 in. larger than each of the duct dimensions. The space between the duct and the sleeve should be packed with mineral fiber and caulked.

- Design final connections to supply air diffusers with flexible ductwork not to exceed 8 ft in length.

- Maintain air velocities within ductwork below the acceptable limits depending upon the location of the ductwork (i.e., in a shaft, above a corridor, above an occupied space). Refer to the Duct Sizing Criteria section in Chap. 6.

- Specify manufacturer-furnished sound attenuating blankets or enclosures for indoor chillers.

- Specify manufacturer-furnished sound attenuating options, such as compressor enclosures or oversized condenser fans, for outdoor chillers.

- Specify manufacturer-furnished sound attenuators for cooling towers.

Vibration Control

All pieces of HVAC equipment that utilize motor-driven rotating elements (such as pumps, fans, and compressors) vibrate, producing a vibratory force that is proportional to the weight of the equipment. This vibration energy will be transmitted to the building structure, piping, and/or ductwork distribution systems if the HVAC equipment is rigidly connected to these elements. The vibration transmitted through these elements may become an annoyance to the building occupants, may affect the operation of sensitive equipment, and may be radiated to certain areas within the building as structure-borne noise.

In order to reduce the transmission of vibration from HVAC equipment to the building structure, piping, and ductwork distribution systems, vibration isolation is required either within the HVAC equipment itself (internal isolation) or at the connection points of the HVAC equipment to these elements (external isolation). Effective vibration isolation is based on the principle that flexible materials transmit vibration energy less efficiently than stiff materials do. Thus, in order to isolate the vibration of HVAC equipment from the (stiff) building structure, piping, and/or ductwork, the HVAC equipment should be connected to these elements with flexible materials.

Vibration isolation is normally described in detail in the project specifications. However, vibration isolation is often shown in the connection details for the various pieces of HVAC equipment included in a project. We will briefly discuss some of the more common types of vibration isolation utilized in commercial buildings.

Equipment Hangers and Supports

Vibration isolation is used in the hangers, floor-mounted supports, and roof-mounted supports for vibrating HVAC equipment and its associated piping. The effectiveness of vibration isolation hangers and supports is measured in terms of isolation efficiency, which is the percentage of the vibratory force that is not transmitted to the support structure. Isolation efficiency is directly proportional to the deflection of the vibration isolator under the operating weight of the HVAC equipment (i.e., isolation efficiency increases with an increase in isolator deflection).

Furthermore, isolation efficiency is increased when the deflection of the vibration isolator is high compared to the deflection of the support structure. This ratio of isolator deflection to the support structure deflection is an important consideration for any piece HVAC equipment installed on a floor (which is not a slab on grade) or roof. But, it is particularly important when determining the appropriate vibration isolation for HVAC equipment that is supported by light construction, such as a roof supported by open-web joists.

The two components that are most often used in vibration isolation for HVAC equipment are:

1. Neoprene
   a. Neoprene's elastic properties provide effective isolation of high-frequency vibration. Neoprene isolators are designed to be fairly stiff, having a deflection of only about ¼ in. Because the deflection is so small, neoprene can be used in conjunction with spring isolators without the adverse effects that can occur due to resonance if two spring isolators are used in series to isolate the same piece of equipment.[8]
   b. Neoprene is often used on the bottom of the base plate of spring isolators to isolate high-frequency vibration.
   c. Separate neoprene vibration isolators are often used as external vibration isolation for internally isolated HVAC equipment, such as chillers and cooling towers, once again, to isolate high-frequency vibration.

2. Steel springs
   a. Spring isolators are used where a higher static deflection, in the range of ¾ to 3½ in., is required to isolate low-frequency vibration.

The following are some common types of vibration isolation hangers and supports and their recommended uses:

1. Vibration isolation hangers
   a. Combination spring and neoprene vibration isolation hangers
      (1) All piping in equipment rooms.
      (2) First 50 ft of piping connected to vibrating HVAC equipment.
      (3) Suspended HVAC equipment that is not internally isolated.
   b. Neoprene vibration isolation hangers
      (1) Suspended HVAC equipment that is internally isolated.

2. Vibration isolation supports
   a. Neoprene pads or molded mounts for equipment bases
      (1) Base-mounted HVAC equipment, such as air handling units, condensing units, heat pump units, chillers, and cooling towers, which is internally isolated.
   b. Spring isolators for equipment bases
      (1) Base-mounted HVAC equipment that is not internally isolated.
   c. Concrete inertia bases with spring isolators
      (1) End-suction pumps.

Flexible Pipe and Duct Connectors

Flexible pipe and duct connectors are used in the piping and ductwork connections to vibrating HVAC equipment to isolate the equipment vibration from the piping and ductwork distribution systems in the building.

As discussed in Chaps. 3 and 6, flexible pipe connectors are either reinforced rubber or stainless steel hose and braid. Flexible pipe connectors should be designed for the piping connections to the following pieces of HVAC equipment:

- End-suction pumps
- Cooling towers
- Water-source heat pump units
- Other pieces of motor-driven equipment where the equipment vibration may be transmitted to the piping system

As discussed in Chap. 6, flexible duct connectors consist of a rubber-coated glass fabric and should be used for the ductwork connections to the following pieces of HVAC equipment:

- Air handling units
- Fan-coil units
- Water-source heat pump units
- Packaged rooftop units
- Fans
- Discharge air duct for fan-powered VAV terminal units
- Other pieces of motor-driven equipment where the equipment vibration may be transmitted to the ductwork system

Additional Resources

For more information, the reader is directed to the Vibration Isolation and Control section in Chap. 48 of the 2011 *ASHRAE Handbook—HVAC Applications*. Table 47 in that chapter lists the recommended vibration isolation for various pieces of HVAC equipment based on the type and size of the equipment and its location within the building. Also, vibration isolation manufacturers are a good source of technical information. The HVAC system designer should consult with a manufacturer's representative or a vibration consultant for any applications that are out of the ordinary or for which further guidance is required.

Endnotes

1. A sound power level of 10^{-12} W is the lowest sound level which a person with excellent hearing can discern.
2. $L_w = 10 \log(1) + 120 = 10(0) + 120 = 120$ dB
3. $L_w = 10 \log(2) + 120 = 10(0.30) + 120 = 123$ dB
4. $L_w = 10 \log(10) + 120 = 10(1) + 120 = 130$ dB

5. A-weighted sound pressure levels can also be measured directly with a sound level meter.

6. Sound attenuators are also referred to as duct silencers.

7. Dynamic insertion loss (DIL) is the difference in sound pressure levels measured within a space before and after the insertion of the sound attenuator. The DIL of a sound attenuator is inversely proportional to the air velocity through the sound attenuator (i.e., the DIL decreases as the air velocity through the sound attenuator increases).

8. Two spring isolators should not be used in series to isolate the same piece of equipment without the expert advice of a vibration consultant because resonance of the unit and excessive vibration can occur at certain rotational speeds.

CHAPTER 9

Automatic Temperature Controls

The automatic temperature control (ATC) system is the part of the HVAC system that causes it to operate according to its design intent. Modern ATC systems are microprocessor-based direct digital control (DDC) systems that have, at their heart, an ATC system controller. The ATC system controller sends output to certain ATC components within the HVAC system commanding them to perform their intended functions, receives input from other ATC system components to determine if the HVAC system is functioning properly, makes any necessary adjustments to the operation of the HVAC system, and signals an alarm when the HVAC system is operating outside of established parameters. Without an ATC system, the HVAC system would either run continuously (out of control) or it would not run at all.

DDC ATC systems are the most widely used systems for medium- and large-sized commercial building projects (projects larger than 10,000 ft²). However, pneumatic[1] and electrical ATC systems may also be utilized to control certain portions of the HVAC systems. Pneumatic ATC systems are commonly found in older buildings; however, some modern buildings use a combination of DDC and pneumatic ATC components. In this case, pneumatic actuators are typically used for the ATC components, which require a high operating torque, such as large control valves and large motor-operated dampers. Electrical ATC systems are commonly used for certain pieces of terminal equipment (such as finned-tube and electric radiators, unit heaters, and cabinet unit heaters) that are not connected to a centralized building automation system (BAS).[2] Electrical ATC systems normally utilize low-voltage[3] (24V ac) ATC components. The 24V ac source for the low-voltage ATC components is derived from the line voltage through a step-down control transformer. Line voltage electrical ATC components are also used where it is not desirable to derive a separate low-voltage control circuit.

Whether they use DDC, pneumatic, or electrical components, ATC systems can be designed to operate in a self-contained mode with no communication to a wider control network, or they can be designed to operate within the framework of a networked control system, often referred to as a BAS. ATC systems that communicate with a BAS are still fully functional, stand-alone systems; that is, they are able to perform their design function even if communication with the BAS is disrupted. However, the advantage of a BAS is that it enables the building operator to monitor and control the various HVAC systems serving a building (or multiple buildings) from a central location, typically a computer workstation. Many building automation systems are also web-based, which enables access to these systems through the Internet from anywhere in the world.

The purpose of this chapter is not to provide an in-depth analysis or detailed description of ATC systems. Rather, it is to provide an overview of the commonly used ATC system components and how these components can work together as a system to enable the various types of commercial HVAC systems they control to perform their intended functions. Where any explanation of the actual operation of the control system is given, it will be in the context of a DDC or electrical ATC system, not a pneumatic ATC system.

Components

ATC systems are composed of various components (referred to as points) that provide input to the ATC system controller or receive output from the controller. The input received by the controller from the points and the output sent by the controller to the points is either analog or digital. Analog information is continuous, or constantly changing. Digital information is in one of two discrete states (such as on or off, opened or closed, voltage or no voltage).

Analog information is communicated through a variable electrical signal [most commonly 0 to 10V dc, 2 to 10V dc, or 4 to 20 milliamperes (mA)] or through a variable resistance. Digital information is communicated through the opening or closing of electrical contacts[4] or the presence or absence of the control voltage. The input and output points for ATC system controllers are referred to as analog input (AI), analog output (AO), digital input (DI), or digital output (DO) points.[5]

Examples of AI points include space temperature, duct static pressure, heating water supply temperature, and chilled water system differential pressure. Examples of AO points include positioning of a modulating control valve, positioning of a variable air volume (VAV) terminal unit primary air damper, modulation of supply fan speed for a VAV air handling unit, and modulation of pump speed for a variable flow pumping system. Examples of DI points include fan status (on/off), smoke detector status (normal/alarm), high-limit heating water temperature (normal/alarm), and low-limit air temperature (normal/alarm). Examples of DO points include fan (on/off), two-position control valve (open/close), smoke damper (open/close), and cooling tower sump heater (on/off).

In the following sections, we will discuss the various types of ATC components that are commonly used to control commercial HVAC systems. Our discussion will not focus on the details of how these components actually work. Instead, it will focus on the types and configurations of the components that are available and how they are commonly integrated into complete ATC systems.

Input

The following are some common ATC system devices that provide input to the ATC system controller.

Analog Input

Temperature Sensors   Temperature is probably the most commonly measured variable in commercial HVAC systems. Temperature sensors used in HVAC systems are either resistance temperature devices (RTDs) or thermistors. Both types of sensors change their resistance with varying temperature; thus the analog input to the ATC system

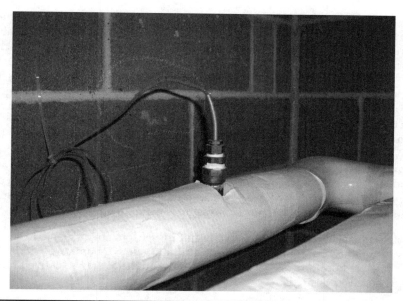

FIGURE 9-1   Photograph of a temperature sensor installed in a pipe.

controller from these sensors is variable resistance. Temperature sensors can measure space temperature, water temperature within pipes, outdoor temperature, and air temperature within ducts. Space temperature sensors can be specified with or without a setpoint adjustment (analog input). They can also be specified with or without an override pushbutton (digital input), which changes the associated air system from unoccupied to occupied mode of operation for a predetermined period of time. Temperature sensors installed in pipes are installed within fluid-filled thermometer wells that extend to the center of the pipe for pipe sizes 4 in. and smaller and a minimum of 2 in. into the fluid for pipes larger than 4 in. Figure 9-1 is a photograph of a temperature sensor installed in a pipe. Outdoor air temperature sensors are normally specified with a sun shield to reduce the effect of the sun's radiant heat on the temperature sensor. Duct temperature sensors are typically 12 in. long and are installed with mounting brackets in the walls of ducts.

Pressure Sensors   Like pressure gauges, pressure sensors measure the difference in pressure between two points. Pressure sensors are equipped with two ports, one that senses the higher pressure and one that senses the lower pressure. It is common for one of these two ports to be open to the atmosphere, in which case the measured pressure is based on the reference of atmospheric pressure and is called gauge pressure. Gauge pressure is measured when there is only one connection to the piping or duct system. Connections of pressure sensors to piping or duct systems are typically made with ¼-in. brass tubing. Air pressure sensors are mounted in the walls of ducts and normally sense only the static pressure of the air within the ducts. Water and steam pressure sensors are mounted in threaded openings in large pipes or in pipe tees for smaller pipes. Water and steam pressure sensors typically sense only the static pressure of the fluid in the pipes.

A duct static pressure sensor is commonly used in the control of the supply fan speed in a VAV air system.[6] The fan speed is modulated through a variable frequency drive (VFD) to maintain the setpoint of the static pressure sensor, which is typically located two-thirds the way down the primary supply air duct system.

Differential pressure is the difference in pressure between two points in a piping or duct system. In order to measure differential pressure, the high- and low-pressure ports of a pressure sensor are piped with tubing from the pressure sensor to the remote points in the pipe or duct. A differential pressure sensor is commonly used in the control of the pump speed in variable flow pumping system. The pump speed is modulated through a VFD to maintain the setpoint of the differential pressure sensor. The differential pressure sensor is connected between the supply and return pipe mains at a point that is indicative of the overall system differential pressure. If the piping system contains multiple branches, it may be necessary to utilize multiple differential pressure sensors to ensure that adequate differential pressure is maintained in all of the branches.

Relative Humidity Sensors   Relative humidity sensors are available to sense space relative humidity, outdoor relative humidity, and relative humidity within ducts. Return air relative humidity is commonly sensed when the average relative humidity of multiple spaces is desired.

Enthalpy Sensors   Enthalpy, which is a measure of the total heat of air (sensible and latent), can also be sensed within spaces, outdoors, and within ducts. Return air and outdoor air enthalpy sensors are commonly used in the control of the airside economizer operation of air systems where the enthalpy of the return air is compared to the enthalpy of the outdoor air. If the air system is in the cooling mode of operation or is required to maintain a constant 55°F discharge air temperature, the ATC system will command the air system utilize up to 100% outdoor air instead of return air when the return air enthalpy is greater than the outdoor air enthalpy.

Carbon Dioxide Sensors   Carbon dioxide (CO_2) sensors are available to sense space CO_2 level, outdoor CO_2 level, and the CO_2 level within ducts. CO_2 sensors are commonly used to accomplish what is called demand controlled ventilation (DCV) for air systems. DCV is an energy-efficient strategy that involves the adjustment of the outdoor airflow delivered by an air system to meet the ventilation needs of the occupants within the spaces served by the air system. Because CO_2 is the gas exhaled by the building occupants, an increase in the space CO_2 level above the outdoor air CO_2 baseline (typically about 450 ppm) indicates the presence of occupants, thus the need for outdoor air ventilation. The sequence of operation for DCV is to adjust the outdoor airflow delivered by the air system as required to maintain the space CO_2 (or return air CO_2) level at a maximum of 1,000 ppm. If the air system serves only one space or several spaces with a similar occupant density and occupancy schedule, the return air CO_2 level can be sensed instead of sensing multiple space CO_2 levels. DCV systems are particularly appropriate for spaces that have infrequent, dense occupancy, such as auditoriums and lecture halls, because the significant outdoor airflow required to ventilate these spaces can be reduced to zero when the spaces are unoccupied, which may be the majority of the time.

One of the requirements of the Leadership in Energy and Environmental Design (LEED) Indoor Environmental Quality (EQ) Credit 1 is to monitor the space CO_2 levels in all densely occupied spaces (greater than or equal to 25 people per 1,000 square feet) and generate an alarm if the conditions vary by 10% or more from the setpoint.

Fluid Flow Measurement Measurement of fluid flow in pipes or ducts requires both a fluid flow sensing element and a converter. The converter receives the input of the sensor (velocity pressure) and calculates the flow rate based on the cross-sectional area of the pipe or duct in which the sensor is installed. The flow rate is then communicated by the converter to the ATC system controller.

Water and Steam Different types of flow sensors are available for measuring water and steam flow. Orifice and venturi flow sensors are inexpensive, have no moving parts, and are the most common types used for measuring fluid flow in commercial HVAC systems. Turbine and vortex-shedding sensors are more costly and also more accurate than orifice and venturi flow sensors. The types of flow sensors should be selected based on the needs of each particular application and the client requirements.

Continuous measurement of water or steam flow in pipes is usually not required for commercial HVAC systems, unless there is a need to meter the heating and cooling energy used and assess a charge for it, as is the case with a district heating and cooling utility and some campus heating and cooling systems. For example, steam flow is continuously measured at the service entrance of the steam utility to each customer in district heating systems. The steam flow rate is totalized and the customer is periodically charged for its steam use, usually in 100-lb increments. If the steam pressure is known and remains constant, the heat of vaporization is also known. Thus, it is only necessary to measure the steam flow in order to determine the steam energy utilized by a customer.

However, if a customer's heating water or chilled water energy use is to be measured and totalized, it is necessary to continuously measure the supply and return temperatures in addition to the water flow. The heating or chilled water energy use can only be measured if the temperature differential is measured simultaneously with the flow rate. Meters that incorporate fluid flow rate and temperature differential are called thermal-energy (or sometimes Btu) meters.

Air Measurement of airflow in ducts is common for commercial HVAC systems, particularly the outdoor airflow required for occupant ventilation in VAV air systems. Section 403 of the 2009 *International Mechanical Code* requires the control system to maintain the minimum required outdoor airflow rates for the entire range of supply airflow.

Airflow is measured in a duct through an airflow measuring station, which consists of a series of airflow sensors (pitot tubes or thermal dispersion elements) that are arranged in a matrix across the duct's cross-sectional area. The airflow sensors measure the average-velocity pressure in the duct and convert this pressure to airflow based on the duct's cross-sectional area. Airflow can also be measured at the inlet of a fan through a fan inlet velocity sensor. Once again, airflow is calculated based on the air velocity pressure and cross-sectional area of the fan inlet. Another example of an airflow measuring device is the velocity sensor mounted in the primary air inlet of single-duct and fan-powered VAV terminal units.

Digital Input

Low-Limit Temperature Sensors A low-limit temperature sensor (also referred to as a freezestat) consists of a continuous temperature sensing element (typically 20 ft long) that is capable of sensing the coldest point along the sensor. The sensor is serpentined across the leaving air side of the first water or steam coil in an air system that is exposed to any mixture of outdoor air. The air system fan motor(s) and outdoor air damper control circuit are wired through the contacts of the low-limit temperature sensor. When the temperature of the coldest point along the sensor drops below the (adjustable) setpoint of the sensor (usually 38°F), the contacts of the sensor open, causing the fan(s) to shut down and the outdoor air damper to close. Low-limit temperature sensors typically require a manual reset because it is important to determine the cause of the shutdown before the air system is reenergized. A low-limit temperature sensor is not required if there is no water or steam coil in the air system.

High-Limit Temperature Sensors   A high-limit temperature sensor is commonly used in the heating water supply connection to a hot water boiler. Its purpose is to shut down the boiler if its (adjustable) setpoint is exceeded. Typically, the setpoint will be 20°F higher than the operating temperature of the boiler. Also, it is common for the high-limit temperature sensor to be an accessory of the boiler itself and to be a factory-wired component of the boiler's operating controls. If not, one must be installed in the heating water supply piping, with the boiler controls wired through the contacts of the high-limit temperature sensor. High-limit temperature sensors require a manual reset because it is important to determine the cause of the high temperature condition before the boiler is reenergized.

Differential Pressure Switches   A differential pressure (DP) switch (Fig. 9-2) is similar to a pressure sensor in that it has high- and low-pressure ports that are piped to remote points in a pipe, duct, or piece of equipment to sense the difference in pressure between

FIGURE 9-2   Photograph of a differential pressure switch.

the two points. DP switches have an adjustable pressure setpoint above which the switch will either open or close its contacts, providing digital input to the ATC system controller. DP switches are commonly piped between the inlet and discharge of pumps and fans to provide the run status (on or off) of the equipment. DP switches are also piped across air filters to provide a dirty filter alarm when the differential pressure across the filters exceeds the setpoint of the DP switch.

Flow Switches   Flow switches are used to identify the presence or absence of fluid flow within pipes and ducts and are sometimes used instead of differential pressure switches to provide the run status of pumps and fans. A flow switch consists of a spring-loaded paddle that is mechanically connected to a switch. The switch is mounted to the wall of a pipe or duct, with the paddle extended through the wall of the pipe or duct into the fluid stream. The position of the paddle changes when there is fluid flow either opening or closing the switch, thereby providing digital input to the ATC system controller.

High Static Pressure Switches   A high static pressure switch is a differential pressure switch that senses the (gauge) static pressure in a duct and opens its contacts if the duct static pressure rises above the high-limit setpoint of the pressure switch. They are commonly used in VAV air systems to shut down the air system fan(s) if the duct static pressure rises too high.

Duct Smoke Detectors   Duct smoke detectors (Fig. 9-3) are mounted to the walls of ducts and are designed to detect the presence of smoke in the air flowing through the ducts. Most types consist of two sampling tubes (one inlet, one return) that extend into the duct and allow air within the duct to circulate through the sensing chamber of the smoke detector (mounted to the wall of the duct) where it is constantly monitored for the presence of smoke.

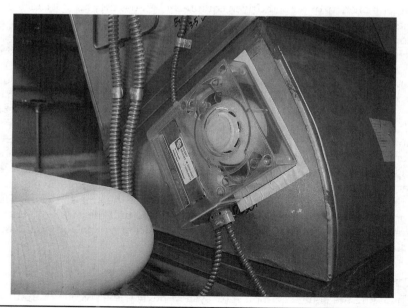

Figure 9-3   Photograph of a duct smoke detector.

The air system fan motor control circuit(s) is wired through the contacts of each smoke detector so that all of the operational capabilities of the air system are shut down when the presence of smoke is detected by any of the smoke detectors in the system. This does not apply if the air system is part of a smoke control system. In this case, the air system would change its mode of operation from normal to smoke control mode. Duct smoke detectors require a manual reset.

Duct smoke detectors are required by the 2009 *International Mechanical Code*, Section 606, to be installed in the return air duct for air systems with a design capacity of 2,000 cfm or greater. They are required in the return air duct connection to the return air riser at each floor for air systems with a design capacity of 15,000 cfm that serve two or more stories. Section 607 of the 2009 *International Mechanical Code* also requires a duct smoke detector for each smoke damper installed within a duct.

End Switches   An end switch is a switch that is mounted on a valve or damper actuator that closes when the actuator reaches its fully stroked opened or closed position.[7] End switches can be separate from, or an integral part of, the valve or damper actuator. End switches are used to provide digital input to the ATC system controller of the (opened or closed) status of automatic control valves and motor-operated dampers. An e1d switch is commonly used to provide the opened status of a motor-operated damper when it is necessary for the damper to be fully opened before a fan can be energized, such as the outdoor air damper for an equipment room ventilation system.

Current Sensing Relays   A current sensing relay consists of an (integral or remote) induction coil and a relay. The induction coil senses electrical current flow through a wire and closes (or opens) its relay contact when current flow is present. Current sensing relays are often used to provide digital input to the ATC system controller of the run status of small (¼ hp or less) direct-drive motors. Current sensing relays should not be used for motors larger than ¼ hp or for belt-driven equipment because they are not capable of detecting a mechanical failure of the driven equipment, such as a broken belt or locked-rotor condition.[8] A differential pressure switch or flow switch should be used to determine the run status of large motors and belt-driven equipment instead of a current sensing relay.

Aquastats   An aquastat is a thermally actuated set of contacts that open or close above a certain temperature. Aquastats are commonly used in 2-pipe systems to keep the control valves for heating-only equipment, such as unit heaters and cabinet unit heaters, closed when the 2-pipe system is in the cooling mode of operation. In this case, the normally closed control valve for the heating-only unit would be wired through the contacts of a strap-on aquastat that is mounted to the dual temperature water supply pipe to the heating-only unit. Whenever the pipe temperature is below 90°F, the contacts of the aquastat open, causing the normally closed control valve to close. When the pipe temperature rises above 90°F, the contacts of the aquastat close. This allows the control valve to function under the control of the unit thermostat. (Figure 9-15, which appears later in this chapter, illustrates the application of an aquastat in the control of a cabinet unit heater in a 2-pipe system.)

Output

Following are some common ATC system devices that receive output from the ATC system controller.

Fluid Flow Control

In this section we will discuss the ATC system devices that are used to automatically control the flow of water, steam, and air in HVAC systems. However, before we begin this discussion, it is important to introduce some general terms associated with this topic.

Fluids are controlled in water and steam systems by control valves, and air is controlled in air systems by motor-operated dampers (MODs). The positions of control valves and MODs are adjusted by actuators, which are electric motors or pneumatic pistons that change the position of control valves and MODs through linear or rotary action. An example of a linear-action actuator is the actuator for a globe-type control valve that is mounted directly to the valve stem and changes the valve position by stroking the valve stem in and out. An example of a rotary-action actuator is a type of actuator for an MOD that rotates a shaft connected to a linkage to change the position of the operable damper blades.

Control valves and MODs can be normally open (N.O.), normally closed (N.C.), or floating, which defines the position of the control valve or MOD when the actuator is de-energized. A spring-return actuator[9] is required to achieve either N.O. or N.C. control valve or MOD operation. Without a spring-return actuator, the control valve or MOD will float; that is, it will not change its position when it is de-energized.

Control valves and MODs can also be either modulating or two-position. Modulating control valves and MODs receive analog output from the ATC system controller and are capable of positioning anywhere between (and including) fully opened and fully closed. Two-position control valves and MODs receive digital output from the ATC system controller and can only be positioned either fully opened or fully closed.

As mentioned above, end switches can be externally mounted to a control valve or MOD actuator or they can be an integral part of the actuator to provide digital input to the ATC system controller of the (opened or closed) status of the control valve or MOD. An actuator can also be equipped with a potentiometer (variable resistor), which provides analog input to the ATC system controller of the position of a modulating control valve or MOD.

Control Valves   As mentioned previously, control valves are used to automatically control the flow of water and steam in commercial HVAC systems. Control valves can have either a 2-way or 3-way configuration.[10] Normally-open control valves are used where fluid flow through the equipment is desired if there is a loss of power to the control valve actuator. Heating and chilled water and steam coils in air handling units typically utilize N.O. control valves as one means to protect against coil freeze-up from outdoor air. Heating coils in fan-coil units, unit heaters, cabinet unit heaters, VAV terminal units, and other heating equipment that is not exposed to outdoor air typically use floating control valves because there is no possibility of freezing these coils and the floating control valve actuators are less costly than spring-return actuators. Normally-closed control valves would only be used where the closed position of the valve is the fail-safe position. An N.C. control valve is commonly used in conjunction with an aquastat for a piece of heating-only equipment, such as a unit heater or cabinet unit heater, that is connected to a 2-pipe system.

Proper sizing of modulating control valves is required to achieve acceptable control of the fluid flow through the equipment served by the control valve. The pressure drop across a modulating control valve in its fully opened position should represent 50 to 70% of the pressure drop through the combined control valve/equipment assembly. More precise control of the fluid flow is achieved when the pressure drop through the control valve represents a significant portion of the combined pressure drop through the control valve/equipment assembly. This is demonstrated qualitatively by considering two control valves for the same application: one is oversized (having a low pressure drop at the design flow rate) and one is properly sized. For the oversized control valve, a small change in the position of the valve stem from the fully closed position to a slightly open position will allow more flow through the valve than the same change in the position of the valve stem for the smaller (properly sized) control valve. Also, once the oversized control valve is opened, further opening of the valve will have less of an effect on the fluid flow than the same change in the position of the valve stem for the properly sized valve. In short, the change in the fluid flow through the control valve is more closely related to the change in the valve stem position for the properly sized control valve than it is for the oversized control valve. This close relation of fluid flow through the control valve to valve stem position is called linearity. Thus, the smaller (properly sized) control valve will function more linearly and control the fluid flow more precisely than the oversized control valve.

The pressure drop through modulating control valves in the fully opened position is defined by the flow coefficient (C_v) of the control valve. The C_v of a control valve is the flow in gallons per minute that causes a 1-psi pressure drop across the control valve. The C_v of a control valve for a water system is defined as follows:

$$C_v = (Q)/(\Delta p)^{1/2}$$

where

C_v = flow coefficient[11]
Q = water flow (gpm)
Δp = pressure drop (psi)

For example, the C_v of a control valve that has a 5-psi pressure drop at 20 gpm is 8.94.[12]

Although the modulating control valves for commercial HVAC systems are normally selected by the ATC contractor, the HVAC system designer must understand the sizing requirements in order to review the ATC submittal prepared by the ATC contractor during construction. Modulating control valves should be sized as described above. However, because a typical pressure drop for modulating control valves is 2 to 5 psi, modulating control valves are commonly one pipe size smaller than the line size of the piping system in which they are installed. Therefore, if a line-sized modulating control valve is submitted by the ATC contractor or if a modulating control valve that is two or more pipe sizes smaller than the line size is submitted, both situations would warrant careful scrutiny to ensure that the control valve is sized properly.

Finally, the pressure drop through a modulating control valve for a hydronic system is normally a significant contributor to the overall system pressure loss and must be included in the pump head calculation for the system pump.

2-Way Control Valves   A 2-way control valve has one inlet and one outlet and is normally of the globe valve design for water and steam applications. However, a butterfly valve may also be used for control valves that are 2½ in. and larger in water systems only. Figure 9-4 illustrates globe-type N.O. and N.C. 2-way control valves. For N.O. 2-way control valves, the spring-return actuator drives the valve stem upward (opening the valve) upon a loss of power. For N.C. 2-way control valves, the spring-return actuator drives the valve stem downward (closing the valve) upon a loss of power.

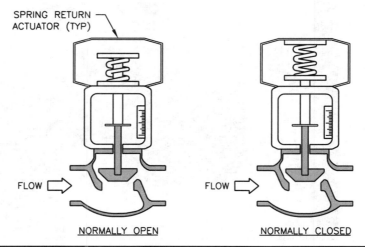

SPRING RETURN —
ACTUATOR (TYP)

FLOW ▭⇨   FLOW ▭⇨

NORMALLY OPEN   NORMALLY CLOSED

FIGURE 9-4   Normally-open and normally-closed 2-way control valves.

Heating water, chilled water, and heat pump water systems with variable-flow pumping utilize 2-way control valves for the air system and terminal unit coils. 2-way control valves should not be used in constant-flow pumping systems; 3-way control valves should be used instead (that is, unless the combined flow rate of all 2-way control valves in the system represents less than 10% of the total system flow or there is a differential pressure-actuated bypass valve installed between the supply and return mains in the system, which ensures that the required minimum flow rate will be circulated by the system pump if a significant number of the 2-way control valves are closed).

Two-position 2-way control valves are used to control the heating water and/or chilled water flow through the coils in small, 100% recirculating air heating and cooling units where modulating control of the fluid flow through the coils is not required. Examples include fan-coil units that do not provide outdoor air ventilation, unit heaters, cabinet unit heaters, and finned-tube radiators. Two-position 2-way control valves are sometimes used to control the condenser water flow through water-source heat pump units in variable flow heat pump water systems. They are also used in the supply, return, and equalizer piping connections to multiple-cell cooling towers or multiple cooling tower installations to isolate the idle cells or idle cooling towers.

Modulating 2-way control valves are used to control the heating water, chilled water, and/or steam flow through the coils in larger heating and cooling units where the mixed air temperature is different from the room temperature and more precise control of the fluid flow through the coils is required. Examples include fan-coil units and air handling units that provide outdoor air ventilation, and VAV terminal units. The steam flow through steam utilization equipment, such as shell and tube heat exchangers and heating coils, will either be controlled by a modulating 2-way control valve or it will not be controlled at all (i.e., there will be no steam control valve and the capacity will be controlled by some other means).

3-Way Control Valves   A 3-way globe-type control valve (Fig. 9-5) has three ports, either two inlets and one outlet (mixing valve) or one inlet and two outlets (diverting valve). Two mechanically interlocked butterfly valves (Fig. 9-6) may also be used to form a 3-way mixing or diverting control valve. This is common for control valves that are 2½ in. and larger.

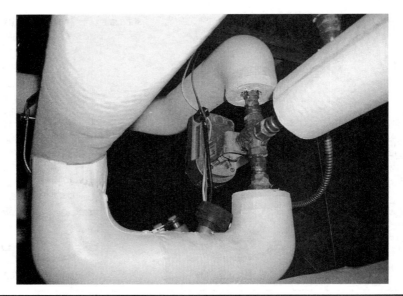

FIGURE 9-5 Photograph of a 3-way globe valve.

FIGURE 9-6 Photograph of a 3-way butterfly valve.

3-way control valves are only used in water systems; they are not used in steam systems. A 3-way mixing valve is a modulating control valve that mixes two inlet streams into one outlet stream. A 3-way diverting valve is a two-position control valve that diverts one inlet stream to one of two outlet ports. The N.O. and N.C. position refers to the straight-through port only (Figs. 9-7 and 9-8). The outlet port of a 3-way mixing valve, which is always the straight-through port, is called the common port. For

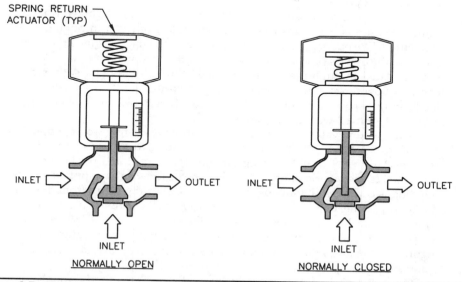

FIGURE 9-7   Normally-open and normally-closed 3-way mixing control valves.

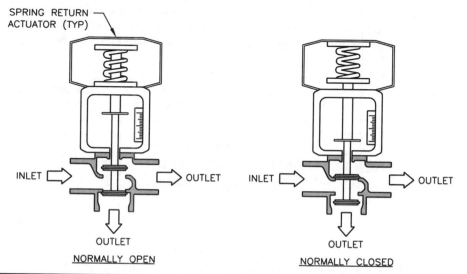

FIGURE 9-8   Normally-open and normally-closed 3-way diverting control valves.

N.O. 3-way mixing control valves, the spring-return actuator drives the valve stem downward (opening the straight-through port) upon a loss of power. For N.C. 3-way mixing control valves, the spring-return actuator drives the valve stem upward (closing the straight-through port) upon a loss of power. For N.O. 3-way diverting control valves, the spring-return actuator drives the valve stem upward (opening the straight-through port) upon a loss of power. For N.C. 3-way diverting control valves, the spring-return actuator drives the valve stem downward (closing the straight-through port) upon a loss of power.

3-way mixing valves are commonly used on heating and chilled water coils in constant flow pumping systems where a constant flow of water through the control valve is required. The outlet pipe of the coil is connected to the straight-through mixing valve inlet and the bypass pipe is connected to the other mixing valve inlet. Whatever flow is not circulated through the coil is circulated through the bypass, thus maintaining a constant flow through the control valve. The position of the control valve is based on the heating or cooling requirement of the coil. [Figure 7-2 shows a heating (or chilled) water coil with a 3-way mixing valve.]

3-way mixing valves are also used to control the heating water supply temperature from a hot water boiler to the building. The heating water supply pipe from the boiler is connected to the straight-through mixing valve inlet and the boiler bypass pipe is connected to the other mixing valve inlet. The heating water supply temperature to the building is controlled by mixing the cool return heating water from the building with the hot supply heating water from the boiler. Often, the temperature of the heating water supplied to the building is reset based on the outdoor air temperature (colder outdoor temperature requires a warmer heating water supply temperature and vice versa). Figure 9-9 illustrates the application of a 3-way mixing valve in the control of heating water supply temperature.

3-way diverting valves are used where full flow is required through one of the two outlet ports. 3-way diverting valves can be used in constant flow pumping systems in the same locations that two-position 2-way control valves are used in variable flow pumping systems. Examples include fan-coil units that do not provide outdoor air ventilation, unit heaters, cabinet unit heaters, finned-tube radiators, and water-source heat pump units. Figure 9-15, which appears later in this chapter, shows the application of a

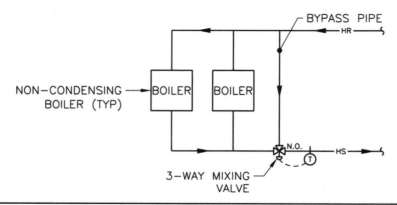

Figure 9-9   Hot water boilers with 3-way mixing valve.

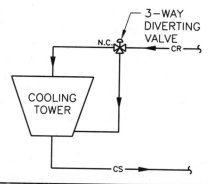

FIGURE 9-10 3-way diverting valve to control bypass around a cooling tower.

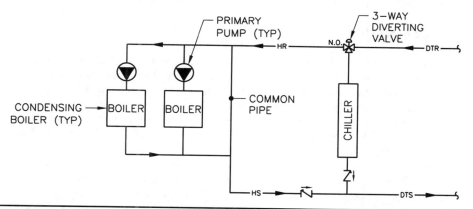

FIGURE 9-11 3-way diverting valve in a 2-pipe system.

3-way diverting valve in the control of a cabinet unit heater in a 2-pipe system. Another application of a 3-way diverting valve is the bypass valve for a cooling tower (Fig. 9-10) where full flow is diverted either through the cooling tower or through the bypass pipe to the cooling tower sump.

3-way diverting valves are also used in 2-pipe hydronic systems where it is necessary to divert full flow either to the boilers or to the chillers, depending on the operating mode of the 2-pipe system (see Fig. 9-11).

Motor-Operated Dampers   Motor-operated dampers (MODs) in an air system are like 2-way control valves in a water or steam system. Their purpose is to open completely to some preset position to allow full airflow, close completely to shutoff airflow, or modulate somewhere in between (and including) fully opened and fully closed.

Two-position (spring-return) N.C. MODs are commonly used to prevent outdoor air from infiltrating through louvers, intake air hoods, and other openings to the outdoors. Examples include the outdoor air intake or relief air openings for equipment room ventilation systems, outdoor air ducts for fan-coil units, and the inlets or outlets of exhaust fans.

Modulating MODs are commonly used in the outdoor air, return air, and relief air ducts associated with air handling units that utilize airside economizer. In this case, it is necessary for the ATC system controller to proportionally position these dampers to meet the needs of the airside economizer mode of operation.

The outdoor air damper for air handling units is typically a (spring-return) N.C. MOD, which provides a measure of protection against coil freeze-up from outdoor air. The relief air damper, which typically tracks with the position of the outdoor air damper, is also a (spring-return) N.C. MOD. The return air damper is typically a (spring-return) N.O. MOD.

It is common for two MODs to be mechanically interlocked by a linkage so that both dampers function together to perform a mixing or diverting operation. As one damper opens, the other damper closes, and vice versa. This is similar to two interlocked butterfly valves functioning together to act as a 3-way mixing or diverting valve in a water system. An example of two MODs performing a mixing function is the outdoor and return air dampers connected to the mixing box of an air handling unit.

Motor Controllers

A motor controller (commonly called a motor starter) is a manual or automatic, motor-rated switching device that starts and stops the motor it controls. It may also control the motor speed. The motor starter also provides thermal overload protection, which de-energizes the motor if the motor draws more than its rated full-load current.[13] Accessories that are available for motor starters include a hand-off-automatic (H-O-A) switch for magnetic motor starters (refer to the Magnetic Motor Starters section below), indicating lights, low-voltage protection, and auxiliary contacts for interlock with other motors or the ATC system.

The motor starters listed below are examples of across-the-line starters that start the motors they control with full voltage. Various types of reduced-voltage starters are available that provide a "soft-start" to reduce the starting (or inrush) current of the motors they control. One example is the wye-start, delta-run (or wye-delta) reduced-voltage starter discussed in the Chillers section of Chap. 4.[14] However, these types of starters are normally recommended for motors that are 100 hp and larger, which are not commonly used in HVAC systems for commercial buildings (with the exception of compressors in large chillers). Thus, they will not be further discussed in this book.

Manual Motor Starters   A manual motor starter consists of a manual motor-rated switch, thermal overload element(s), and optional accessories. The use of manual motor starters is limited to motors that do not require automatic control. For example, a manual motor starter is commonly used to control the operation of the exhaust fan serving a kitchen exhaust hood. In this case, the manual motor starter would be mounted near the hood and it would be equipped with auxiliary contacts that close when the exhaust fan is (manually) started and open when the exhaust fan is (manually) stopped. The control circuit of the makeup air unit would be wired through the auxiliary contacts in the exhaust fan starter so that the makeup air unit is energized through its magnetic motor starter whenever the exhaust fan is energized. Thus, the operation of the make-up air unit (which runs only when the exhaust fan is running) is said to be interlocked with the operation of the exhaust fan. There are other applications of manual motor starters. However, because they have no automatic capability, we will discuss them no further in this chapter.

Magnetic Motor Starters   A magnetic motor starter is a magnetic switching device that uses a reduced-voltage control circuit[15] to open and close the load-carrying contacts through a solenoid coil. Overload protection is provided for the motor by thermal overload elements connected in series with the load contacts. When the setting of any of the overload elements is exceeded, its associated contact in the control circuit opens, deenergizing the motor. A common accessory for a magnetic motor starter is an H-O-A switch. The three-position H-O-A switch controls whether the starter operates in the hand mode (manual on), off mode (manual off), or automatic mode (under control of the ATC system). Other common accessories for magnetic motor starters include auxiliary contacts and indicating lights. Figure 9-12 is a schematic of a magnetic motor starter for a three-phase motor with a 120V/1Ø control transformer, thermal overload elements, and H-O-A switch.

A combination magnetic motor starter has a disconnect switch built into the starter enclosure. Figure 9-13 shows a combination magnetic motor starter. The three-pole disconnect switch is located in the upper right-hand corner of the enclosure, the magnetic motor starter is located in the lower left-hand corner, and the control transformer and associated primary and secondary fusing is located in the lower right-hand corner.

Magnetic motor starters are used to control single-speed and multispeed motors. For single-speed motors, full voltage is applied to the motor windings whenever the load-carrying contacts are closed. Magnetic motor starters for multispeed motors contain a separate set of load-carrying contacts for each set of motor windings (i.e., a two-speed motor would have two sets of windings and the starter would have two sets of load-carrying contacts).

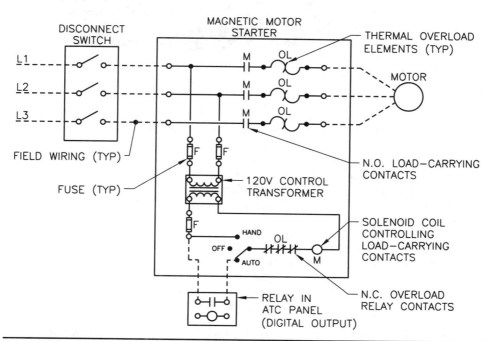

FIGURE 9-12   Magnetic motor starter with accessories.

FIGURE 9-13   Photograph of a combination magnetic motor starter.

Magnetic motor starters for HVAC equipment typically receive digital output from the ATC system controller. If the starter controls a multispeed fan, a separate digital output is required for each speed. Magnetic motor starters can provide digital input to the ATC system controller through an auxiliary contact. However, like the current sensing relay discussed earlier, the auxiliary contact in the control circuit will not detect a mechanical failure of the driven equipment. Magnetic motor starters are used for almost all three-phase motors in HVAC systems and for single-phase motors that require automatic control.

Variable Frequency Drives   A VFD[16] is an automatic motor starter with the same motor protection features as a magnetic motor starter, but its mode of operation is much different. A VFD does not open and close contacts to start and stop the motor. Rather, it uses solid-state components to change the line-side alternating current waveform entering the VFD into a variable-voltage, variable-frequency load-side waveform that is sent to the motor. By varying the frequency of the load-side waveform, the VFD is able to modulate the speed of the motor anywhere from completely off to full speed.

FIGURE 9-14   Photograph of a variable frequency drive with a hand-off-automatic switch and a bypass contactor.

A VFD (Fig. 9-14) can also be equipped with an H-O-A switch, which enables hand, off, or automatic operation (through the ATC system) of the VFD. Also, if the motor must remain in service at all times, electrically interlocked isolation and bypass contactors, a bypass magnetic motor starter, and a VFD-off-bypass switch can be specified to allow the VFD to be isolated for maintenance or replacement while the motor is controlled by the bypass magnetic motor starter. However, these bypass components are not required if a full-size, standby piece of equipment, controlled by a separate VFD or starter, is available.

VFDs are frequently used in variable-flow pumping and VAV air systems to control the system pumps and fans. The use of VFDs greatly improves the energy efficiency of the systems in which they are utilized because the input power to the motor is proportional to the cube of the motor speed. Thus, when the speed of a pump or fan motor is decreased to one-half of full speed, the input power is decreased to one-eighth (plus losses) of the full-load input power.

VFDs receive analog output from the ATC system controller and can also provide analog input to the ATC system controller of the actual operating frequency of the VFD.

Step Controllers

Step controllers, which consist of a series of contactors that close individually, are commonly used to stage electric heating elements and compressors. Step controllers require a separate digital output from the ATC system controller for each stage.

Connections

Electrical and DDC ATC systems require a line voltage source of electrical power (typically 120V/1Ø) from which the ATC system controllers can derive the low-voltage control power that is required by the ATC system components. For major pieces of equipment, such as air handling units, chillers, and boilers, a separate 120V/1Ø power connection is required to each ATC system controller. However, for smaller pieces of equipment, particularly VAV terminal units, it is common for a line voltage power connection to be made at the ATC system controller for one unit, from which low-voltage control power is fed to the ATC system controllers on several other units, thus eliminating the need for separate line voltage power connections to all of the units. For equipment that has a line voltage electrical connection, such as fan-powered VAV terminal units, electric heating coils, or fan-coil units, a separate 120V/1Ø electrical power connection is not required for the ATC system. The low-voltage control power can be derived through a control transformer from the line voltage feed to the equipment.

Line voltage electrical systems are normally provided by the electrical contractor, and low-voltage electrical systems are normally provided by the ATC contractor. Therefore, it is necessary for the HVAC system designer to inform the project electrical engineer of the locations where line voltage electrical power connections are required for the ATC system in order for these connections to be incorporated into the design of the electrical power distribution system for the building.

Control valves typically have the same ends (soldered, threaded, or flanged) for connection to the piping systems in which they are installed as other similar-sized valves in the system. MODs are constructed with a frame that mounts inside of the ducts in which they are installed.

Systems

The best way to demonstrate the control functions of the ATC components we have discussed is to show how these components are configured to control certain HVAC systems commonly found in commercial buildings. The following are selected ATC diagrams, sequences of operation, and DDC system point lists as they would appear on the construction drawings.[17] The purpose of the ATC diagram is to provide the ATC contractor with an understanding of the components that are required to accomplish the intended function of the HVAC system. Each HVAC system that is controlled by an ATC system should have an ATC diagram and a sequence of operation. The sequence of operation describes the logic of the ATC system. It should start with a general statement of the type of ATC system, and then describe its various modes of operation, alarms, and safeties. The DDC point list describes the various input and output points for the DDC ATC system controller associated with the HVAC system.

In developing the ATC diagram, it is not necessary to show all of the ancillary systems, such as transformers and relays, that are required to enable the system to work, nor is it necessary to give a detailed explanation in the sequence of operation of how the ATC system components actually perform their required functions. The purpose of the ATC diagram and sequence of operation is to show the intent of the design. The DDC point list is supplemental to the ATC diagram and lists the input that the DDC system controller receives from each ATC component and the output that the DDC system controller sends to each ATC component. For example, the operation of an MOD shown on the ATC diagram will be described in the sequence of operation. However, specific information, such as the requirement for an analog input to indicate the position of the MOD, will be given in the DDC point list.

The project specifications should require the ATC contractor to submit detailed shop drawings showing exactly how the ATC system will be configured. The ATC shop drawings should include wiring diagrams for each HVAC system, as well as product data for all of the ATC system components. During the review of these shop drawings, the HVAC system designer will have the opportunity to ensure that the ATC systems proposed by the ATC contractor are suitable to accomplish the design intent of each HVAC system.

Hot Water Cabinet Unit Heater

We will start off with a simple ATC diagram and sequence of operation for a hot water cabinet unit heater that is installed in a 2-pipe system (Fig. 9-15). The ATC system is electric with no connection to the BAS; thus there is no DDC point list associated with this system.

Sequence of Operation

The hot water cabinet unit heater shall be controlled by the unit-mounted electric thermostat.

The 3-way control valve shall open to the coil and the unit fan shall be energized when the space temperature drops below the 70°F (adjustable) setpoint of the thermostat. Upon a rise in space temperature above setpoint, the reverse shall occur.

The strap-on aquastat shall keep the control valve closed to the coil whenever the pipe temperature is below 90°F to prevent chilled water flow through the cabinet unit heater.

Notes to Reader

It is not necessary to specify exactly how the strap-on aquastat is to keep the control valve closed. The ATC contractor will know that it is necessary to wire the control valve actuator through the contacts of the aquastat and that the aquastat contacts need to be open below the setpoint temperature in order to keep the control valve closed. A 3-way diverting control valve is used to allow dual-temperature water to circulate either through the cabinet unit heater or through the bypass pipe so the aquastat continuously senses the temperature of the dual-temperature water. If a two-position 2-way control valve were used, there would be no flow through the control valve when it was in the closed position. Thus, the aquastat would not accurately sense the temperature of the water circulating through the 2-pipe system. Finally, whether the ATC components

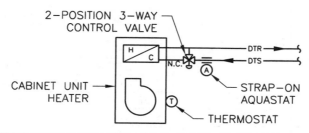

Figure 9-15 Hot water cabinet unit heater ATC diagram.

are line voltage or low voltage will be described in the project specifications; the ATC diagram and sequence of operation is the same regardless of the voltage of the electrical ATC components.

Parallel Fan-Powered Variable Air Volume Terminal Unit

Figures 9-16 and 9-17 illustrate an ATC diagram and DDC point list for a parallel fan-powered VAV terminal unit with hot water heat.

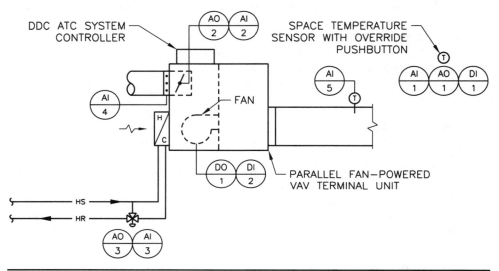

FIGURE 9-16 Parallel fan-powered VAV terminal unit with hot water heat ATC diagram.

DDC POINT LIST

POINT TYPE	POINT #	DESCRIPTION
AI	AI-1	SPACE TEMPERATURE
	AI-2	PRIMARY AIR DAMPER POSITION
	AI-3	HOT WATER COIL VALVE POSITION
	AI-4	PRIMARY AIRFLOW
	AI-5	DISCHARGE AIR TEMPERATURE
DI	DI-1	OVERRIDE
	DI-2	FAN STATUS
AO	AO-1	SPACE TEMPERATURE SETPOINT
	AO-2	PRIMARY AIR DAMPER POSITION
	AO-3	HOT WATER COIL VALVE POSITION
DO	DO-1	FAN ON/OFF

FIGURE 9-17 Parallel fan-powered VAV terminal unit with hot water heat DDC point list.

Sequence of Operation

General   The VAV terminal unit shall be controlled by the DDC ATC system controller. Occupied and unoccupied modes of operation shall be determined by the time schedule of the BAS. Graphics shall be provided in the BAS for all components of the system, identifying the current mode of operation, and the setpoints and current values of all points.

Morning Warm-up Mode   The primary air damper shall open completely. The VAV box fan shall be energized and the heating coil control valve shall open completely. Once the space temperature is within 2°F of the occupied setpoint, the VAV terminal unit shall function in the occupied mode.

Occupied Cooling Mode   The terminal unit fan shall not run in the occupied cooling mode. The primary air damper shall modulate between its minimum and maximum positions to maintain the cooling setpoint of the space temperature sensor.

Occupied Heating Mode   After the primary air damper has reached its minimum position, upon a further drop in space temperature below the heating setpoint of the space temperature sensor, the terminal unit fan shall be energized. Upon a further drop in space temperature, the heating coil control valve shall modulate open to the coil as required to maintain the space heating temperature setpoint. Upon a rise in space temperature above setpoint, the reverse shall occur.

Unoccupied Mode   The terminal unit fan shall be energized and the heating coil control valve shall modulate open to the coil as required to maintain the night setback temperature setpoint. Upon a rise in space temperature above setpoint, the terminal unit fan shall be de-energized.

Override Mode   The VAV terminal unit and its associated air system shall be indexed from the unoccupied mode to the occupied mode of operation for a predetermined period of time (adjustable) either manually through the override pushbutton located on each space temperature sensor or through the BAS.

Notes to Reader

This sequence of operation assumes that all of the spaces in the building that require heat during unoccupied periods are served by fan-powered VAV terminal units. Thus, there is no need for the air handling unit fan(s) to operate during unoccupied periods. The VAV air handling unit sequence of operation (see below) is consistent with this type of system. However, if there are certain areas within the building that are served by single-duct VAV terminal units with heating coils, it will be necessary for the air handling unit fan(s) to operate during unoccupied periods in order for these VAV terminal units to be able to supply heat to the areas they serve. In this case, the fan-powered VAV terminal unit sequence of operation would need to be modified to add a description of the primary air damper operation during the unoccupied mode of operation.

VAV Air Handling Unit

Figures 9-18 and 9-19 illustrate an ATC diagram and DDC point list for a VAV air handling unit with supply and return fans, hot water preheat and chilled water cooling coils, enthalpy economizer, and continuous outdoor airflow measurement.

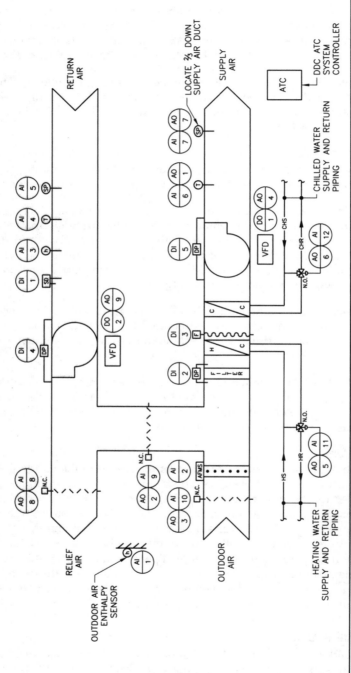

Figure 9-18 VAV air handling unit ATC diagram.

DDC POINT LIST

POINT TYPE	POINT #	DESCRIPTION	ALARM
ANALOG INPUT	AI-1	OUTDOOR AIR RELATIVE HUMIDITY	
	AI-2	OUTDOOR AIRFLOW	
	AI-3	RETURN AIR RELATIVE HUMIDITY	
	AI-4	RETURN AIR TEMPERATURE	
	AI-5	RETURN AIR STATIC PRESSURE	
	AI-6	DISCHARGE AIR TEMPERATURE	
	AI-7	SUPPLY AIR STATIC PRESSURE	
	AI-8	RELIEF AIR DAMPER POSITION	
	AI-9	RETURN AIR DAMPER POSITION	
	AI-10	OUTDOOR AIR DAMPER POSITION	
	AI-11	HEATING COIL CONTROL VALVE POS.	
	AI-12	COOLING COIL CONTROL VALVE POS.	
DIGITAL INPUT	DI-1	RETURN AIR SMOKE DETECTION	YES
	DI-2	FILTER DIFFERENTIAL PRESSURE	YES
	DI-3	LOW LIMIT TEMPERATURE	YES
	DI-4	RETURN FAN STATUS	YES
	DI-5	SUPPLY FAN STATUS	YES
ANALOG OUTPUT	AO-1	DISCHARGE AIR TEMP. SETPOINT	
	AO-2	RETURN AIR DAMPER POSITION	
	AO-3	OUTDOOR AIR DAMPER POSITION	
	AO-4	SUPPLY FAN VFD SPEED	
	AO-5	HEATING COIL CONTROL VALVE POS.	
	AO-6	COOLING COIL CONTROL VALVE POS.	
	AO-7	SUPPLY AIR STATIC PRESS. SETPOINT	
	AO-8	RELIEF AIR DAMPER POSITION	
	AO-9	RETURN FAN VFD SPEED	
DIGITAL OUTPUT	DO-1	SUPPLY FAN START/STOP	
	DO-2	RETURN FAN START/STOP	

FIGURE 9-19   VAV air handling unit DDC point list.

Sequence of Operation

General   The air handling unit shall be controlled by the DDC ATC system controller. Occupied and unoccupied modes of operation shall be determined by the time schedule of the BAS. Graphics shall be provided in the BAS for all components of the system, identifying the current mode of operation, setpoints and current values of all points, and the status of all alarms and safeties.

Morning Warm-up Mode   When indexed to the occupied mode by the time schedule of the BAS, the supply and return fans shall be energized and run continuously.

The outdoor and relief air dampers shall be fully closed and the return air damper shall be fully open.

The supply fan VFD shall modulate the supply fan speed to maintain the 1.50-in. w.c. (adjustable) setpoint of the supply air static pressure sensor. The return fan VFD shall modulate the return fan speed to maintain the −0.25-in. w.c. (adjustable) setpoint of the return air static pressure sensor.

The preheat coil control valve shall modulate to maintain a 105°F discharge air temperature.

The morning warm-up mode of operation shall continue as long as the return air temperature is below 68°F, as sensed by the return air temperature sensor.

When the return air temperature reaches 68°F, the morning warm-up mode shall be completed and the unit shall function in the occupied mode.

Occupied Mode   When the morning warm-up mode is complete, the outdoor, return, and relief air dampers shall position to deliver the minimum outdoor airflow, as measured by the outdoor airflow measuring station.

The supply fan VFD shall modulate the supply fan speed to maintain the 1.50-in. w.c. setpoint of the supply air static pressure sensor. The return fan VFD shall modulate the return fan speed to maintain the −0.25-in. w.c. setpoint of the return air static pressure sensor.

Enthalpy Economizer Mode   The DDC ATC system controller shall continuously compare the return air enthalpy to the outdoor air enthalpy.

When the return air enthalpy is greater than outdoor air enthalpy, the preheat coil control valve shall modulate to maintain a minimum of 55°F (adjustable) discharge air temperature, as sensed by the discharge air temperature sensor. Upon a rise in discharge air temperature above 55°F, the outdoor and relief air dampers shall modulate open and the return air damper shall modulate closed proportionally to maintain the 55°F discharge air temperature setpoint. Once the outdoor and relief air dampers reach their fully open positions, upon a further rise in discharge air temperature above setpoint, the chilled water coil shall modulate open to the coil to maintain the discharge air temperature setpoint.

Upon a drop in discharge air temperature below setpoint, the reverse shall occur.

Minimum Outdoor Air Mode   When the return air enthalpy is less than outdoor air enthalpy, the outdoor, return, and relief air dampers shall position to deliver the minimum outdoor airflow, as measured by the outdoor airflow measuring station. The outdoor, relief, and return air dampers shall modulate as required to ensure that the minimum outdoor airflow is continuously delivered for the entire range of supply airflow. Upon a rise in discharge air temperature above 55°F, the chilled water coil shall modulate open to the coil to maintain the discharge air temperature setpoint.

Upon a drop in discharge air temperature below setpoint, the reverse shall occur.

Unoccupied Mode   When indexed to the unoccupied mode by the time schedule of the BAS, the outdoor and relief air dampers shall close completely and the return air damper shall open completely.

The supply and return fans shall be de-energized.

Override Mode   The unit shall be indexed from the unoccupied mode to the occupied mode of operation for a predetermined period of time (adjustable) either manually through the override pushbutton located on any VAV terminal unit space temperature sensor or through the BAS.

Fan Failure Alarm   A fan failure alarm shall signal at the BAS when either the supply or the return fan fails to operate, as sensed by the differential pressure switch piped between the inlet and discharge of each fan.

Dirty Filter Alarm   A dirty filter alarm shall signal at the BAS when the differential pressure across the filter section exceeds the 0.70-in. w.c. (adjustable) setpoint.

Safeties   Upon a drop in the preheat coil leaving air temperature below 38°F, as sensed by the low-limit temperature sensor serpentined across the downstream face of the coil, the supply and return fans shall be de-energized, the outdoor and relief air dampers shall close completely, the return air damper shall open completely, the preheat coil control valve shall open fully to the coil, and an alarm shall signal at the BAS. The low-limit temperature sensor shall require a manual reset for the unit to be restarted.

The return air duct smoke detector shall be hard-wired through the supply and return fan VFD control circuits. When the return air duct smoke detector detects the presence of smoke, the supply and return fan shall be de-energized, the outdoor and relief air dampers shall close completely, the return air damper shall open completely, and an alarm shall signal at the fire alarm control panel and at the BAS. The duct smoke detector shall require a manual reset for the unit to be restarted.

Notes to Reader

Morning Warm-up Mode   The morning warm-up mode of operation is intended to utilize the heating capability of the air handling unit to assist the other heating terminal equipment in the building (such as the VAV terminal units) to warm the building from its night setback temperature of approximately 60°F to within 2°F of the occupied heating setpoint as quickly as possible. The morning warm-up mode of operation stops short of the occupied heating setpoint so that the air handling unit does not overshoot the occupied temperature. Because the morning warm-up mode of operation occurs before the occupied hours of operation, it is not necessary to provide outdoor air ventilation until the morning warm-up mode of operation is complete and the unit begins to operate in the occupied mode of operation. The supply and return fans operate during the morning warm-up mode of operation the same way they operate during the occupied mode of operation; that is, their fan speed is modulated to maintain the setpoint of the supply and return air static pressure sensors. Static pressure setpoints should be given for these sensors, but their actual settings will be finalized by the testing, adjusting, and balancing contractor during start-up. The supply air static pressure setpoint should be as low as possible, yet it should be high enough to enable the supply fan to deliver the maximum cooling airflow to the majority of the VAV terminal units in order to simulate the design cooling airflow condition. The return air static pressure setpoint should be adjusted to maintain a positive building static pressure. In fact, a building differential pressure sensor can be used instead of a return air static pressure sensor to control the return fan speed. In this case, the low-pressure port of the building differential pressure sensor would be connected by ¼-in. tubing to a point outdoors and the high-pressure port would be connected by ¼-in. tubing to an area indoors that is judged to be representative of the overall building pressure. It is important that the termination of the low-pressure tubing outdoors not be affected by wind pressure; otherwise, an erroneous outdoor pressure measurement could result at certain times.

Occupied Mode   During the occupied mode of operation, the supply and return fans continue to operate as in the morning warm-up mode of operation. Outdoor air ventilation is introduced to the building to maintain the required minimum outdoor airflow, as measured by the outdoor airflow measuring station.

Enthalpy Economizer Mode   Because the unit has an enthalpy (airside) economizer mode of operation, the enthalpy of the return air is constantly compared to the enthalpy of the outdoor air. When cooling of the mixed air is required, the ATC system controller will position the outdoor and return air dampers to favor the air with the lower enthalpy because it requires less energy to cool to the 55°F discharge air temperature.

Thus, when the return air enthalpy is greater than the outdoor air enthalpy, and cooling of the mixed air is required, the outdoor air damper will modulate beyond its minimum outdoor airflow position to allow a greater percentage of outdoor airflow into the mixing box of the air handling unit. The outdoor air damper is allowed to modulate as far as its fully opened position if that is what is necessary to meet the 55°F setpoint of the discharge air temperature sensor. In this mode of operation, the return air damper closes proportionally as the outdoor air damper opens, eventually to its fully closed position if necessary. The relief air damper position tracks with the position of the outdoor air damper so that the excess outdoor air delivered by the air handling unit is given a path to be relieved from the building in order to prevent overpressurization of the building.

Once the outdoor air damper reaches its fully open position, further cooling from the outdoor air is no longer possible. As the discharge air temperature rises above the 55°F setpoint of the discharge air temperature sensor, mechanical cooling must be provided by the chilled water cooling coil in order to maintain the 55°F setpoint of the discharge air temperature sensor. The outdoor air damper will remain at its fully open position (relief air damper also fully open and the return air damper fully closed) as long as the return air enthalpy is greater than the outdoor air enthalpy.

When the outdoor air enthalpy rises above the return air enthalpy, less energy is required to cool the return air than is required to cool the outdoor air. Therefore, the outdoor air damper will be positioned to deliver the minimum outdoor airflow, and the relief and return air dampers will reposition accordingly. In this mode of operation, minimum outdoor airflow will be delivered during the occupied mode of operation and mechanical cooling will be provided by the chilled water cooling coil in order to maintain the 55°F setpoint of the discharge air temperature sensor.

As the discharge air temperature drops below the 55°F setpoint of the discharge air temperature sensor, the process reverses. If the unit is in the economizer mode of operation, mechanical cooling will be reduced to zero. The outdoor air damper will then begin to close all the way to its minimum position if that is what is required. The relief and return air dampers will also reposition accordingly. Once the outdoor air damper reaches its minimum position, upon a further drop in the discharge air temperature, heating will be provided by the preheat coil to maintain the 55°F setpoint of the discharge air temperature sensor.

Unoccupied Mode   During the unoccupied mode of operation, the supply and return fans shut down and the outdoor air damper closes to keep the unit from freezing during the winter. The return air damper is positioned to the fully open position and the relief air damper (like the outdoor air damper) is closed. This sequence of operation assumes that all of the spaces in the building requiring heat during unoccupied periods are served by fan-powered VAV terminal units. Thus, there is no need for the air handling unit fans to operate during unoccupied periods. However, if there are certain areas within the building that are served by single-duct VAV terminal units with heating coils, it will be necessary for the air handling unit fans to operate during unoccupied periods in order for these VAV terminal units to be able to supply heat to the areas they serve.

Building Automation Systems

A modern BAS is a computerized network of devices that communicate over a local area network (LAN) that, in some ways, is similar to the LAN used in many businesses. The BAS is often capable of managing not only the HVAC system controls but also security and access control, fire and life safety, closed-circuit television, asset tracking, and other systems as well.

For smaller projects, a BAS is not necessary. All of the HVAC systems operate in a stand-alone fashion with no communication to a wider control network. Also, for projects with a BAS, it is not necessary for all ATC systems to communicate with the BAS. Some pieces of HVAC equipment in the building, particularly the small pieces of terminal equipment, such as finned-tube and electric radiators, unit heaters, and cabinet unit heaters, do not require central monitoring and control and are, therefore, not connected to the BAS.

BAS Architecture

The architecture of the HVAC system controls portion of the BAS often consists of a two-tiered network (Fig. 9-20).

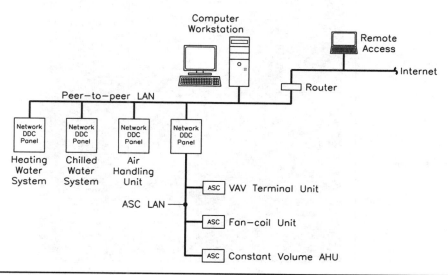

FIGURE 9-20   Typical BAS architecture for HVAC system controls.

Peer-to-Peer LAN

The first tier is the primary LAN, called the peer-to-peer LAN. The devices on the peer-to-peer LAN communicate directly with each other and include the central computer workstation (or server), the network DDC panels that control major pieces of HVAC equipment and communicate with the second tier LAN, and a router that enables access to the BAS from the Internet for web-based systems. Various network technologies are utilized for the peer-to-peer LAN, but Ethernet is the most common.

The application software developed by the BAS manufacturer is loaded onto the central computer workstation. This software is customized by the local BAS vendor for the project to include graphics for each of the HVAC systems connected to the BAS. The network DDC panels (Fig. 9-21) are fully programmable and have expanded capabilities for monitoring and controlling HVAC systems with many points, such as central heating and cooling plants and air handling units.

Application-Specific LAN

The second tier is the application-specific LAN, which functions under a network DDC panel. This LAN uses a different data communication protocol than the peer-to-peer LAN. HVAC systems with fewer points, such as VAV terminal units and fan-coil

FIGURE 9-21   Photograph of a network DDC panel.

units, are connected to the application-specific LAN through their individual application-specific controllers (ASCs). ASCs are designed to perform specific control functions for a particular type of HVAC equipment, such as a VAV terminal unit or a fan-coil unit. Therefore, ASCs are not programmable, cannot connect to the peer-to-peer LAN, and have a limited number of points through which to perform their intended control functions.

Communication

All network DDC panels and ASCs function in a stand-alone mode; that is, they do not require communication with the BAS to perform their control functions. Communication with the BAS is for sharing information between devices and for providing the capability of monitoring and controlling the HVAC systems from a central or remote location.

Operators can communicate with any device on the BAS through the central computer workstation, through the Internet for web-based systems, or through a portable operator terminal, which may be a laptop computer, personal digital assistant (PDA), or other network-compatible device supported by the BAS software. The portable operator terminal can be connected by a network data cable to any network DDC panel, ASC, or room temperature sensor. The ability of the portable operator terminal to communicate with the BAS through the room temperature sensors is often utilized by the testing, adjusting, and balancing contractor during start-up and commissioning of the HVAC systems.

Interoperability

It is important for the HVAC system designer to understand that the HVAC system controls portion of the BAS may be just one part of an overall BAS that manages other systems in the building (and possibly other buildings) as well. Also, the manufacturer of the overall BAS may be different than the manufacturer of the HVAC system controls portion of the BAS. In this case, the HVAC system controls portion of the BAS must be able to communicate with the overall BAS.

The ability of one manufacturer of BAS devices to communicate with other manufacturers' BAS devices is called interoperability. Interoperability of BAS devices is a fairly recent development. Prior to 1995, BAS devices could only communicate with a BAS of the same manufacturer. This inability to communicate among system manufacturers meant that once a BAS manufactured by a particular company was established in a facility, the BAS devices connected to that BAS for all future projects had to be manufactured by that company. This eliminated competition for the BAS devices on future projects, requiring the BAS portion of a project to be sole-source to the company whose BAS was established in the facility.

This lack of competition left many building owners dissatisfied. As a result, many BAS manufacturers and ASHRAE cooperated to develop an open data communication protocol (or language) that allows the BAS devices manufactured by various companies to communicate with each other. This open data communication protocol is called BACnet®. BACnet is the language that the BAS devices use to communicate with each other. However, not all BAS devices, especially older BAS devices, utilize BACnet as their native language. In this case, a "gateway" is required in order for these devices to communicate with a BACnet BAS. Many BAS manufacturers now offer BACnet as the native language, thereby eliminating the need for gateways.

A second well-known open communication protocol is called LonWorks®, which was developed by the Echelon Corporation. LonWorks not only describes the data communication protocol (called LonTalk®) but also describes the devices themselves. All LonWorks devices contain a Neuron® core processor, which is manufactured by the Echelon Corporation. The presence of this processor in the LonWorks devices enables the communication of these devices over the LonTalk LAN. Many BAS manufacturers offer a product line that utilizes LonWorks devices. Thus, there will be competition among local BAS vendors for a BAS that is specified to be LonWorks. Although Lon-Works enables communication between BAS devices of different manufacturers, the devices themselves require the proprietary LonWorks Neuron core processor and the LAN must be a LonTalk LAN. Thus, there is still a level of propriety in a LonWorks BAS. Another difference between LonWorks and other building automation systems is in the BAS architecture. A LonWorks BAS has a flat peer-to-peer architecture instead of the two-tiered architecture described above. Therefore, all LonWorks devices (even the ASCs) communicate directly with each other over the LonTalk LAN.

Unfortunately, BACnet and LonWorks are two completely different languages. Thus, devices using one language cannot communicate directly with devices using the other language. Therefore, if the BAS is to utilize an open communication protocol (as it should), the building owner will need to decide early in the design process for a new building whether the BAS will be BACnet or LonWorks. Also, for renovation projects, the BAS devices must utilize the same communication protocol as the existing BAS, which may be BACnet, LonWorks, or a proprietary communication protocol of a particular BAS manufacturer.

Endnotes

1. Pneumatic ATC systems utilize compressed air to perform the control functions.
2. A BAS can also be referred to as a building management system (BMS), a facility management and control system (FMCS), or other customer-specific names, all of which mean the same thing.
3. Low voltage refers to wiring and electrical devices using 30V ac or less.
4. Contacts are the switch portion of a relay. A relay is an electromagnetic switch whose contacts are opened and closed by the presence or absence of current flow through a solenoid coil. Normally open contacts are open when the solenoid coil is de-energized and closed when the solenoid coil is energized. Normally closed contacts are closed when the solenoid coil is de-energized and open when the solenoid coil is energized.
5. Digital input and digital output points can also be referred to as binary input (BI) and binary output (BO) points, respectively.
6. A duct static pressure sensor may also be used in the control of the return fan speed if a return fan is required.
7. A separate end switch is required for each position (opened and closed).
8. A locked-rotor condition occurs when the motor is energized, but the shaft is either not turning or is requiring higher than the rated torque of the motor. This adverse condition, which causes the motor to draw more than its rated full-load current, can potentially damage the motor. However, a locked-rotor condition will not be detected by a current sensing relay because the motor is still drawing electrical current.

9. A spring-return actuator contains an internal spring that returns the actuator to its normal position when it is de-energized.

10. A 4-way control valve configuration is also available but is not used in commercial HVAC systems.

11. Although C_v has units of gpm/psi$^{1/2}$, it is usually expressed as a dimensionless term.

12. $C_v = (20)/(5)^{1/2} = 8.94$.

13. The thermal overload elements (one per phase conductor) must be matched to the full-load current of the specific motor.

14. Wye-delta starters are commonly used to start the compressors in chillers that are 100 tons and larger.

15. Power for the control circuit is typically derived from a control transformer mounted within the starter enclosure. The control circuit voltage is normally 120V/1∅.

16. VFDs are also called variable speed drives (VSDs), adjustable frequency drives (AFDs), adjustable speed drives (ASDs), and pulse-width-modulated drives (PWM drives).

17. Sometimes the sequences of operation and DDC point lists are included in the project specifications. However, it is convenient for the ATC contractor if they are shown on the drawings adjacent to the ATC diagrams.

CHAPTER 10

Construction Drawings

Construction drawings are used to graphically convey the intent of the HVAC system design to the contractor who will be installing the HVAC systems. They also become a record of the HVAC systems for the building owner once the project is completed. The project specifications, which define additional requirements of the HVAC system design, also accompany the construction drawings and together form the construction documents. However, the project specifications will not be discussed in this book. In this chapter, we will describe the process of preparing HVAC construction drawings and the conventions commonly employed to clearly present the design intent.

Legend and Abbreviations

Symbols and abbreviations are used extensively in HVAC construction drawings. Thus, it is necessary to explain the symbols and abbreviations that are used in a legend, which also includes a list of abbreviations. The legend and abbreviations are usually presented on the first drawing of the HVAC drawing set. Figures 10-1 and 10-2 illustrate some of the symbols and abbreviations that are commonly used in the preparation of HVAC construction drawings. Although the symbols vary from one company to another, the actual symbols used are not as important as the need to explain what the symbols represent in order to ensure a clear communication of the HVAC system design.

Terminology

Though it may seem a bit tedious, it is important to discuss the proper use of the following terms commonly used in the preparation of construction drawings.

Shall: The word "shall" should be used instead of the word "will" in the notes on the construction drawings. The word "shall" carries with it a stronger connotation, meaning there are no exceptions or alternatives to the stated directive. It is acceptable for the word "will" to be used when describing the actions of the building owner or other parties that are not bound by the requirements of the HVAC construction drawings.

New: Use of the word "new" should be avoided on construction drawings wherever possible. All work shown in a bold line weight, scheduled, or detailed on the construction drawings is, by definition, new and is required to be provided by the contractor unless noted otherwise. This eliminates any confusion that may be caused if the word "new" is used in some instances to describe the work but is not used universally on the drawings for all new work. For example, if the word "new" is only sometimes used to refer to the new work, the contractor may misinterpret something that is

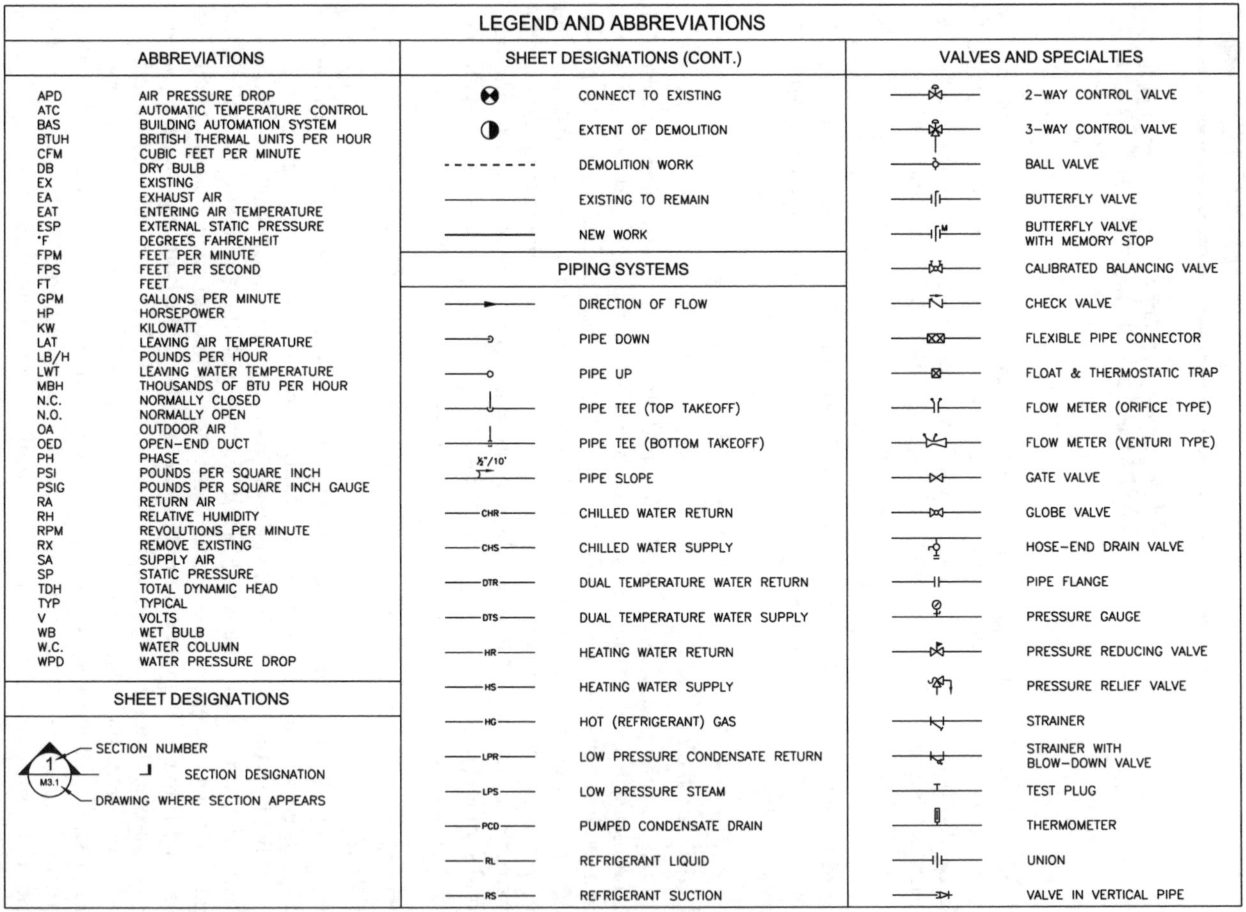

LEGEND AND ABBREVIATIONS

ABBREVIATIONS

APD	AIR PRESSURE DROP
ATC	AUTOMATIC TEMPERATURE CONTROL
BAS	BUILDING AUTOMATION SYSTEM
BTUH	BRITISH THERMAL UNITS PER HOUR
CFM	CUBIC FEET PER MINUTE
DB	DRY BULB
EX	EXISTING
EA	EXHAUST AIR
EAT	ENTERING AIR TEMPERATURE
ESP	EXTERNAL STATIC PRESSURE
°F	DEGREES FAHRENHEIT
FPM	FEET PER MINUTE
FPS	FEET PER SECOND
FT	FEET
GPM	GALLONS PER MINUTE
HP	HORSEPOWER
KW	KILOWATT
LAT	LEAVING AIR TEMPERATURE
LB/H	POUNDS PER HOUR
LWT	LEAVING WATER TEMPERATURE
MBH	THOUSANDS OF BTU PER HOUR
N.C.	NORMALLY CLOSED
N.O.	NORMALLY OPEN
OA	OUTDOOR AIR
OED	OPEN-END DUCT
PH	PHASE
PSI	POUNDS PER SQUARE INCH
PSIG	POUNDS PER SQUARE INCH GAUGE
RA	RETURN AIR
RH	RELATIVE HUMIDITY
RPM	REVOLUTIONS PER MINUTE
RX	REMOVE EXISTING
SA	SUPPLY AIR
SP	STATIC PRESSURE
TDH	TOTAL DYNAMIC HEAD
TYP	TYPICAL
V	VOLTS
WB	WET BULB
W.C.	WATER COLUMN
WPD	WATER PRESSURE DROP

SHEET DESIGNATIONS

SECTION NUMBER

SECTION DESIGNATION

DRAWING WHERE SECTION APPEARS

1

M3.1

SHEET DESIGNATIONS (CONT.)

	CONNECT TO EXISTING
	EXTENT OF DEMOLITION
	DEMOLITION WORK
	EXISTING TO REMAIN
	NEW WORK

PIPING SYSTEMS

	DIRECTION OF FLOW
	PIPE DOWN
	PIPE UP
	PIPE TEE (TOP TAKEOFF)
	PIPE TEE (BOTTOM TAKEOFF)
X/10'	PIPE SLOPE
CHR	CHILLED WATER RETURN
CHS	CHILLED WATER SUPPLY
DTR	DUAL TEMPERATURE WATER RETURN
DTS	DUAL TEMPERATURE WATER SUPPLY
HR	HEATING WATER RETURN
HS	HEATING WATER SUPPLY
HG	HOT (REFRIGERANT) GAS
LPR	LOW PRESSURE CONDENSATE RETURN
LPS	LOW PRESSURE STEAM
PCD	PUMPED CONDENSATE DRAIN
RL	REFRIGERANT LIQUID
RS	REFRIGERANT SUCTION

VALVES AND SPECIALTIES

	2-WAY CONTROL VALVE
	3-WAY CONTROL VALVE
	BALL VALVE
	BUTTERFLY VALVE
	BUTTERFLY VALVE WITH MEMORY STOP
	CALIBRATED BALANCING VALVE
	CHECK VALVE
	FLEXIBLE PIPE CONNECTOR
	FLOAT & THERMOSTATIC TRAP
	FLOW METER (ORIFICE TYPE)
	FLOW METER (VENTURI TYPE)
	GATE VALVE
	GLOBE VALVE
	HOSE-END DRAIN VALVE
	PIPE FLANGE
	PRESSURE GAUGE
	PRESSURE REDUCING VALVE
	PRESSURE RELIEF VALVE
	STRAINER
	STRAINER WITH BLOW-DOWN VALVE
	TEST PLUG
	THERMOMETER
	UNION
	VALVE IN VERTICAL PIPE

Figure 10-1 Abbreviations, sheet designations, and piping, valve, and specialty symbols.

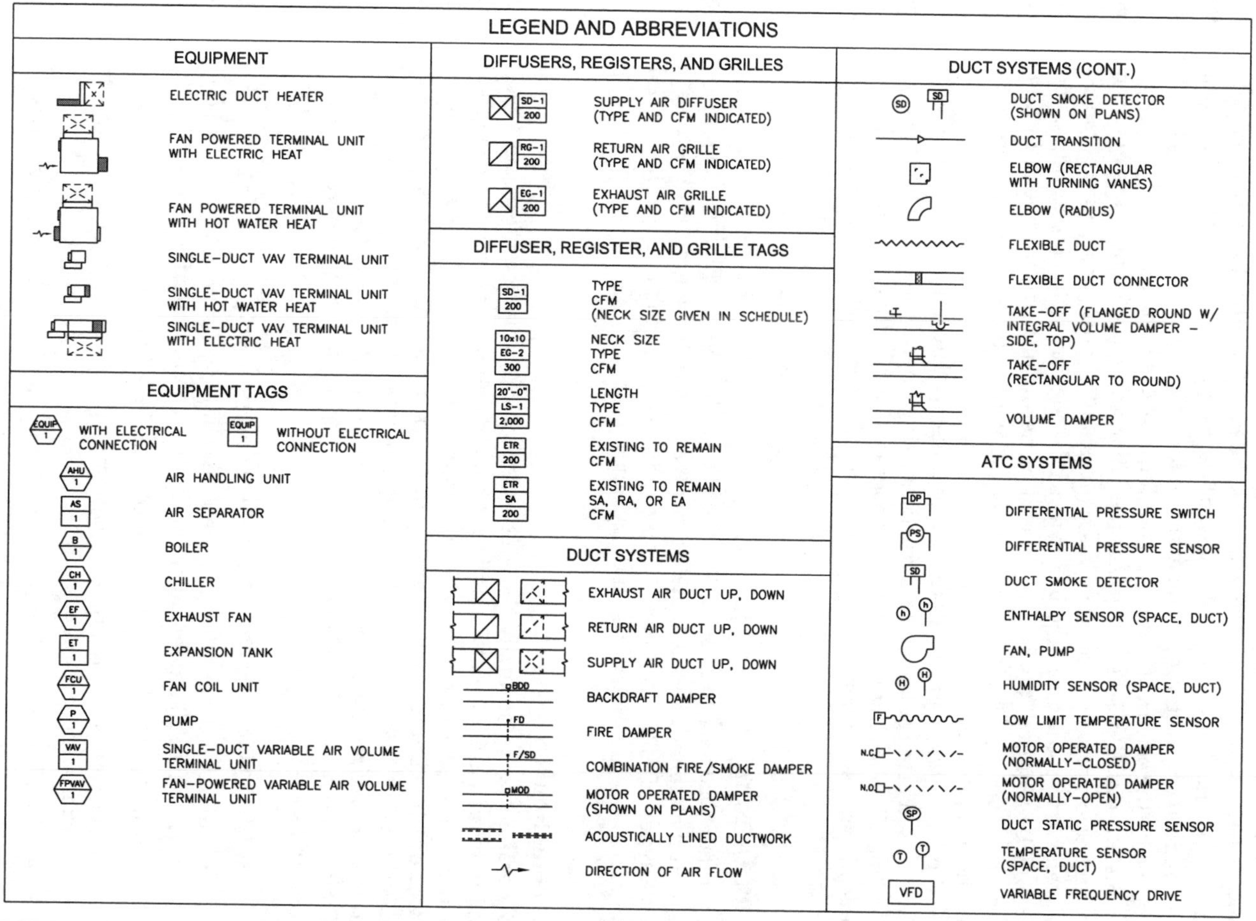

LEGEND AND ABBREVIATIONS

EQUIPMENT

Symbol	Description
	ELECTRIC DUCT HEATER
	FAN POWERED TERMINAL UNIT WITH ELECTRIC HEAT
	FAN POWERED TERMINAL UNIT WITH HOT WATER HEAT
	SINGLE-DUCT VAV TERMINAL UNIT
	SINGLE-DUCT VAV TERMINAL UNIT WITH HOT WATER HEAT
	SINGLE-DUCT VAV TERMINAL UNIT WITH ELECTRIC HEAT

EQUIPMENT TAGS

Symbol	Description
EQUIP 1	WITH ELECTRICAL CONNECTION
	WITHOUT ELECTRICAL CONNECTION
AHU 1	AIR HANDLING UNIT
AS 1	AIR SEPARATOR
B 1	BOILER
CH 1	CHILLER
EF 1	EXHAUST FAN
ET 1	EXPANSION TANK
FCU 1	FAN COIL UNIT
P 1	PUMP
VAV 1	SINGLE-DUCT VARIABLE AIR VOLUME TERMINAL UNIT
FPVAV 1	FAN-POWERED VARIABLE AIR VOLUME TERMINAL UNIT

DIFFUSERS, REGISTERS, AND GRILLES

Symbol	Description
SD-1 / 200	SUPPLY AIR DIFFUSER (TYPE AND CFM INDICATED)
RG-1 / 200	RETURN AIR GRILLE (TYPE AND CFM INDICATED)
EG-1 / 200	EXHAUST AIR GRILLE (TYPE AND CFM INDICATED)

DIFFUSER, REGISTER, AND GRILLE TAGS

Symbol	Description
SD-1 / 200	TYPE / CFM (NECK SIZE GIVEN IN SCHEDULE)
10x10 / EG-2 / 300	NECK SIZE / TYPE / CFM
20'-0" / LS-1 / 2,000	LENGTH / TYPE / CFM
ETR / 200	EXISTING TO REMAIN / CFM
ETR / SA / 200	EXISTING TO REMAIN / SA, RA, OR EA / CFM

DUCT SYSTEMS

Symbol	Description
	EXHAUST AIR DUCT UP, DOWN
	RETURN AIR DUCT UP, DOWN
	SUPPLY AIR DUCT UP, DOWN
▭800	BACKDRAFT DAMPER
FD	FIRE DAMPER
F/SD	COMBINATION FIRE/SMOKE DAMPER
▭400	MOTOR OPERATED DAMPER (SHOWN ON PLANS)
	ACOUSTICALLY LINED DUCTWORK
	DIRECTION OF AIR FLOW

DUCT SYSTEMS (CONT.)

Symbol	Description
SD	DUCT SMOKE DETECTOR (SHOWN ON PLANS)
	DUCT TRANSITION
	ELBOW (RECTANGULAR WITH TURNING VANES)
	ELBOW (RADIUS)
	FLEXIBLE DUCT
	FLEXIBLE DUCT CONNECTOR
	TAKE-OFF (FLANGED ROUND W/ INTEGRAL VOLUME DAMPER – SIDE, TOP)
	TAKE-OFF (RECTANGULAR TO ROUND)
	VOLUME DAMPER

ATC SYSTEMS

Symbol	Description
DP	DIFFERENTIAL PRESSURE SWITCH
PS	DIFFERENTIAL PRESSURE SENSOR
SD	DUCT SMOKE DETECTOR
E	ENTHALPY SENSOR (SPACE, DUCT)
FP	FAN, PUMP
H	HUMIDITY SENSOR (SPACE, DUCT)
	LOW LIMIT TEMPERATURE SENSOR
NC	MOTOR OPERATED DAMPER (NORMALLY-CLOSED)
NO	MOTOR OPERATED DAMPER (NORMALLY-OPEN)
SP	DUCT STATIC PRESSURE SENSOR
T	TEMPERATURE SENSOR (SPACE, DUCT)
VFD	VARIABLE FREQUENCY DRIVE

FIGURE 10-2 Equipment, duct, and automatic temperature control system symbols.

not labeled as "new" as being provided by others or as being shown for informational purposes only.

Furnish: *Furnish* means to purchase a product and deliver it to the jobsite or another construction team member. For example, the HVAC contractor may be required by the project specifications to furnish the starters for all motor-driven HVAC equipment to the electrical contractor for installation. In this case, the cost of installing the equipment is borne by the electrical contractor, not the HVAC contractor.

Install: *Install* means to furnish the labor and equipment necessary to connect a product and prepare it for use. This word should be used carefully because it clearly indicates the product is not to be furnished by the contractor. For example, the contractor may be required to install a piece of owner-furnished equipment. In this case, the cost of purchasing and delivering the equipment to the jobsite is borne by the owner, not the installing contractor.

Provide: *Provide* means to furnish and install. Thus, the word "provide" is almost always used in the notes on the construction drawings. From the definitions of "furnish" and "install" above, it is obvious that confusion can arise when the words "furnish" or "install" are used when the word "provide" is what is actually intended.

Notes on Drawings

It is often necessary to supplement the HVAC drawings with notes that provide information that cannot be adequately portrayed through the graphical entities. There are three types of notes commonly employed in engineering and architectural drawings: general notes, drawing notes, and key notes.

General Notes

General notes are normally shown on the first drawing of each design discipline and apply to all work of that discipline. As the name implies, they are general in nature and give the contractor the basic requirements related to the work of that discipline. More detailed information regarding the work of each discipline is given in the project specifications. The following is a list of general notes that may be used for the HVAC construction drawings. These general notes would be used for a renovation project. They would have to be edited for a new project to delete any references to the existing conditions.

- General notes shall apply to all HVAC drawings.
- All key notes indicated on the drawings as "typical" are to be considered as shown at all other similar conditions whether noted or not.
- All HVAC work shall be complete and ready for satisfactory service.
- The contract drawings are diagrammatic and are intended to convey the general arrangement of the work.
- The contractor is responsible for the means, methods, and work scheduling associated with the installation of the HVAC systems.
- Examine the site and observe the conditions under which the work will be installed. No allowances will be made for errors or omissions resulting from the failure to completely examine the site.

- Verify the size and location of all existing services. Notify the engineer of all discrepancies that exist between the contract documents and the existing services before making any connections to the existing services.

- Coordinate the size and location of roof penetrations and flashing requirements with the work of other trades.

- Route piping and duct systems parallel and perpendicular to the building lines. Mount as close as possible to the underside of the building structure.

- Coordinate the installation of the HVAC systems with the work of other trades. Provide offsets in piping and ductwork as required at no additional cost to avoid obstructions.

- Mount room sensors and switches at 4 ft, 0 in. above finished floor unless noted otherwise.

- Support all equipment from the building structure to provide a vibration-free installation.

- Ductwork dimensions shown on the drawings are internal airflow dimensions. Increase the sheet metal ductwork dimensions by 2 in. to accommodate 1-in. duct liner where required.

- Provide flexible duct connectors on all ductwork connections to fans or air handling units.

- Provide ½-in. mesh aluminum screen over the opening of all open-ended ductwork.

- Provide manual air vents at all high points and drains at all low points of hydronic piping systems.

- Pitch all hydronic piping ¼ in. in 10 ft in the direction of terminal equipment to enable the system to be drained.

- Pitch all steam and steam condensate return piping ½ in. in 10 ft in the direction of the steam traps to enable steam condensate to be removed from the system.

- Ensure that adequate clearance exists for the installation and maintenance of all work shown on the drawings and described in the specifications.

- Provide access panels (installed in walls or ceilings) and/or access doors (installed in ductwork) that are indicated or required for access to concealed HVAC devices that may require future inspection, repair, or adjustment.

- Identify all HVAC piping and equipment as to its function and equipment number indicated on the drawings.

- Identify all piping systems with cylindrical self-coiling plastic sheet that snaps over piping insulation and is held tightly in place without the use of adhesive tape or straps. Pipe identification shall be provided with flow arrows and lettering that is at least 1 in. high.

- Identify all HVAC equipment with engraved, color-coded laminated plastic markers with contact-type, permanent adhesive. Match equipment schedules on the drawings as closely as possible for equipment designations.

- Provide sleeves and caulk all piping penetrations through walls and floors and patch to match the adjacent construction. Provide chrome-plated escutcheons[1] on all piping penetrations in exposed locations.

- Provide sleeves and patch all duct penetrations through walls and floors to match the existing construction. Sleeve dimensions shall be 1 in. larger than insulated duct dimensions. The space between the duct and the sleeve shall be packed with mineral fiber and caulked.

- Firestop all penetrations through fire-resistance-rated walls, floors, or assemblies in accordance with the applicable codes and standards.

- Seal all penetrations through waterproof construction in accordance with the waterproofing manufacturer's instructions. All work shall be performed by approved contractors if required by the manufacturer to maintain the warranty on the material.

For renovation projects, it is helpful to add general demolition notes on the first drawing in the HVAC drawing set. These notes describe the general requirements of the demolition work, eliminating the need for repetitive instructions in the demolition key notes such as "remove all associated hangers and supports" or "dispose of all equipment and materials to be removed in accordance with federal, state, and local regulations." The following is a list of general demolition notes that may be used for the HVAC construction drawings for a renovation project:

- Demolition notes shall apply to all HVAC demolition drawings.

- Demolition shall be performed as neatly as practical and with the minimum disruption to the building activities and occupants.

- Remove all existing hangers and supports associated with the demolition work.

- Where a portion of existing piping or ductwork is indicated to be removed, the remaining piping or ductwork shall be capped and reinsulated to match existing.

- All equipment and materials being removed, and not indicated to be given to the owner, shall be disposed of by the contractor in accordance with all federal, state, and local laws, ordinances, rules, and regulations.

- All equipment and materials indicated to be reused or given to the owner shall be carefully removed so as not to damage the equipment or material or affect its reuse. Any such equipment and materials damaged by the contractor shall be replaced with new by the contractor at no expense to the owner.

Drawing Notes

Drawing notes apply only to the drawing on which they appear. Examples of drawing notes are those that describe the process, or phases, in which the work on a particular drawing must be performed. The following is an example of drawing notes that describe the work scheduling of a kitchen and dining room renovation where the kitchen must remain in service throughout the renovation:

- The dining room will be shut down for a 2-week period to allow for the work in the dining room to be performed. The kitchen will remain operable during the 2-week shutdown of the dining room. Provisions shall be made to allow the kitchen staff free access to the kitchen at all times.

- All demolition and new work in the dining room and kitchen, including work associated with air handling unit AHU-1, air-cooled condensing unit ACCU-1, exhaust fans, variable air volume (VAV) terminal units, supply air ductwork, air devices (air inlets and outlets), piping, electrical, and automatic temperature control (ATC) systems shall be performed during this 2-week period.

- All work in the kitchen shall be performed between 9:00 p.m. and 11:00 a.m. The kitchen shall be cleaned up and open areas in the ceiling covered with plastic sheeting upon completion of each day's work.

- One of the two exhaust hoods in the kitchen must be operable at all times throughout the course of the construction.

Key Notes

Key notes are used when important information that cannot be adequately portrayed graphically must be communicated. Sometimes called specific notes, key notes consist of a numbered list of notes that make specific references to items on a drawing through the use of callouts. The callout is usually represented by a leader with a numeric note "bubble" attached to it.

Key notes may have multiple applications on the same drawing, in which case one or two instances should be identified on the drawing and the key note should contain the word "typical," which means it has multiple applications to similar conditions on the drawing (which are obvious to the reader of the drawing). Key notes should not be used where adequate information is portrayed graphically on the drawing. An example of an unnecessary key note would be "Supply air ductwork above ceiling." Unless there are portions of the building shown on the drawing that do not have finished ceilings, it is obvious that the ductwork is to be installed above the ceiling; thus, this key note would provide no useful information to the contractor.

It is important for key notes to be direct, complete, and unambiguous. Key notes should be stated as a directive (or command) using the active voice, not the passive voice (e.g., "Remove existing air handling unit" instead of "Existing air handling unit to be removed"). The passive voice is ambiguous and may cause the contractor to ask, "Who is removing the existing air handling unit?" Key notes should be complete and include all necessary instructions that are not already explained in the general notes or drawing notes. Finally, key notes must be unambiguous; that is, they must clearly explain the work that the HVAC contractor must perform and the work that will be performed by others. The following is an example of a direct, complete, unambiguous key note:

"Remove existing air handling unit and all associated ductwork and piping to point indicated. Remove all associated ATC components and control wiring. Remove existing concrete housekeeping pad. Patch all openings in existing walls and floors where the ductwork and piping are removed to match existing. Existing electrical power wiring shall be removed by the electrical contractor."

Linework

The weight (or thickness) and color (percent gray scale) of the linework on HVAC drawings is a useful tool in making a distinction between the architectural floor plan, reflected ceiling plan, existing HVAC work to remain, HVAC demolition work, and HVAC new work. The following conventions should be used to make the HVAC demolition and new work stand out from the existing HVAC work and the architectural floor plan and reflected ceiling plan:

1. Architectural floor plan
   a. Line weight—0.15 mm
   b. Line type—continuous
   c. Line color—50% gray scale

2. Reflected ceiling plan
   a. Line weight—0.15 mm
   b. Line type—continuous
   c. Line color—20% gray scale

3. Existing HVAC work to remain
   a. Line weight—0.20 mm
   b. Line type—continuous
   c. Line color—black

4. HVAC demolition work
   a. Line weight—0.50 mm
   b. Line type—dashed
   c. Line color—black

5. HVAC new work
   a. Line weight—0.50 mm
   b. Line type—continuous
   c. Line color—black

Another important point regarding the linework for HVAC drawings is that of consistent spacing between the single-line representations of parallel pipes. The space between parallel pipes should be 6 in. in the model file for pipe sizes 3 in. and smaller. Thus, for a drawing that is presented at 1/8 in. = 1 ft scale, the space between the pipes will be 1/16 in. and the space between parallel pipes for a drawing that is presented at ¼ in. = 1 ft scale will be 1/8 in. The space between parallel pipes should be 12 in. in the model file for pipe sizes 4 to 6 in. For pipes that are 8 in. and larger, the pipe spacing in the model file should be approximately equal to the actual distance between the pipe centerlines (accounting for approximately 2 in. of insulation on the pipes and 2 in. of spacing between the facing of the insulation on adjacent pipes). For example, the pipe spacing for two parallel 12-in. pipes would be 18 in. in the model file.

Computer-Aided Design Considerations

It is not the purpose of this book to describe in detail the standards that a particular company should use in preparing computer-aided design (CAD) drawings. However, there are certain conventions that aid in both the process of preparing CAD drawings and in sharing CAD information with other members of the design and construction teams.

CAD Standard

The most important thing is to have a standard for CAD drawings that defines the conventions for layering, file naming, symbols, line weights and line types, fonts, and so on. Many of these conventions are defined in detail in the *U.S. National CAD Standard*, which is available online. Although this standard need not be adopted verbatim as a company CAD standard, it does provide reasonable guidelines that should be followed if there are no compelling reasons to digress from its recommendations. One process that is defined in the *U.S. National CAD Standard* is the use of model files and sheet files in the preparation of CAD drawings.

Although it may seem a little awkward at first, the use of model files and sheet files in preparing HVAC floor plans, large-scale floor plans, and sections is a helpful convention. This method allows each aspect of the HVAC work to be drawn in only one file, while enabling this same information to be presented in different ways on multiple drawings. For example, the HVAC existing conditions and demolition work are drawn in only the HVAC demolition model file, yet the demolition work can be presented using a bold dashed line weight on the HVAC demolition drawings and not be shown at all on the HVAC new work drawings. At the same time, the HVAC existing conditions are available for presentation in both the HVAC demolition and new work drawings. The process, in brief, is described below.

Model Files

The model file contains the HVAC work drawn in the "model space" at full size (e.g., a 36-in.-wide duct is drawn 36 in. wide). There will be separate model files for both the HVAC demolition (and HVAC existing conditions) and the HVAC new work for each floor of the building. The first step in creating an HVAC model file is to assign the appropriate values to the CAD software variables in the drawing, such as the dimension scale, line type scale, and text size. The values for these variables depend upon the scale in which the information will ultimately be presented on the plotted drawings. Next, the respective architectural floor plan is attached as an external reference file. The HVAC demolition model files contain the existing HVAC work drawn on "Existing" layers and the demolition work drawn on "Demolition" layers.[2] Each HVAC new work model file will attach the respective architectural floor plan and the respective HVAC demolition model file as external reference files. The "Demolition" layers in the referenced HVAC demolition drawing are turned off and the new work is drawn in the HVAC new work model file on appropriately named layers for the HVAC new work. Equipment numbers, duct sizes, pipe sizes, and pipe designations within the piping linework are drawn in the model files.

Sheet Files

The sheet files are the files that are plotted. There will be one sheet file for each drawing in the HVAC drawing set. For HVAC demolition and new work floor plan drawings, large-scale floor plan drawings, and sections, the appropriate demolition and new work model files are referenced into the "model space" of the sheet file that will present the work. The title block is referenced into the "layout" of the sheet file. A viewport is defined in the layout where the work is to be presented on the plotted drawing and the viewport is assigned the appropriate scale factor (e.g., 1/8 in. = 1 ft). All annotations, such as plan titles, key notes, and their associated callouts, are drawn in the layout of the sheet file. The layout of the sheet file is plotted at a 1-to-1 scale factor; thus, the page setup of the layout matches the actual sheet size of the plotted drawing.

Other Conventions

Other conventions such as layer names within each type of CAD drawing and file naming conventions should also be used consistently in order to increase the efficiency of CAD drawing production and simplify the transfer of information with other members of the design team. These conventions are defined in detail in the *U.S. National CAD Standard.*

It should also be noted that certain clients may have their own CAD standards that must be used in the preparation of CAD drawings for their projects. It is very important to obtain these standards at the beginning of the project before any design work commences so that all HVAC CAD drawings are set up and completed in accordance with the client's CAD standard. It is very difficult to go back and change a set of CAD drawings to meet a different CAD standard once the CAD work is complete.

Regarding the numbering of the HVAC drawings, it is common for the project architect, who is typically the design team leader, to define the drawing numbering convention. This convention should be followed by all design disciplines in order for the entire architectural and engineering drawing set to be consistently numbered.

Demolition Plans

For renovation projects, it is common for the first drawings in the HVAC drawing set to describe the demolition of existing HVAC systems that is required. There are various conventions that may be employed to describe the demolition of the existing HVAC systems. However, the conventions employed are not as important as the need to clearly describe the extent of demolition that is required and the work that is to remain.

If all of the HVAC services in a particular area, such as an equipment room, are to be removed and only the services outside of the area are to remain, the area can be surrounded by a boundary on the demolition drawing and a key note (and possibly drawing notes) can be used to verbally describe the demolition work within the area. This convention is acceptable because the contractor's estimate for the demolition work will normally be based on floor area or weight of the services to be removed, and the contractor will have to visit the project site before submitting the bid for the work. Design time is better spent detailing the locations and sizes of the services to remain outside of the demolition area so that the new work connections to these existing services can be fully coordinated.

However, when the demolition work is selective (only a portion of the existing services in a particular area is to be removed), it is necessary to show the extent of the demolition work in more detail by using graphical entities and key notes. A symbol is commonly used to show the limit where the demolition work ends and the existing services remain (Fig. 10-3). It is only necessary to show the amount of existing services that will remain to the extent that is required to clearly portray the design intent and provide useful information for coordination.

It is necessary for the HVAC system designer to conduct a field survey for every renovation project in order to determine the sizes and locations of the existing HVAC systems that will be affected by the work of the project. The existing building drawings

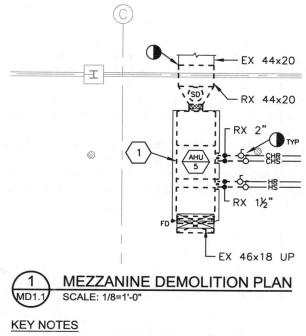

MEZZANINE DEMOLITION PLAN
SCALE: 1/8=1'-0"

KEY NOTES

⟨1⟩ REMOVE EXISTING AIR HANDLING UNIT AND HEATING WATER PUMP. REMOVE DUCTWORK AND PIPING TO POINTS INDICATED. EXISTING SHUTOFF VALVES AND HOUSEKEEPING PAD TO REMAIN. REMOVE ALL ASSOCIATED ATC COMPONENTS, SMOKE DETECTOR, AND CONTROL WIRING. EXISTING ELECTRICAL POWER WIRING SHALL BE REMOVED BY THE ELECTRICAL CONTRACTOR.

FIGURE 10-3 Demolition plan.

should never be depended upon as an accurate reflection of the existing conditions. The existing drawings are, however, helpful in providing the general arrangement of the existing services and should be brought to the field and marked up with the actual conditions. The provision of accurate information regarding the sizes and locations of the existing services is invaluable in preparing a well-coordinated set of construction drawings for renovation projects.

Figure 10-3 is an example of a drawing that shows the selective demolition of an air handling unit located within an equipment room. The drawing is presented at a 1/8 in. = 1 ft scale[3] and utilizes a bold, dashed line type to identify the demolition work. Existing work to remain is drawn with a thin, continuous line type. A symbol is used to show the extent of the demolition work and a key note is given to provide supplemental information. The air handling unit, housekeeping pad, and ductwork are drawn to scale in their actual locations; piping is represented with single lines. Only the portions of the existing services that affect the work are shown.

New Work Plans

Figure 10-4 is the HVAC new work drawing associated with the demolition drawing shown if Fig. 10-3. Note that the words "new work" are not required in the plan title. The drawing utilizes a bold, continuous line type to identify the new work. Existing work to remain is drawn with a thin, continuous line type. Lines are broken and/or shown in a hidden line type where the particular service is routed beneath another service that is also shown on the plan. This convention is necessary in order to provide a clear presentation of the work.

This particular drawing is presented at a 1/8 in. = 1 ft scale because it is possible (though somewhat crowded) to show the necessary information on the plan. A section is also provided to present supplemental information that cannot be shown in the plan view (Fig. 10-5). The location where the section is cut is designated by a symbol on the

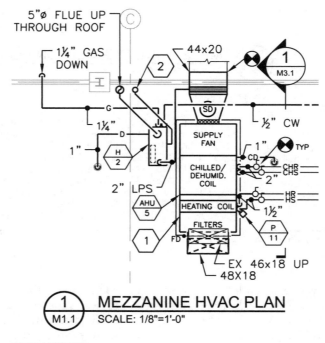

KEY NOTES

1   AIR HANDLING UNIT MOUNTED ON 4" HOUSEKEEPING PAD. ENLARGE EXISTING HOUSEKEEPING PAD TO BE 4" LARGER THAN UNIT ON ALL SIDES.

2   4" PVC INTAKE PIPE UP THROUGH ROOF. PROVIDE ½"x½" MESH SCREEN OVER OPENING.

FIGURE 10-4   HVAC plan.

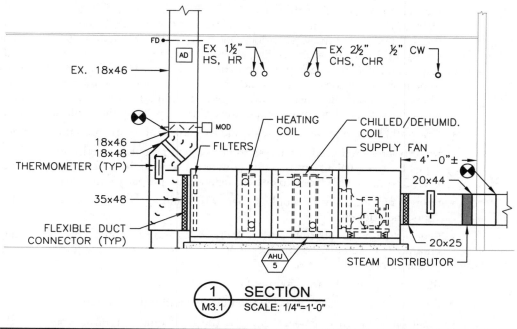

FIGURE 10-5 Section drawing of an air handling unit.

drawing along with the section number and the drawing where the section appears. Symbols are used to show the connections to the existing services and key notes are given to provide supplemental information.

Simplicity and clarity are the keys to presenting information on either a demolition or new work floor plan. It is not necessary to show every valve, fitting, and specialty because these items will be shown in the connection details for each piece of equipment. Each piece of equipment is given a designation (shown in the equipment tag) on the floor plan, which is used in scheduling the equipment capacities.

Common designations for HVAC equipment are as follows:

- B-1, etc.   Boilers
- CH-1, etc.   Chillers
- P-1, etc.   Pumps
- AS-1   Air separator

All designations for equipment should be shown in the legend.

Other conventions that should be followed in the preparation of new work plans are as follows:

- Sometimes it is helpful to present the HVAC equipment and ductwork on one plan and the HVAC equipment and piping on a separate plan. This keeps the drawings from becoming too crowded and aids in clear presentation of the work.

- It is helpful to show the reflected ceiling plan for areas with finished ceilings on HVAC plans. Because the reflected ceiling plan can usually be referenced from the architectural drawings, it does not need to be drawn from scratch. The ceiling grid and all of the ceiling devices should be shown on the reflected ceiling plan in a thin, light gray color. The HVAC work, including the air devices to be installed in the ceiling, should be drawn in a bold, continuous line type. This convention will assist the contractor in coordinating the location of ceiling air devices with other equipment in the ceiling, especially the lighting fixtures.
- Draw all HVAC equipment to scale and provide a minimum amount of detail.
- Identify areas required for access to HVAC equipment in a light, hidden line type. This is particularly true for crowded installations where access to the HVAC equipment is likely to be obstructed if not properly coordinated.
- Show main shutoff valves only. Other valves and specialties are shown in the connection details.
- Shade duct-mounted coils and coils in VAV terminal units. This assists the contractor in identifying the locations of HVAC equipment requiring piping connections.
- Show all volume dampers without exception.
- Show turning vanes in all rectangular elbows without exception.
- Break pipes that cross under other pipes. The gap should be 1/32 in. when plotted.
- Use a hidden line type for all work that is underneath other work.
- Use proper conventions for supply, return, and exhaust air ducts up and down (refer to the sample legend in Fig. 10-2).
- Use proper conventions for pipe rises and drops (refer to the sample legend in Fig. 10-1).

Section Drawings

Section drawings should be prepared for areas requiring close coordination with the work of other trades. Examples include:

- Breakouts of ductwork and piping from risers in shafts
- Corridors where the ductwork is the largest
- Areas requiring close coordination with unusual architectural and structural elements
- Crowded areas in equipment rooms
- Areas around large pieces of HVAC equipment in equipment rooms

The same drawing conventions that apply to new work plans apply to section drawings. All entities shown on the sections should be drawn to scale. Section drawings are normally presented at ¼ in. = 1 ft scale, although larger scales can be used if necessary to show detail. The architectural and structural building components and pertinent work of other trades should be drawn in order to show how the HVAC systems are coordinated with these elements. See Figs. 10-5 and 10-6 for a section drawing and photograph of the air handling unit shown in the sample HVAC plan in Fig. 10-4.

FIGURE **10-6**   Photograph of the air handling unit shown in the section drawing (Fig. 10-5).

Large-Scale Plans

Like section drawings, large-scale plans should be prepared for areas requiring close coordination with the work of other trades. The same drawing conventions that apply to new work plans apply to large-scale drawings. All entities shown on the large-scale drawings should be drawn to scale. Large-scale drawings are normally presented at ¼ in. = 1 ft scale.

The use of the larger scale usually makes it possible to show the HVAC equipment, ductwork, and piping on the same drawing (Fig. 10-7). The larger scale also allows for more detail to be shown, such as flexible duct connectors for all duct connections to fans or air handling units, housekeeping pads, floor drains, and portions of work by other trades.

Details

Details are an important part of HVAC construction drawings because much of what is shown in them is repeated multiple times throughout the project. For example, the specialties shown in the hot water heating coil detail for a VAV terminal unit could apply to more than a hundred VAV terminal units for a large project. Thus, if the balancing valve is omitted from the detail, the cost to incorporate it through a change order to the contract could be thousands of dollars.

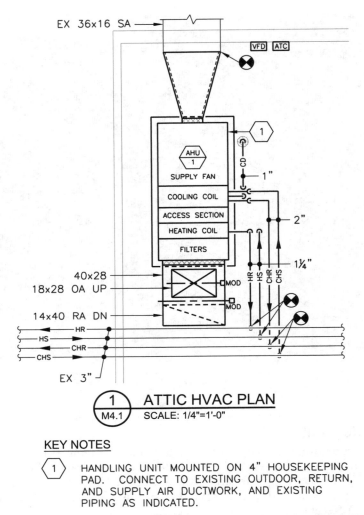

EX 36x16 SA

VFD ATC

AHU 1

SUPPLY FAN

CD

1"

COOLING COIL

ACCESS SECTION

2"

HEATING COIL

FILTERS

1¼"

40x28
18x28 OA UP
MOD

HR HS CHR CHS

14x40 RA DN
MOD

HR
HS
CHR
CHS

EX 3"

① ATTIC HVAC PLAN
M4.1 SCALE: 1/4"=1'-0"

KEY NOTES

⟨1⟩ HANDLING UNIT MOUNTED ON 4" HOUSEKEEPING PAD. CONNECT TO EXISTING OUTDOOR, RETURN, AND SUPPLY AIR DUCTWORK, AND EXISTING PIPING AS INDICATED.

FIGURE 10-7 Large-scale HVAC plan, presented at ¼ in. = 1 ft.

However, some judgment needs to be used when preparing the details for a project. If a piece of equipment has fairly simple connections that are spelled out in the project specification, a detail is not required. One example is a supply air diffuser that is to be mounted within a ceiling grid and has a piece of flexible ductwork connected to it. However, for pieces of equipment that have more components and the orientation of the components with respect to each other is important, a detail should be developed. One example is a heating or cooling coil. Details are also provided for components that are not necessarily related to any particular piece of HVAC equipment. Examples include pipe hangers and pipe penetrations through various architectural components such as walls, roofs, and floors. Finally, a detail may be required in the form of large-scale plan and section views of a piece of HVAC equipment, piping, or ductwork where close coordination with a particular architectural feature is required. An example may be the installation of a fan coil unit below a window or recessed above a ceiling.

Details describing HVAC equipment connections should show every pipe, valve, specialty, duct, duct accessory, hanger, support, and ATC component required to properly install the equipment in accordance with the design intent. Notes can be added to describe pertinent information that may be difficult to show graphically. The notes can also be used to describe coordination with other trades that may not be described elsewhere in the drawings or specifications. The HVAC equipment itself need not be shown in great detail unless it is required to convey the design intent. With few exceptions, information that should be omitted from the details includes pipe and duct sizes, dimensions of the equipment, product specifications, and equipment quantities. Unless this information is the same for each piece of equipment, as in the case of a ¾-in. hose-end drain valve, it should be shown elsewhere in the drawings and specifications in order to avoid the potential for conflicts between the detail and other parts of the construction documents.

Most consulting engineering companies have a standard library of details that they have developed. The HVAC system designer should become familiar with these details because they will usually apply to 80 to 90% of the details required for a project.

The details that would apply to the air handling unit replacement shown in the sample demolition plan, HVAC plan, and section above include:

- Heating water coil detail (Fig. 10-8)
- Chilled water coil detail (Fig. 10-9)
- Condensate drain detail (Fig. 10-10)
- Gas-fired steam humidifier detail (Fig. 10-11)
- Gas vent and combustion air intake detail (Fig. 10-12)

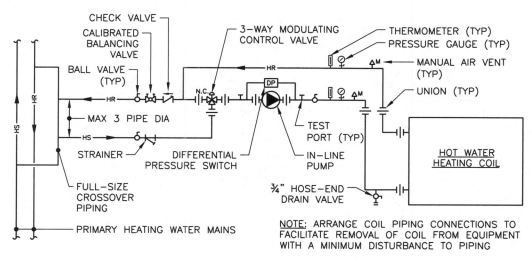

FIGURE 10-8 Heating water coil detail.

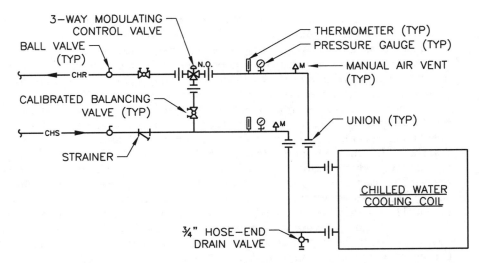

NOTE: ARRANGE COIL PIPING CONNECTIONS TO FACILITATE REMOVAL OF COIL FROM EQUIPMENT WITH A MINIMUM DISTURBANCE TO PIPING

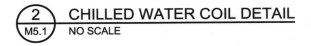

CHILLED WATER COIL DETAIL

2 M5.1   NO SCALE

FIGURE 10-9 Chilled water coil detail.

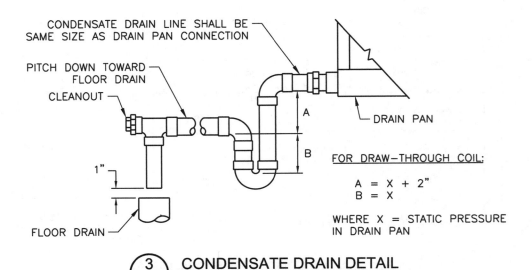

FOR DRAW-THROUGH COIL:

A = X + 2"
B = X

WHERE X = STATIC PRESSURE IN DRAIN PAN

CONDENSATE DRAIN DETAIL

3 M5.1   NO SCALE

FIGURE 10-10 Condensate drain detail.

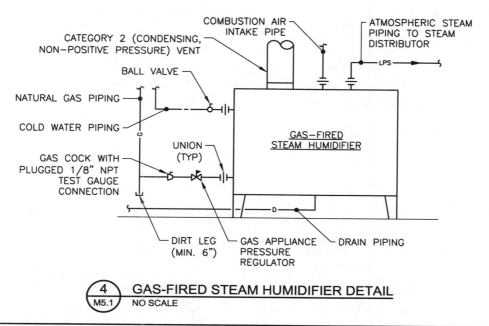

COMBUSTION AIR INTAKE PIPE

ATMOSPHERIC STEAM PIPING TO STEAM DISTRIBUTOR

CATEGORY 2 (CONDENSING, NON-POSITIVE PRESSURE) VENT

BALL VALVE

NATURAL GAS PIPING

COLD WATER PIPING

LPS

GAS-FIRED STEAM HUMIDIFIER

UNION (TYP)

GAS COCK WITH PLUGGED 1/8" NPT TEST GAUGE CONNECTION

DIRT LEG (MIN. 6")

GAS APPLIANCE PRESSURE REGULATOR

DRAIN PIPING

D

4 / M5.1   GAS-FIRED STEAM HUMIDIFIER DETAIL   NO SCALE

FIGURE 10-11   Gas-fired steam humidifier detail.

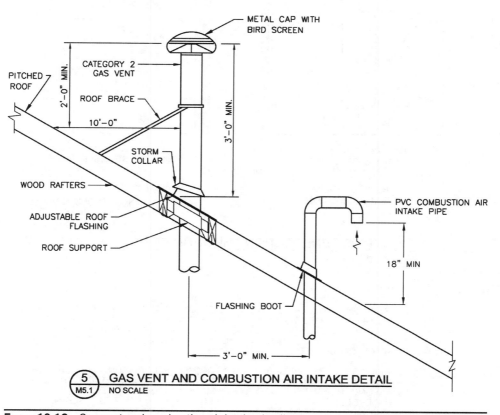

METAL CAP WITH BIRD SCREEN

CATEGORY 2 GAS VENT

PITCHED ROOF

2'-0" MIN.

3'-0" MIN.

ROOF BRACE

10'-0"

STORM COLLAR

WOOD RAFTERS

ADJUSTABLE ROOF FLASHING

ROOF SUPPORT

PVC COMBUSTION AIR INTAKE PIPE

18" MIN

FLASHING BOOT

3'-0" MIN.

5 / M5.1   GAS VENT AND COMBUSTION AIR INTAKE DETAIL   NO SCALE

FIGURE 10-12   Gas vent and combustion air intake detail.

- Pipe penetration through precast plank detail (Fig. 10-13)
- Pipe penetration through interior wall detail (Fig. 10-14)
- Pipe hanger detail (Fig. 10-15)

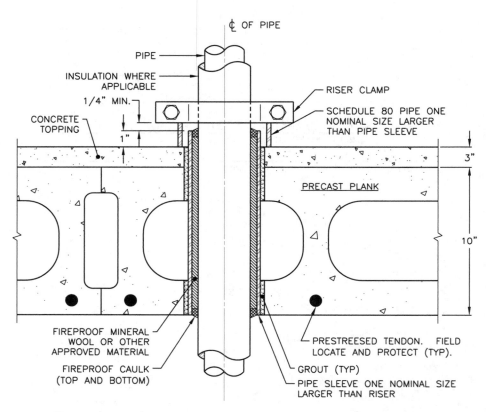

⌀ OF PIPE

PIPE

INSULATION WHERE APPLICABLE

1/4" MIN.

CONCRETE TOPPING

1"

RISER CLAMP

SCHEDULE 80 PIPE ONE NOMINAL SIZE LARGER THAN PIPE SLEEVE

3"

PRECAST PLANK

10"

FIREPROOF MINERAL WOOL OR OTHER APPROVED MATERIAL

FIREPROOF CAULK (TOP AND BOTTOM)

PRESTREESED TENDON. FIELD LOCATE AND PROTECT (TYP).

GROUT (TYP)

PIPE SLEEVE ONE NOMINAL SIZE LARGER THAN RISER

⑥ **PIPE PENETRATION THROUGH PRECAST PLANK DETAIL**
M5.1 NO SCALE

NOTES:

1. PLAN LOCATIONS OF ALL DECK PENETRATIONS ARE APPROXIMATE AND SHALL BE FIELD AJUSTED AS REQUIRED TO NOT INTERFERE WITH OR DAMAGE EXISTING PRESTRESSED TENDONS. IN GENERAL, PENETRATIONS ARE EXPECTED TO PASS THROUGH THE HOLLOW CORES OF THE PRECAST PLANKS, WHICH OCCUR BETWEEN THE TENDONS.

2. PRIOR TO CORE DRILLING, EACH PROPOSED PENETRATION LOCATION SHALL BE VERIFIED BY DRILLING A ¾" DIAMETER PILOT HOLE COMPLETELY THROUGH THE PRECAST PLANK DECK, AND VERIFYING WITH A METAL DETECTOR (FROM THE UNDERSIDE) THAT THE EXISTING TENDONS ARE AT LEAST 1½" BEYOND THE LIMITS OF THE PROPOSED HOLE.

FIGURE 10-13 Pipe penetration through precast plank detail.

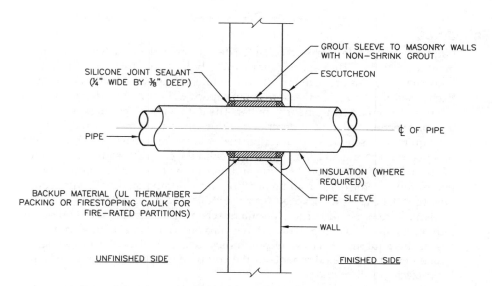

SILICONE JOINT SEALANT
(¼" WIDE BY ⅜" DEEP)

GROUT SLEEVE TO MASONRY WALLS
WITH NON–SHRINK GROUT

ESCUTCHEON

PIPE

℄ OF PIPE

INSULATION (WHERE
REQUIRED)

PIPE SLEEVE

BACKUP MATERIAL (UL THERMAFIBER
PACKING OR FIRESTOPPING CAULK FOR
FIRE–RATED PARTITIONS)

WALL

UNFINISHED SIDE

FINISHED SIDE

⑦ **PIPE PENETRATION THROUGH INTERIOR WALL DETAIL**
M5.1 NO SCALE

NOTE:

1.   AT THE CONTRACTOR'S OPTION, A UL LISTED FIRESTOPPING PIPE SLEEVE
ASSEMBLY MAY BE UTILIZED.

FIGURE 10-14   Pipe penetration through interior wall detail.

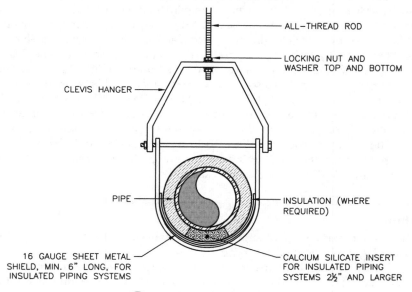

ALL–THREAD ROD

LOCKING NUT AND
WASHER TOP AND BOTTOM

CLEVIS HANGER

PIPE

INSULATION (WHERE
REQUIRED)

16 GAUGE SHEET METAL
SHIELD, MIN. 6" LONG, FOR
INSULATED PIPING SYSTEMS

CALCIUM SILICATE INSERT
FOR INSULATED PIPING
SYSTEMS 2½" AND LARGER

⑧ **PIPE HANGER DETAIL**
M5.1 NO SCALE

FIGURE 10-15   Pipe hanger detail.

Schedules

Equipment schedules are used to list the design capacities for each major piece of equipment for a project. The capacities for similar types of equipment are listed in the same schedule (e.g., fans are listed in a fan schedule; pumps are listed in a pump schedule). Each schedule should list the equipment capacities in sufficient detail for the contractor to order it from an equipment supplier. The data used by manufacturers in their catalogs or selection software to describe their products is often the same data presented in the equipment schedule to describe the equipment capacities. The criteria used to describe the equipment capacities are given in the column headings. It is good practice to give the criterion in each heading and to list the units of that criterion in parenthesis. For example, some of the column headings for a pump schedule would read as follows: flow (gpm), head (ft w.c.), speed (rpm), impeller diameter (in.).

Notes are often added to equipment schedules to list the required accessories for individual pieces of equipment and may also describe some distinguishing characteristics of the equipment. However, the notes do not replace the project specifications, which describe the requirements of the equipment in detail; they merely supplement the project specifications.

Often, the particular manufacturer and model number of the equipment that formed the basis for the design are listed in the schedule for each piece of equipment as well. This is not done to show any preference of one manufacturer over another. Rather, it is done to give the contractor a better understanding of the configuration of the particular piece of equipment. Other manufacturers listed as acceptable manufacturers in the project specifications can also submit their equipment for use as long as it meets the scheduled capacities and the requirements of the project specifications.

It is helpful to develop standard equipment schedules using a worksheet program, such as Microsoft Excel®. Worksheets are particularly helpful in presenting information in tabular format because they enable data to be easily presented in a consistent format. Also, it is easy to add or delete columns or rows and modify the data in each cell. Third-party software is available that can import the data from a worksheet into the CAD software and convert the worksheet entities into CAD lines and text. The CAD text can be edited if changes are required. It is also possible to establish a link between the worksheet and the drawing to update the schedule in the CAD drawing each time the data in the worksheet is modified. This process of linking worksheets to CAD drawings is much more efficient than creating schedules with the CAD software by using only lines and text.

The schedules shown in Figs. 10-16 through 10-18 would be used to describe the equipment capacities for the air handling unit replacement shown in the sample demolition plan, HVAC plan, and section above.

Diagrams

There are three types of diagrams commonly used in HVAC drawings: flow diagrams, riser diagrams, and ATC diagrams.

Flow and riser diagrams are intended to supplement the information shown on the floor plan drawings and connection details for the HVAC equipment. Information that should be shown on the flow and riser diagrams includes the pipe and duct sizes and flow rates, main shutoff valves, main dampers, flow meters, airflow measuring stations, and main ATC devices. These components may also be shown on the floor plan drawings

AIR HANDLING UNIT SCHEDULE

DESIG.	SERVICE	SUPPLY FAN						
		SUPPLY AIR (CFM)	OUTDOOR AIR (CFM)	ESP (IN. W.C.)	TSP (IN. W.C.)	FAN SPEED (RPM)	BRAKE (HP)	MOTOR (HP)
AHU-5	SERVICE AREA	3,600	3,600	1.60	4.07	2,693	4.37	5

AIR HANDLING UNIT SCHEDULE (CONT.)

CAPACITY (MBH)	AREA (SF)	FACE VEL (FPM)	EAT (°F)	LAT (°F)	HOT WATER HEATING COIL							
					APD (IN. W.C.)	ROWS	FPI	FLUID	FLOW (GPM)	EWT (°F)	LWT (°F)	WPD (FT W.C.)
272.2	7.50	480	0.0	70.0	0.15	2	8	WATER	13.6	180	140	5.4

AIR HANDLING UNIT SCHEDULE (CONT.)

HEAT PIPE COIL 1		TOTAL (MBH)	SENSIBLE (MBH)	AREA (SF)	FACE VEL (FPM)	CHILLED WATER COOLING COIL										HEAT PIPE COIL 2	
EAT DB/WB (°F)	LAT DB/WB (°F)					EAT DB/WB (°F)	LAT DB/WB (°F)	APD (IN. W.C.)	ROWS	FPI	FLUID	FLOW (GPM)	EWT (°F)	LWT (°F)	WPD (FT W.C.)	EAT DB/WB (°F)	LAT DB/WB (°F)
95.0/78.0	81.2/74.4	233.0	98.4	7.88	457	81.2/74.4	55.9/55.6	1.64	8	12	WATER	23.3	44	64	10.6	55.9/55.6	69.7/60.8

AIR HANDLING UNIT SCHEDULE (CONT.)

FILTERS			ELEC. (VOLT./PH)	DIMENSIONS (L x W x H)	ELEC. (VOLT./PH)	WEIGHT (LBS)	BASIS OF DESIGN	NOTES
THICKNESS	TYPE	EFFICIENCY						
2"	PLEATED	MERV 8	480/3	132" x 66" x 48"	480/3	4,950	CIRCUL-AIRE TMP-3600-HPD	1,2,3

NOTES:
1. PLENUM SUPPLY FAN.
2. DOUBLE-WALL CONSTRUCTION.
3. STAINLESS STEEL DRAIN PAN.

FIGURE 10-16 Air handling unit schedule.

GAS-FIRED STEAM HUMIDIFIER SCHEDULE

DESIG.	SERVICE	AIRFLOW (CFM)	ENT AIR (°F/%RH)	LVG AIR (°F/%RH)	LOAD (LB/H)	PRESSURE (PSIG)	WATER TYPE	DUCT SIZE (W x H)	MAX VAPOR TRAIL (FT)	ELEC. (VOLT./PH)	BASIS OF DESIGN	NOTES
H-2	AHU-5	3,600	0/0	70/35	87.3	0	POTABLE	44 x 20	2.5	120/1	DRISTEEM GTS-200	1,2

NOTES:
1. SIZE AND NUMBER OF DISPERSION TUBES SHALL BE DETERMINED BY THE MANUFACTURER.
2. PROVIDE COMPLETE GAS TRAIN AND FULLY-MODULATING NATURAL GAS BURNER.

Figure 10-17 Gas-fired steam humidifier schedule.

PUMP SCHEDULE

DESIG.	SERVICE	TYPE	FLUID	FLOW (GPM)	HEAD (FT W.C.)	SPEED (RPM)	IMPELLER DIA (IN.)	MOTOR (HP)	ELEC. (VOLT./PH)	BASIS OF DESIGN
P-11	AHU-5 HEATING COIL	IN-LINE	WATER	14	20	1,750	4.50	1/3	120/1	TACO MODEL NO. 1611

Figure 10-18 Pump schedule.

or the connection details for the HVAC equipment. Consequently, there is some redundancy in the presentation of this information, making it important to ensure that the flow and riser diagrams are consistent with the floor plans and equipment connection details. It is not necessary to show all of the valves and specialties for the HVAC equipment on the flow and riser diagrams because it unnecessarily clutters the diagrams and is a potential source of conflict with the equipment connection details. These components are more clearly shown in the connection details for the HVAC equipment.

ATC diagrams illustrate the complete ATC systems for the central plant,[4] air systems, and terminal equipment. ATC diagrams require a sequence of operation and a direct digital control (DDC) point list (for DDC systems). ATC diagrams are discussed in detail in Chap. 9.

Flow Diagrams

Flow diagrams illustrate the major components of the central plant and include equipment designations, pipe sizes, and flow rates. Figure 10-19 illustrates a sample flow diagram for a four-pipe heating and cooling plant. One helpful device in preparing the

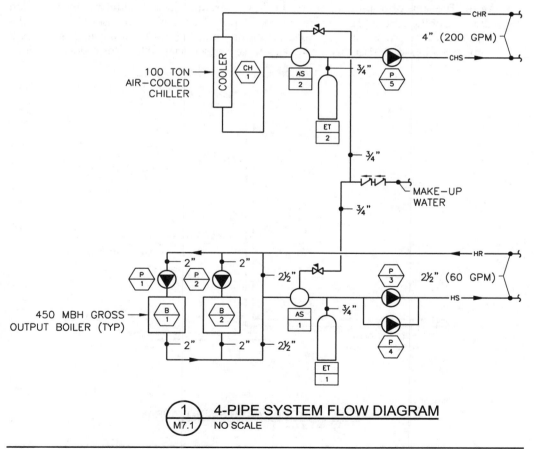

FIGURE 10-19 4-pipe system flow diagram.

flow diagram for a central plant is to orient the equipment on the diagram in a fashion similar to the floor plan drawing of the central plant equipment room. For example, if the floor plan drawing of the equipment room is oriented with north at the top of the sheet, the boilers are on the west side of the equipment room, the secondary heating water pumps are on the east side of the equipment room, and the main heating water piping exits the east wall of the equipment room, the flow diagram should be oriented with the boilers on the left-hand side of the flow diagram, the secondary pumps on the right-hand side of the flow diagram, and the main heating water piping to the building also on the right-hand side of the flow diagram. This is not a hard and fast rule, but it does aid in providing a consistent presentation of the information and helps the reader of the drawings to compare the flow diagram to the floor plan drawings without having to reorient the diagram to the plan.

Riser Diagrams

Riser diagrams illustrate the major components of the piping and ductwork distribution systems, including the piping and duct risers and the connections to these risers at each floor. Riser diagrams also show the equipment designations, pipe sizes, duct sizes, and flow rates.[5] The same principle of orienting the flow diagram in a fashion similar to the floor plan drawing of the central plant equipment room applies to orienting the riser diagram to the piping or ductwork distribution system in the building. Figures 10-20 and 10-21 illustrate a sample heating water riser diagram and a sample air riser diagram for a three-story building.

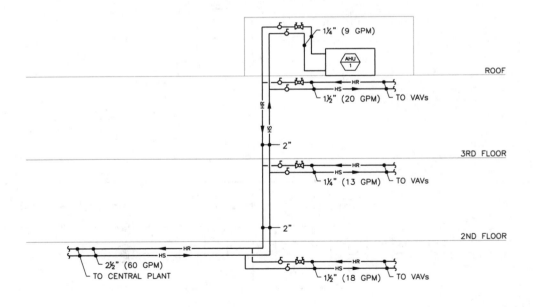

FIGURE 10-20 Heating water riser diagram.

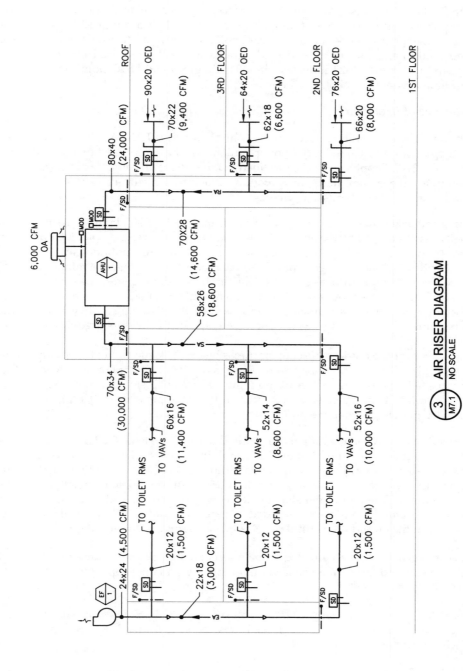

Figure 10-21 Air riser diagram.

355

ATC Diagrams

The final components of the construction drawings for the air handling unit replacement illustrated in Figs. 10-3 through 10-6 and Figs. 10-8 through 10-18 are the ATC diagram, sequence of operation, and DDC point list for the air handling unit and associated equipment. (No flow or riser diagrams would be required for this project.) Figures 10-22 and 10-23 illustrate the ATC diagram and DDC point list for the air handling unit replacement.

Sequence of Operation

General   The air handling unit shall be controlled by the DDC ATC system controller. Occupied and unoccupied modes of operation shall be determined by the time schedule of the BAS. Graphics shall be provided in the BAS for all components of the system, identifying the current mode of operation, setpoints and current values of all points, and the status of all alarms and safeties.

Occupied Mode   When the unit is energized, the outdoor air damper shall open and shall be proved open by the end switch mounted on the damper actuator. Upon closure of the end switch, the supply fan shall be energized and run continuously.

The heating water coil control valve shall modulate to maintain a minimum of 68°F discharge air temperature as sensed by the discharge air temperature sensor.

The chilled water coil control valve shall modulate to maintain a maximum of 55°F discharge air dew point temperature as sensed by the discharge air hygrometer, and a maximum of 75°F discharge air temperature as sensed by the discharge air temperature sensor.

Whenever the outdoor air temperature is below 70°F, the circulating pump shall be energized and run continuously. Upon a rise in outdoor air temperature above 75°F, the pump shall be de-energized. A pump failure alarm shall signal at the BAS when the circulator fails to operate, as sensed by the differential pressure switch piped between the inlet and discharge of the pump.

Humidification   When the air handling unit is started, the BAS shall send a start signal to the gas-fired steam humidifier. The humidifier shall be energized upon receiving proof of airflow from the duct mounted airflow switch. The humidifier shall operate under its packaged controls to maintain a 35% (adjustable) space relative humidity (RH) setpoint. The humidifier shall be de-energized if the 50% RH (adjustable) setting of the high-limit humidistat is exceeded. A general humidifier alarm shall signal at the BAS when the humidifier control panel is in any alarm condition.

Unoccupied Mode   When indexed to the unoccupied mode by the time schedule of the BAS, the outdoor air damper shall close completely, and the supply fan and humidifier shall be de-energized.

Fan Failure Alarm   A fan failure alarm shall signal at the BAS when the supply fan fails to operate, as sensed by the differential pressure switch piped between the inlet and discharge of the fan.

Dirty Filter Alarm   A dirty filter alarm shall signal at the BAS when the differential pressure across the filter section exceeds the 0.70-in. w.c. (adjustable) setpoint.

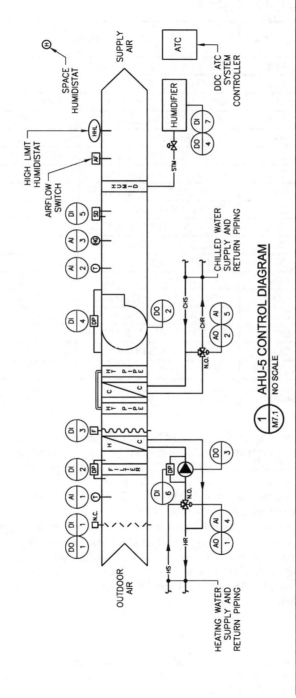

FIGURE 10-22 Air handling unit ATC diagram.

DDC POINT LIST

POINT TYPE	POINT #	DESCRIPTION	ALARM
ANALOG INPUT	AI-1	OUTDOOR AIR TEMPERATURE	
	AI-2	DISCHARGE AIR TEMPERATURE	
	AI-3	DISCHARGE AIR AIR DEW POINT	
	AI-4	HEATING COIL CONTROL VALVE POS.	
	AI-5	COOLING COIL CONTROL VALVE POS.	
DIGITAL INPUT	DI-1	OUTDOOR AIR DAMPER POSITION	YES
	DI-2	FILTER DIFFERENTIAL PRESSURE	YES
	DI-3	LOW LIMIT TEMPERATURE	YES
	DI-4	SUPPLY FAN STATUS	YES
	DI-5	SUPPLY AIR SMOKE DETECTION	YES
	DI-6	CIRCULATING PUMP STATUS	YES
	DI-7	GENERAL HUMIDIFIER ALARM	YES
ANALOG OUTPUT	AO-1	HEATING COIL CONTROL VALVE POS.	
	AO-2	COOLING COIL CONTROL VALVE POS.	
DIGITAL OUTPUT	DO-1	OUTDOOR AIR DAMPER POSITION	
	DO-2	SUPPLY FAN START/STOP	
	DO-3	CIRCULATING PUMP START/STOP	
	DO-4	HUMIDIFIER START/STOP	

FIGURE 10-23   Air handling unit DDC point list.

Safeties   Upon a drop in the heating coil leaving air temperature below 38°F, as sensed by the low-limit temperature sensor serpentined across the downstream face of the coil, the supply fan shall be de-energized, the outdoor air damper shall close completely, the heating coil control valve shall open fully to the coil, the humidifier shall be de-energized, and an alarm shall signal at the BAS. The low-limit temperature sensor shall require a manual reset for the unit to be restarted.

The supply air duct smoke detector shall be hard-wired through the supply fan starter control circuit. When the supply air duct smoke detector detects the presence of smoke, the supply fan shall be de-energized, the outdoor air damper shall close completely, the humidifier shall be de-energized, and an alarm shall signal at the fire alarm control panel and at the BAS. The duct smoke detector shall require a manual reset for the unit to be restarted.

Endnotes

1. An escutcheon is a doughnut-shaped flat plate that is installed around a pipe penetration through a wall. Its purpose is to cover the rough opening in the wall around the pipe and provide a finished appearance.

2. Refer to the *U.S. National CAD Standard* for recommended layer names for each file type.

3. A 1/8 in. = 1 ft scale is commonly employed for demolition drawings because this scale is usually adequate to present the detail that is required.

4. The flow diagram for the central plant can often be used as the basis for developing the central plant ATC diagram.

5. Riser diagrams are typically used for buildings that have three or more stories. Riser diagrams are not necessary for buildings that have fewer than three stories because the information can be clearly portrayed on the floor plan drawings for the piping and ductwork distribution systems.

Index

Note: Page numbers followed by *f* denote figures; page numbers followed by *n* denote notes; page numbers followed by *t* denote tables.